Disaster Risk Reduction

Methods, Approaches and Practices

Series Editor

Rajib Shaw, Keio University, Shonan Fujisawa Campus, Fujisawa, Japan

Disaster risk reduction is a process that leads to the safety of communities and nations. After the 2005 World Conference on Disaster Reduction, held in Kobe, Japan, the Hyogo Framework for Action (HFA) was adopted as a framework for risk reduction. The academic research and higher education in disaster risk reduction has made, and continues to make, a gradual shift from pure basic research to applied, implementation-oriented research. More emphasis is being given to multi-stakeholder collaboration and multi-disciplinary research. Emerging university networks in Asia, Europe, Africa, and the Americas have urged process-oriented research in the disaster risk reduction field. With this in mind, this new series will promote the output of action research on disaster risk reduction, which will be useful for a wide range of stakeholders including academicians, professionals, practitioners, and students and researchers in related fields. The series will focus on emerging needs in the risk reduction field, starting from climate change adaptation, urban ecosystem, coastal risk reduction, education for sustainable development, community-based practices, risk communication, and human security, among other areas. Through academic review, this series will encourage young researchers and practitioners to analyze field practices and link them to theory and policies with logic, data, and evidence. In this way, the series will emphasize evidence-based risk reduction methods, approaches, and practices.

All submitted proposals are carefully reviewed by the series editor.

Publication Ethics and Publication Malpractice Statement: https://www.springernature.com/gp/policies/book-publishing-policies.

Ethical, Conflict-of-Interest, and Editorial Policy Statement: https://www.springer.com/us/authors-editors/journal-author/journal-author-helpdesk/before-you-start/before-you-start/1330#c14214.

Rajib Shaw · Takako Izumi · Riyanti Djalante ·
Fumihiko Imamura
Editors

Two Decades from the Indian Ocean Tsunami

Key Challenges and Advancements

Springer

Editors
Rajib Shaw
Graduate School of Media and Governance
Keio University
Fujisawa, Kanagawa, Japan

Riyanti Djalante
ASEAN Secretariat
Jakarta, Indonesia

Takako Izumi
International Research Institute of Disaster Science
Tohoku University
Sendai, Miyagi, Japan

Fumihiko Imamura
International Research Institute of Disaster Science
Tohoku University
Sendai, Miyagi, Japan

ISSN 2196-4106 ISSN 2196-4114 (electronic)
Disaster Risk Reduction
ISBN 978-981-96-2668-7 ISBN 978-981-96-2669-4 (eBook)
https://doi.org/10.1007/978-981-96-2669-4

This Springer imprint is published by the registered company Springer Nature Singapore Pte Ltd.
The registered company address is: 152 Beach Road, #21-01/04 Gateway East, Singapore 189721, Singapore

Preface

The Indian Ocean Tsunami of December 26, 2004, was a watershed moment in the history of disaster events. With its catastrophic impact stretching across 14 countries and causing the tragic loss of over 230,000 lives, it brought the world face-to-face with the tremendous power of disasters. It also revealed the vulnerabilities in disaster preparedness and response systems on a global scale, particularly in coastal regions. Yet, amid the destruction and despair, the world witnessed an unprecedented outpouring of humanitarian support, international cooperation, and a strengthened resolve to prevent future tragedies of such magnitude.

As we reflect on the two decades that have passed since this fateful event, "*Two Decades from the Indian Ocean Tsunami: Key Challenges and Advancements*" offers a comprehensive review of the progress made and the ongoing challenges that remain in disaster risk reduction (DRR). This volume provides a platform for experts, practitioners, and policymakers to explore the lessons learned from the tsunami, evaluate the advancements in disaster management strategies, and highlight the complexities of building resilience in a world where disasters are increasing in frequency and intensity due to climate change.

The contributions in this book focus on several key areas: the evolution of early warning systems, improvements in global and regional cooperation, community resilience-building initiatives, and the integration of science, technology, and policy in mitigating disaster risk. The chapters also address regional and country experiences over the past two decades. In total, we have 22 chapters spread into two parts.

We dedicate this book to the memory of those who perished in the 2004 tsunami, to the survivors who rebuilt their lives, and to the countless individuals and organizations who have worked tirelessly over the last two decades to improve disaster resilience and response globally.

We will be very happy if the readers find this book useful for their work.

Fujisawa, Japan	Rajib Shaw
Sendai, Japan	Takako Izumi
Jakarta, Indonesia	Riyanti Djalante
Sendai, Japan	Fumihiko Imamura

About This Book

At the threshold of 20 years from Indian Ocean Tsunami, the disaster risk reduction field evolved significantly in the affected countries as well as globally. Several mega disasters happened in the last 20 years, the risk landscape has become more complex, and technology, governance, and people's resilience have gained positive momentum. We had two world conferences on disaster risk reduction, resulting in two global disaster risk reduction frameworks, which were aligned to the development frameworks as well as climate change agreements. The need for synergy of disaster, development, and climate change are increasingly felt. COVID-19 and post COVID risk landscape have brought a new dimension of compound and complex disasters and the need for All Hazards Approach.

With this backdrop, the book will provide some glimpses of advancement in the disaster risk reduction and resilience-building efforts in the affected countries and will focus on few cross-cutting issues related to governance, technology, and communities. The book will have two major parts: cross-cutting issues and country experiences.

This book deliberates on the aspects as mentioned above related to disaster risk reduction, preparedness, and management. Apart from the introduction, the book has two parts: (1) *Cross Cutting Issues*, and (2) *Regional and Country Experiences.*

Contents

Editors and Contributors

About the Editors

Rajib Shaw is a professor in Graduate School of Media and Governance of Keio University, Japan. He is the co-founder of a Delhi (India) based social entrepreneur startup Resilience Innovation Knowledge Academy (RIKA), and RIKA Institute. He is the Co-chair of the United Nations Asia Pacific Science Technology Advisory Group (AP-STAG) and was CLA for IPCC's 6th Assessment Report. Professor Shaw is the recipient of the prestigious "Pravasi Bharatiya Samman Award (PBSA)" in 2021 for his contribution in education sector. He also received United Nations Sasakawa Award in 2022 for his lifetime contributions in the field of disaster risk reduction. He published 81 books and over 450 research papers. More about his work can be found in: www.rajibshaw.org.

Takako Izumi is a Professor at the International Research Institute of Disaster Science (IRIDeS), Tohoku University, Japan since 2013. She also serves as Program Director of the Multi Hazards Program under the Association of Pacific Lim Universities (APRU), which comprises 60 universities and academic institutes in the Pacific Rim. Her research interests include international and regional frameworks/strategies for disaster risk reduction, international humanitarian assistance, and DRR initiatives at the local and community levels. Previously, she has almost 15-year experience with an international NGO and the UN agencies including UN Habitat, UNOCHA, UNDRR and so on for disaster risk reduction, response and recovery especially in Asia. In May 2015, she has been appointed as a member of the UNDRR Asian Pacific Science Technology Advisory Group (APSTAG) to support the efforts of governments and stakeholders in the implementation of the Sendai Framework for DRR. She holds Ph.D. in Global Environmental Studies from Kyoto University, Japan.

Riyanti Djalante is the Assistant Director/Head of Disaster Management and Humanitarian Assistance Division at the Secretariat of the Association of Southeast Asian Nations (ASEAN). She has vast experience related to disaster risk reduction and climate change adaptation research and policy. She is the Lead Author of IPCC Assessment Report 6 Working Group II, IPCC Special Report on impacts of 1.5° warming, and the Global Environmental Outlook 6 of the UN Environment Program. She is the editor of the *Journal of Progress in Disaster Science*. Her former appointments include a member of the Scientific Committee of the IRDR, research fellow of the Earth System Governance Network, and Social Science Fellow of the International Council for Science. She worked as a Research Associate at the UNU Institute of Environment and Human Security (UNU-EHS) in Germany and at the Institute of Advance Science and Sustainability in Japan (UNU-IAS). She also worked for the government in Indonesia in the field of disaster risk management and development planning.

Fumihiko Imamura is an expert on tsunami modeling for warning, mitigation planning and education/awareness and was a member of a study group of the reconstruction design council in response to the 2011 Tohoku earthquake at cabinet office, and the committee for technical investigation on Countermeasures for Earthquakes and Tsunamis of the Central Disaster Management Council. He is also a member of Science Council of Japan, Science member of the Central Disaster Management Council in Japan and was the president of Japan Society for Natural Disaster Science in 2008–2011. He has conducted several field surveys as a leader in earthquakes and tsunami damage investigations since 1992 in Nicaragua and Indonesia. He was the Director of International Research Institute of Disaster Science (IRIDeS), Tohoku University during 2014–2023. He has guided many graduate students and has an impressive list of publications.

Contributors

Tofael Ahamed Institute of Life and Environmental Sciences, University of Tsukuba, Tsukuba, Japan

Iftekhar Ahmed University of Newcastle, Newcastle, Australia

Daniel P. Aldrich Northeastern University, Boston, USA

Dilanthi Amaratunga Global Disaster Resilience Centre, School of Applied Sciences, University of Huddersfield, Queensgate, Huddersfield, UK

Ariyaningsih Graduate School of Media and Governance, Keio University, Minato, Japan;
Department of Civil Engineering and Planning, Institut Teknologi Kalimantan, Balikpapan, Indonesia

Ayush Baskota Disaster Preparedness, Mitigation and Management, Asian Institute of Technology, Bangkok, Thailand

Jeeten Kumar Bhagnani Disaster Preparedness, Mitigation and Management, Asian Institute of Technology, Bangkok, Thailand

A. K. M. Abdul Ahad Biswas Department of Disaster Risk Management, Faculty of Environmental Science and Disaster Management, Patuakhali Science and Technology University, Dumki, Bangladesh

Ranit Chatterjee RIKA Institute, New Delhi, India

Ambika Dabral RIKA Institute, New Delhi, India

Promi Debnath Department of Disaster Risk Management, Faculty of Environmental Science and Disaster Management, Patuakhali Science and Technology University, Dumki, Bangladesh

Riyanti Djalante The ASEAN Secretariat, Jakarta, Indonesia

Md. Faisal Department of Disaster Resilience and Engineering, Faculty of Environmental Science and Disaster Management, Patuakhali Science and Technology University, Dumki, Bangladesh

Ardiansyah Fauzi Department of Mathematics, Physics and Electrical Engineering, Northumbria University, Newcastle upon Tyne, UK

Nishara Fernando Department of Sociology, University of Colombo, Colombo, Sri Lanka

Yeshani Fernando Federation of Sri Lankan Local Government Authorities, Colombo, Sri Lanka

Yinan Gao Graduate School of Media and Governance, Keio University, Fujisawa, Japan

Krishnakali Ghosh RIKA Institute, New Delhi, India

Rajan Chandra Ghosh Department of Emergency Management, Faculty of Environmental Science and Disaster Management, Patuakhali Science and Technology University, Dumki, Bangladesh

Subhajit Ghosh Disaster Preparedness, Mitigation and Management, Asian Institute of Technology, Bangkok, Thailand

Richard Haigh Global Disaster Resilience Centre, School of Applied Sciences, University of Huddersfield, Queensgate, Huddersfield, UK

Md. Fuad Hasan Department of Emergency Management, Faculty of Environmental Science and Disaster Management, Patuakhali Science and Technology University, Dumki, Patuakhali, Bangladesh;
Department of Environment Protection, German University Bangladesh, Gazipur, Bangladesh

Md. Mahmudul Hasan Department of Geo-Information Science and Earth Observation, Faculty of Environmental Science and Disaster Management, Patuakhali Science and Technology University, Dumki, Bangladesh

Papri Hazra Department of Environmental Science, Faculty of Environmental Science and Disaster Management, Patuakhali Science and Technology University, Dumki, Bangladesh

Fumihiko Imamura International Research Institute of Disaster Science (IRIDeS), Tohoku University, Sendai, Japan

Mikio Ishiwatari Meiji University, Tokyo, Japan

Ismeti Research Center for Law, National Research and Innovation Agency, Jakarta, Indonesia

Takako Izumi International Research Institute of Disaster Science (IRIDeS), Tohoku University, Sendai, Tohoku, Japan

Sakiko Kanbara Kobe City College of Nursing, Kobe, Japan

Sujeewa Kumari University of Colombo, Colombo, Sri Lanka

Kamaluddin Latief College of Public Health, Program in Global Health and Health Security, Taipei Medical University, Taipei, Taiwan

Arunabh Mitra Graduate School of Media and Governance, Keio University, Fujisawa, Japan

Eriko Motoyama Servis EDA Pvt. Ltd., Kuala Lumpur, Malaysia

Rekha Nianthi University of Peradeniya, Peradeniya, Sri Lanka

Laely Nurhidayah Research Center for Law, National Research and Innovation Agency, Jakarta, Indonesia

Indrajit Pal Disaster Preparedness, Mitigation and Management, Asian Institute of Technology, Bangkok, Thailand

Ma. Theresa P. Pamaong Graduate School of Media and Governance, Keio University, Fujisawa, Japan

U. T. G. Perera Department of Geography, Faculty of Arts, University of Colombo, Colombo, Sri Lanka

H. Rahayu The School of Architecture, Planning and Policy Development, ITB, Bandung, Indonesia

Harkunti Pertiwi Rahayu Chair of WG1, Indian Ocean Tsunami Warning and Mitigation System (2015–2024), Hyderabad, India;
The School of Architecture, Planning and Policy Development, ITB, Bandung, Indonesia

Md. Abdur Rahman Office of the Senior Consultant, Knowledge Craft Bangladesh (KCB), Dhaka, Bangladesh

Md. Rasheduzzaman Department of Emergency Management, Faculty of Environmental Science and Disaster Management, Patuakhali Science and Technology University, Dumki, Patuakhali, Bangladesh

Sujata Saunik Graduate School of Media and Governance, Keio University, Fujisawa, Japan

Md. Shamsuzzoha Department of Emergency Management, Faculty of Environmental Science and Disaster Management, Patuakhali Science and Technology University, Dumki, Patuakhali, Bangladesh;
Graduate School of Life and Environmental Sciences, University of Tsukuba, Tsukuba, Japan

Rajib Shaw RIKA Institute, New Delhi, India;
Graduate School of Media and Governance, Keio University, Minato, Japan

Riyan Benny Sukmara Department of Civil Engineering and Planning, Institut Teknologi Kalimantan, Balikpapan, Indonesia;
Department of Civil Engineering, National Central University, Taoyuan City, Taiwan

Agus Suntoro Research Center for Law, National Research and Innovation Agency, Jakarta, Indonesia

Anawat Suppasri International Research Institute of Disaster Science (IRIDeS), Tohoku University, Sendai, Japan

Divya Suresh Graduate School of Media and Governance, Keio University, Fujisawa, Japan

Md. Foisal Talukder Department of Emergency Management, Faculty of Environmental Science and Disaster Management, Patuakhali Science and Technology University, Dumki, Patuakhali, Bangladesh

Ratchaneekorn Thongthip Disaster Preparedness, Mitigation and Management, Asian Institute of Technology, Bangkok, Thailand

Henny Warsilah Centre for Society and Culture, National Research and Innovation Agency, Jakarta, Indonesia

Yafang Wen Graduate School of Media and Governance, Keio University, Fujisawa, Japan

Deepthi Wickramasinghe University of Colombo, Colombo, Sri Lanka

Nurrohman Wijaya Department of Urban and Regional Planning, Bandung Institute of Technology, Bandung, Indonesia

Ray-Shyan Wu Department of Civil Engineering, National Central University, Taoyuan City, Taiwan

Ayse Yildiz University of Leicester, Leicester, UK

Yanwu Zhang Graduate School of Media and Governance, Keio University, Fujisawa, Japan

Chapter 1
Introduction and Overview of Disaster Risk Landscape: Perspectives of Twenty Years from the Indian Ocean Tsunami of 2004

Rajib Shaw, Takako Izumi, Riyanti Djalante, and Fumihiko Imamura

Abstract The Indian Ocean Tsunami of 2004 was one of the devastating disasters in the recent past, which changed significantly the global risk reduction approaches. Two major frameworks Hyogo Framework for Action and Sendai Framework for Disaster Risk Reduction have evolved. Sendai Framework is aligned with other global frameworks, like SDGs (Sustainable Development Goals) and Paris Agreement. Over the last 20 years, several mega disasters happened in the Asia Pacific region as well as in the world, causing significant impacts on lives and properties. The GET (governance education and technology) components of disaster risk reduction have also evolved significantly with the incorporation of disaster laws, focusing on localization, incorporation of disaster education in schools and higher education systems, and evolution of new technologies. Several new concepts also emerged over the last 20 years: systemic risk approach, from risk to resilience, ecosystem base resilience, resilient infrastructure, pre-disaster recovery planning, impact-based investment, urban rural collective resilience, all hazards approach, Early warning for all, and enhanced roles of stakeholders. Future risk reduction needs to be more inclusive, and technologies need to be affordable.

Keywords Indian Ocean tsunami · Mega disaster · Adaptive governance · Risk reduction education · Emerging technologies

R. Shaw (✉)
Graduate School of Media and Governance, Keio University, Fujisawa, Japan
e-mail: shaw@sfc.keio.ac.jp

T. Izumi · F. Imamura
International Research Institute of Disaster Science (IRIDeS), Tohoku University, Sendai, Japan

R. Djalante
ASEAN Secretariat, Jakarta, Indonesia

R. Shaw et al. (eds.), *Two Decades from the Indian Ocean Tsunami*, Disaster Risk Reduction, https://doi.org/10.1007/978-981-96-2669-4_1

1.1 Introduction

The Indian Ocean Tsunami of December 26, 2004, is one of the deadliest disasters in modern history. Triggered by a massive undersea earthquake with a magnitude of 9.1–9.3 off the west coast of northern Sumatra, Indonesia, the tsunami devastated coastal areas across 14 countries. The earthquake, caused by the subduction of the Indian plate beneath the Burma plate, generated powerful waves up to 30 m high, which struck countries like Indonesia, Thailand, Sri Lanka, India, and even as far as the east coast of Africa (Shaw 2014). The tsunami caused an estimated 230,000–250,000 deaths with more than 57 nationalities, displacing millions of people and leading to widespread destruction of infrastructure and livelihoods. Indonesia was the hardest hit, particularly the province of Aceh, where entire communities were wiped out. The disaster highlighted the lack of an effective early warning system in the Indian Ocean at that time. As a result, a significant global effort followed to establish the Indian Ocean Tsunami Warning System, launched in 2006. The 2004 tsunami also prompted greater international collaboration on disaster risk reduction, with a focus on early warning systems, preparedness, and resilient infrastructure.

Over the past 20 years, disaster risk reduction (DRR) has evolved significantly, driven by an increasing awareness of the complex and interconnected nature of disasters, technological advancements, and the need for more inclusive, sustainable approaches. Although the International Decade of Natural Disaster Reduction (IDNDR: 1990–1999) has been able to make certain awareness raising on the need of pre-disaster risk reduction and preparedness activities, DRR efforts were primarily reactive, focusing on responding to disasters after they occurred. However, with the adoption of the Hyogo Framework for Action (2005–2015, UNISDR 2005) and later the Sendai Framework for Disaster Risk Reduction (2015–2030, UNDRR 2015), the emphasis shifted toward a more proactive approach. Governments and organizations now prioritize risk reduction through preparedness, prevention, and resilience building rather than focusing solely on response and recovery.

Technological advancements have revolutionized DRR. Real-time data collection, predictive modeling, remote sensing, and Geographic Information Systems (GIS) have improved risk assessments and early warning systems. For example, early warning systems for earthquakes, tsunamis, and storms have been significantly enhanced, allowing for timely evacuations and minimizing casualties. Moreover, big data and artificial intelligence (AI) have enabled better analysis of disaster patterns and the development of more precise disaster response strategies.

There has been a growing recognition of the importance of localization in DRR. Community-based disaster risk reduction (CBDRR) has become a core principle, recognizing that local communities are often the first responders and have valuable knowledge about local risks. DRR has increasingly focused on building the capacity of local actors and fostering community resilience through education, training, and inclusive participation. DRR strategies have also become more inclusive, with a greater focus on addressing the vulnerabilities of marginalized groups, including women, children, the elderly, and people with disabilities. The Sendai

Framework emphasizes reducing vulnerability and promoting equality, ensuring that DRR efforts reach all segments of society.

2015 was a landmark year where three major global frameworks were adopted (UN 2015a, 2015b, 2015c): Sendai Framework for Disaster Risk Reduction (SFDRR: 2015–203), Sustainable Development Goals (SDGs: 2015–2030) and Paris Agreement on Climate Change (2015–2030). An analysis of these frameworks shows synergies in terms of mutual references, while SFDRR has referred 20 times the word "sustainable development", and 15 times "climate change". SDG document has referred 12 times "disaster risk" and 20 times "climate change", and Paris Agreement has referred 16 times "sustainable development" and 1 time "disaster risk". The lack of mentioning "disaster risk" in Paris Agreement is due to the use of different terminology of "loss and damages." Another interesting aspect of these three frameworks is the use of the term "local", which has been used 10 times in 35 page document of SDGs, 48 times in 25 page document of SFDRR, and 9 times in 32 page document in Paris Agreement. The term "local" has been used in reference to local authorities, communities, culture, materials, planning, governance, knowledge, priorities and adaptation, etc. This shows that all these three global frameworks have a strong local focus. While the frameworks provide larger global perspectives and targets, their implementation needs to be rooted in local (Shaw et al. 2016).

In summary, DRR has transitioned from a reactive, top-down approach to a proactive, inclusive, and technology-driven model. This shift has led to more effective, equitable, and sustainable disaster management worldwide. This chapter tries to provide an overview of the last 20 years, as well as it provides an overview of the book's structures and contents.

1.2 Ten Major Disasters Over Last 20 years

While each and every life loss is important, and the disaster risk reduction measures need to protect every life, following are some of the mega disasters which happened in the last 20 years:

1. **Hurricane Katrina (2005)**: A devastating hurricane that hit New Orleans and the U.S. Gulf Coast, causing over 1,800 deaths and US$125 billion in damage.
2. **Cyclone Nargis (2008):** A major category 3 cyclone struck in the coastal area of Myanmar, causing more than 120,000 deaths and wiping out several coastal villages.
3. **Sichuan Earthquake, China (2008)**: A powerful 7.9 magnitude earthquake struck China's Sichuan province, resulting in around 87,000 deaths and widespread destruction.
4. **Haiti Earthquake (2010)**: A 7.0 magnitude earthquake hit Haiti, leading to around 230,000 deaths and a severe humanitarian crisis due to widespread infrastructure collapse.

5. **Tohoku Earthquake and Tsunami, Japan (2011)**: A 9.0 magnitude earthquake and resulting tsunami killed over 18,500 people, triggered the Fukushima nuclear disaster, and caused vast destruction in north-eastern Japan.
6. **Typhoon Haiyan, Philippines (2013)**: One of the strongest tropical cyclones ever recorded, it caused over 6,000 deaths and widespread devastation in the Philippines.
7. **Nepal Earthquake (2015)**: A 7.8 magnitude earthquake struck Nepal, causing nearly 9,000 deaths and severe damage to buildings, infrastructure, and cultural heritage sites.
8. **Sulawesi Earthquake and Tsunami, Indonesia (2018)**: A 7.5 magnitude earthquake triggered a tsunami and liquefaction in Indonesia, resulting in over 4,300 deaths and significant damage.
9. **COVID-19 Pandemic (2020–2022)**: The global pandemic has caused over 6 million deaths worldwide and had far-reaching economic, social, and health impacts across the globe.
10. **Turkey-Syria Earthquake (2023)**: A 7.8 magnitude earthquake struck south-eastern Turkey and northern Syria, causing over 50,000 deaths and widespread destruction of infrastructure.

The key issues emerging from the above list can be summarized as follows: (1) the frequency and severity of natural hazards are increasing, (2) due to high at risk population, the exposure become very high, resulting in high number of casualty, (3) financial losses are also increasing due to high damages of buildings and infrastructures, (4) disasters are becoming diversified, and need multi hazard/all hazard approaches, and (5) COVID-19 (as a biological hazard), as well as other natural hazards induced disasters, are posing key challenges to development pathways of many countries.

Needless to say that, apart from these mega and visible/widely media covered disasters, there are many new types of disasters, which are often local and/or invisible. Heat wave is one of them, and it is affecting people and livelihoods across the globe, be it in the developing or developed countries. Due to prolonged dry days, we are seeing increasing numbers of bush fires. Local cloud-burst and inundation flooding in the cities are getting more common, causing significant disruptions in the urban lives and livelihoods. Also, COVID-19 has pointed the importance of focusing on all types of hazards, not just natural hazards. Biological, radiological, technical, and cyber hazards are becoming more common, and we are seeing more complex and compound disasters.

1.3 Evolution of Governance, Education, and Technology

This section describes briefly the evolution of GET (Governance Education and Technology) domains in disaster risk reduction.

1.3.1 Governance

In post Indian Ocean Tsunami, different countries have developed and enacted Disaster Law at national level (WB 2024). For example, India passed the Disaster Management Act in 2005, creating a formal legal framework for disaster risk reduction and management. This legislation established a National Disaster Management Authority (NDMA) and a disaster fund, focusing on preparedness and capacity building. Sri Lanka established its Disaster Management Law in 2005–2006, followed by Pakistan in 2006 as an Ordinance (turned into a Law in 2010), followed by China and Indonesia in 2007, Philippines in 2010, Bangladesh in 2012, and Vietnam in 2013. All these Laws provide strong legislative and governance measures for different countries. DRR governance has evolved to become more inclusive, focusing on vulnerable populations such as women, children, the elderly, and people with disabilities. The Sendai Framework highlights the importance of reducing vulnerability and promoting equity, ensuring that DRR strategies are inclusive and sensitive to the needs of all segments of society.

A major shift in DRR governance has been the increased focus on localization. National governments have recognized the importance of empowering local communities, as they are often the first responders to disasters. The focus on community-based disaster risk reduction (CBDRR) and local-level planning has been strengthened to ensure that policies and strategies are tailored to the specific needs and vulnerabilities of local populations.

As urbanization has accelerated, particularly in developing countries, DRR governance has increasingly focused on building resilient cities. Programs like the Making Cities Resilient Campaign have encouraged urban planners to incorporate disaster resilience into infrastructure development, land use planning, and housing. The role of the private sector in DRR governance has expanded significantly, particularly in building resilient infrastructure, providing insurance solutions, and developing risk financing mechanisms. Governments have increasingly collaborated with businesses to reduce economic vulnerability to disasters through public–private partnerships (Izumi and Shaw 2015).

The intersection between disaster risk reduction and climate change adaptation has gained prominence. DRR governance now incorporates climate risk assessments, with strategies designed to address the increasing frequency and intensity of weather-related disasters, such as storms, floods, and droughts, which are exacerbated by climate change. In summary, DRR governance has shifted from a response-centric approach to one that emphasizes proactive risk management, scientific integration, local engagement, inclusivity, and international collaboration. This evolution reflects the growing recognition that reducing disaster risks requires coordinated, multi-sectoral, and multi-level efforts to ensure sustainable and resilient development.

1.3.2 Education

The Hyogo Framework emphasized the importance of public awareness and education to foster a culture of safety and resilience at all levels. Education during this period focused on building awareness about disaster risks and preparedness, with governments encouraged to integrate DRR into school curricula. Over the years, many countries have incorporated DRR into formal education curricula, recognizing that children and youth are critical to building a culture of resilience. By 2024, subjects related to DRR were being taught in primary and secondary schools in various regions, with lessons on emergency preparedness, climate change, and sustainable development. Governments and international organizations like UNESCO and UNICEF developed specific curricular materials and guides for teachers to ensure that students are educated on disaster preparedness, safety measures, and environmental stewardship (Shaw et al. 2021).

Beyond schools, DRR education became embedded in informal education settings like media campaigns, public seminars, and digital platforms. Social media, mobile apps, and online platforms increasingly served as tools to disseminate DRR knowledge, reaching a wider audience. Initiatives like the UNISDR's Global Campaign on Disaster Risk Reduction and World Tsunami Awareness Day promoted community-wide learning through events, storytelling, and information dissemination. These campaigns were designed to create a culture of preparedness beyond the classroom, making disaster literacy accessible to the general public.

The importance of local knowledge and indigenous practices in disaster management became more widely acknowledged in DRR education. Educational programs increasingly integrated traditional wisdom with modern science, teaching communities how to blend indigenous coping mechanisms with formal risk reduction strategies. This shift promoted cultural sensitivity and acknowledged the role of indigenous communities in managing risks through sustainable practices, fostering resilience in the face of natural hazards.

The post Indian Ocean Tsunami period saw the expansion of higher education programs focused on disaster management and risk reduction. Universities around the world established programs in disaster science, risk management, and emergency response, training a new generation of professionals equipped with the skills to reduce disaster risks and manage emergencies (Shaw et al. 2017). Many countries introduced certification programs for disaster risk managers, emergency responders, and urban planners. This professionalization of DRR education ensured that specialists across various sectors, from health to engineering, have the knowledge to incorporate risk reduction principles into their work. Post COVID-19 has also changed significantly the ways universities and higher education institutions perceive the risk with significant changes in pre-disaster risk management approaches (Izumi and Shaw 2015).

1.3.3 Technology

There has been significant advancement in technology in the last 20 years. Satellite technology advanced greatly between 2005 and 2024, offering high-resolution imagery and real-time data to monitor hazards such as earthquakes, floods, hurricanes, and wildfires. Earth observation satellites like Landsat, Copernicus, and NOAA satellites have become key tools for tracking environmental changes and disaster impacts. This has improved the Early Warning System globally and locally (Sakurai and Shaw 2021). Unmanned Aerial Vehicles (UAVs) or drones revolutionized post-disaster assessment by providing aerial imagery in real-time, especially in areas that are hard to access after disasters. Drones were increasingly used for disaster mapping, search and rescue, and delivering supplies in emergency situations. The 2011 East Japan Earthquake and Tsunami was one of the turning points of the drone technology and its usage.

The ability to gather and analyze large datasets in real-time has improved disaster forecasting and risk assessment. Technologies like big data analytics helped identify patterns in weather events, seismic activity, and human behavior, aiding early warning systems and preparedness strategies. Advanced computational models, powered by big data, have allowed researchers to simulate disaster scenarios and predict their impacts with greater accuracy. This has enabled governments and agencies to develop more effective risk reduction and response plans.

Artificial Intelligence (AI) has played a transformative role in analyzing vast amounts of data to identify early warning signs of disasters. For example, AI models can detect early seismic activities or climate patterns that may indicate approaching storms, allowing for faster, more accurate alerts. Machine Learning for Risk Assessment: Machine learning algorithms are used to predict the likelihood of hazards in specific regions by analyzing historical data and real-time environmental information. This has helped improve flood prediction models, wildfire risk maps, and earthquake impact assessments (Kanbara et al. 2022).

The advancement of technology from 2005 to 2024 revolutionized DRR education also. Interactive digital platforms, online simulations, virtual reality (VR), and Geographic Information Systems (GIS) became widely used to teach DRR concepts, offering immersive and practical learning experiences. E-learning platforms and mobile apps provide accessible DRR education to remote communities and professionals in need of continuous learning. Mobile alerts, interactive games, and digital learning modules facilitated the rapid dissemination of disaster information and training.

1.4 Evolution of Different Other Concepts

In this complex and dynamic risk landscape, different new and emerging risks will evolve, and to cope with these, many new concepts and approaches are also evolving. Following are the ten key concepts which evolved strongly over the last 20 years.

Systemic risk: The systemic risk approach in disaster risk reduction focuses on understanding how interconnected risks across various sectors and systems can create cascading effects during disasters. It emphasizes the need to address complex vulnerabilities, interdependencies, and feedback loops, ensuring a comprehensive, multi-dimensional strategy for managing risks and reducing their widespread impacts.

From risk to resilience: While 20 years hack, we used to focus on disaster reduction, the concept changed to disaster risk reduction. Focus on risk has been one of the key aspects over several years. In the current context, it has evolved to resilience building.

Ecosystem-based resilience: Ecosystem-based resilience for disaster risk reduction involves using natural ecosystems, like wetlands, forests, and mangroves, to mitigate disaster risks. These ecosystems act as buffers against hazards, such as floods and storms, while enhancing biodiversity and supporting livelihoods. This approach promotes sustainable, nature-based solutions to build community resilience against disasters.

Resilient infrastructure (hard and soft): Significant focus has been made on resilient infrastructures, especially to protect the critical infrastructures like schools, hospitals, local government offices, and critical transport and road networks, and vital infrastructures like electricity, water support and telecommunication system. Soft infrastructure has also got significant attention over the years, which is critical to support the hard infrastructures.

Pre-disaster recovery planning: Disasters are often considered as used or misused development opportunities. However, in post-disaster situations needs a balance between speed and quality. To enhance this balance, the new concept of pre-disaster recovery planning has evolved, which prepares recovery plans through scenario-based approaches of disaster damages. This helps significantly in streamlining recovery plans in the post-disaster situation.

Impact based investment: Disaster risk reduction investment is critical to reduce loss and damages due to disasters. A new concept of impact-based investment has evolved over the past several years, where the investment is directly related to tangible impacts, and has been practiced by many development agencies globally.

Urban rural collective resilience: While urban resilience has been a common concept during early 2000, the resource dependence of urban and rural areas, the breaking of the supply chain during the disaster has prompted the evolution of the new concept of urban rural collective resilience. The concept has promoted the use of urban resources to enhance rural ecosystem and resources to ensure the supply chain resilience.

All Hazards Approach: All Hazards Approach (AHA) has become popular in recent years with increasing complex and compound disasters, where two or three types of disasters like natural, biological, radiological, cyber, etc., are happening together. The concept of NATECH (natural hazard induced technological disaster) has become also critical and specific guidelines are currently made for addressing these types of complex/cascading emergencies.

Early Warning for All: While there has been significant development in the early warning system (EWS), however, the last mile connectivity has always been a challenge. To make early warning inclusive and accessible to everyone, the United Nations has started the concept of "*no one should be left behind*." Early Warning For All is the United Nations Secretary General's key message and appeal to make everyone covered in the EWS.

Stakeholder's roles: There have been significant changes in the roles of the stakeholders from advising to co-designing and co-implementing. Private sector has become a core stakeholder from their CSR (corporate social responsibility) to making risk reduction as their core business. Civil society and science and technology stakeholders have also been part of the decision making, and co-delivery process.

1.5 About the Book

The book is a modest attempt to make an overview of post Indian Tsunami development over the last 20 years. The book consists of 22 chapters. After this introduction chapter. Part 1 covers eleven key cross-cutting issues, and their evolution over the last 20 years. The section covers risk governance, all hazards approach, early warning system, systemic risk consideration, planetary to individual health connectivity, mechanism of single/multi hazards, early warning system, science technology and innovation, ecosystem-based approaches, stakeholders' roles, housing issues, and community-based disaster risk reduction. The chapters focus not only on the 2004 tsunami affected areas but cover regional and global overviews. Part 2 covers the evolution of the disaster risk reduction regimes in affected countries with specific relation to legislation, plans and policies, early warning systems, and related technology development in the field of disaster risk reduction, response and recovery process over the last 20 years. This section also covers regional approaches in ASEAN region.

1.6 Postscript

The year 2024 marks 20 years from the Indian Ocean Tsunami. The year 2025 marks 30 years from the Great Hanshin Awaji Earthquake (popularly known as Kobe Earthquake) of Japan. The year 2025 also marks 10 years from the Sendai Framework on Disaster Risk Reduction. There will be another 5 years to go toward the end of the current global frameworks, including SFDRR, SDGs, and Paris Agreement. Prolonged three years of COVID-19 (2020, 2021, and 2022) have posed significant development challenges, and most of the Sendai, Paris, and SDGs targets need to be rethought, since many of them cannot be achieved in the prescribed time frame. We also do not know what are the new risks waiting in future in the complex risk landscape. Thus, adaptive governance is the key to address the risk reduction process. Science and technology play an important role in the adaptive governance. The key to future science and technology would be inclusive as well as affordable. As mentioned earlier, the future risk reduction needs to have one slogan "*no one should be left behind*".

References

Izumi T, Shaw R (2015) Disaster management and private sectors: challenges and potentials. Springer, p 342

Kanbara S, Shaw R, Kato N, Miyazaki H, Morita A (2022) Society 5.0, digital transformation and disasters: Past, present and future. Springer, p 217

Sakurai M, Shaw R (2021) Emerging technologies for disaster resilience: Practical cases and theories. Springer, p 260

Shaw R, Prabhakar SVRK, Chiba Y (2016) Policy brief: SDGs, DRR and CCA: Potential for strengthening inter-linkages. Available from and accessed on 24th of September 2024. https://pub.iges.or.jp/pub/sdgs-drr-and-cca-potential-strengthening-inter

Shaw R, Shiwaku K, Izumi T (2017) Science and technology in disaster risk reduction: Potentials and challenges. Elsevier Academic Press, p 525

Shaw R, Sakurai A, Oikawa Y (2021) New realization of disaster risk reduction education in the context of a global pandemic: Lessons from Japan. Int J Disaster Risk Sci. https://doi.org/10.1007/s13753-021-00337-7

Shaw R (2014) Disaster recovery: Used or mis-used development opportunities. Springer Publisher, p 431

UN (2015c) Adoption of the Paris agreement, UNFCCC COP 21st Session, Paris, France, p 32

UN (2015a) Sendai framework for disaster risk reduction 2015–2030, UN ISDR document, 25 pages, Sendai, Japan

UN (2015b) Transforming our world: the 2030 agenda for sustainable development, UN General Assembly Resolution, 35 pages, NY, USA

UNDRR (2015) United Nations Office for Disaster Risk Reduction (UNDRR). Sendai framework for disaster risk reduction 2015–2030. Available from accessed on 23rd of September 2024. https://www.undrr.org/publication/sendai-framework-disaster-risk-reduction-2015-2030

UNISDR (2005) Hyogo framework for action, United nations international strategies for disaster reduction. Available from accessed on 23rd of September 2024. https://www.unisdr.org/2005/wcdr/intergover/official-doc/L-docs/Hyogo-framework-for-action-english.pdf

WB (2024) World bank open data. Available from accessed on 23rd of September 2024. https://data.worldbank.org

Part I
Cross Cutting Issues

Chapter 2
Transforming Disaster Risk Governance: Towards More Collaborative and Inclusive Approach for Tsunami Risk Reduction

Laely Nurhidayah, Riyanti Djalante, Agus Suntoro, and Ismeti

Abstract Two decades after the 2004 Indian Ocean Earthquake and Tsunami, one of the deadliest in modern history, the commitment and global response to address disaster risk reduction and building resilience have been rapidly developed through two major international instruments, The Hyogo Framework of Action (HFA) 2005–2015 and the Sendai Framework for Disaster Risk Reduction 2015–2030. One of the priority actions in the Sendai Framework is strengthening disaster risk governance at global regional, national, and local levels. Key lessons learned at global and regional levels are: a greater need to foster and transform risk governance towards more cooperation and collaboration; and working across silos—especially in regional and local contexts—to foster more efficient planning; address climate change; protect biodiversity; eradicate poverty; improve food, nutrition, health, and education; implement ecosystem-based approaches in disaster risk reduction; and linking technology to people. In addition, it is urgently needed to promote initiatives towards more inclusive disaster risk management policies for tsunami preparedness and response at global and regional levels. In the ASEAN region, it is reported based on pre-disaster analysis of existing data that women facing discrimination, such as the poorest women of minority ethnicities and those living in rural areas face the greatest obstacles to achieve the SDGs by 2030. ASEAN has established a regional framework on protection gender and inclusion in disaster management 2021–2025. However, not all ASEAN member states have integrated protection, gender, and inclusion in Disaster Management including tsunami preparedness. The aim of this chapter is to examine global and regional initiatives for risk governance toward strengthened tsunami risk reduction. The chapter will employ grey critical literature analysis. This chapter will provide insight on strengthening collaborative and inclusive tsunami

L. Nurhidayah (✉) · A. Suntoro · Ismeti
Research Center for Law, National Research and Innovation Agency, Jakarta, Indonesia
e-mail: lael003@brin.go.id

R. Djalante
The ASEAN Secretariat, Jakarta, Indonesia

R. Shaw et al. (eds.), *Two Decades from the Indian Ocean Tsunami*, Disaster Risk Reduction, https://doi.org/10.1007/978-981-96-2669-4_2

risk management and institutions globally and regionally and its implications at the local level.

Keywords Risk governance · Tsunami · Inclusive DRR · Multi-stakeholder · Collaboration

2.1 Introduction

Two decades after the 2004 Indian Ocean Earthquake and Tsunami, one of the deadliest in modern history, the commitment and global response to address disaster risk reduction and building resilience have been rapidly developed through two major international instruments, The Hyogo Framework of Action (HFA) 2005–2015 and the Sendai Framework for Disaster Risk Reduction 2015–2030. The expected outcome of Hyogo Framework of Action (HFA) 2005–2015 is the substantial reduction of disaster losses in lives and in social, economic, and environmental assets of communities and countries (UNISDR 2005). To achieve these outcomes countries need to have strategic goals to integrate disaster risk reduction into sustainable development policies and planning. In addition, countries need to develop and strengthen institutional mechanisms and capacities to build resilience to hazards, as well as incorporations of risk reduction approaches into the implementation of emergency preparedness, response, and recovery programs. The cross-cutting issues that have been highlighted in HFA include: multi-hazards approach, gender perspective and cultural diversity, community and volunteer participation and capacity building, and technology transfer. Similarly with HFA, the Sendai Framework goals are to achieve resilience and reduce disaster risk with priorities of action including: understanding risk, strengthening risk governance, investing for resilience, and enhancing preparedness and recovery. To achieve these goals, the Sendai framework calls for the implementation of integrated and inclusive approaches of all sectors include: "economic, structural, legal, social, health, cultural, educational, environmental, technological, political and institutional measures that prevent and reduce hazard exposure and vulnerability to disaster, increase preparedness for response and recovery, and thus strengthen resilience" (United Nation Office for Disaster Risk Reduction (UNISDR) 2023). The Sendai framework highlights the need to focus actions at all levels including: local, national, regional, and global levels.

Strengthening Disaster Risk Governance is crucial——particularly as measures to reduce the damage and losses from the impacts of disaster include institutions and their relevant systems, mechanism, policy, and legal frameworks. Other arrangements to guide, coordinate, and oversee, disaster risk reduction and related areas are needed (UNDRR 2024). At the domestic level, disaster instruments supported by a network of laws, policies, and procedures play an important role in providing architecture for DRM. Disaster risk reduction has shifted the paradigm from response and recovery to prevention and preparedness (Jones et al. 2014). This shift has been recognized in ASEAN. The ASEAN Framework on Anticipatory Action in Disaster Management represents a significant shift in how the Association of Southeast Asian Nations (ASEAN) plans to tackle natural hazard-related disasters. The region is moving towards mechanisms for preparedness and response that can operate based on the One ASEAN One Response principles of speed, scale, and solidarity (ASEAN 2022). ASEAN recognizes that planning and prevention need to be multi-hazards based and multi-sectoral. Furthermore, land use law, building codes, and environmental protection are prevention measures in disaster management (Srinivas and Nakagawa 2008). However, in low-income countries, environmental protection receives comparatively low priority in national policies and development plans. According to scientists, coastal vegetation has been widely recognized as a natural method to reduce the energy of tsunami waves (Tanaka 2009). The concept of using nature in disaster risk reduction has been promoted around the world. Ecosystem Disaster Risk Reduction (Eco-DRR) is the concept of reducing the risk to natural hazards by avoiding development in disaster-prone areas, as well as by using healthy ecosystems as buffers to protect people's lives and properties (Ministry of the Environment 2016).

Furthermore, strengthening measures that reduce poverty, increase access to food security, and improve health and education are critical policy options that can significantly reduce the vulnerability of communities to disaster. The poor are also usually the most vulnerable to disasters and have the lowest capacity to recover from them. In addition, people affected by the tsunami face severe food security problems in the short and long-term because parents and relatives have been lost, livelihood assets have been destroyed, and sources of income no longer exist. It is evident that tsunami disasters hit fisheries and agriculture sectors in a country affected. For example, in Aceh tsunami has long-term effects and damage which threatened the continued use of coastal and fisheries resources.

While global and regional initiatives on DRR are crucial as policy guidance, effective DRR relies on clear and actionable communication at a local level. Djalante et al. suggested that four characteristics of influence disaster resilience include: polycentric and multi-layers institutions, participation and collaboration, self-organization and networks, and learning and innovation (Djalante and Holley 2013). In addition, DRR has recognized the importance of full, equal, and effective inclusion in risk management and recovery planning. Specifically, the Sendai Framework has called for special attention to "women and girls, people with disabilities rural populations, Indigenous peoples, ethnic and linguistic minorities, migrants, displaced individuals, gender and sexual minorities, youth, and older persons". The United Nations International Strategy for Disaster Reduction (UNISDR) explicitly proposed the establishment of a social system living with risk and stressed the promotion of regional sustainable development by building the capacity of communities to mitigate risk (Shi et al. 2010).

This chapter will examine risk governance initiatives at global and regional ASEAN. The aim is to strengthen tsunami disaster risk governance in the future at all levels. Risk governance is key to establish social resilience. This chapter is divided into five sections. First section is highlighted the introduction, second is exploring the risk governance theory, third section is discussing the Transforming

Risk Governance: collaboration and cooperation and working across silos as important measures to effectively strengthen disaster risk reduction and fourth, is inclusive tsunami disaster management as to increase effectively policy to integrate those most at risk of being left behind into the decision-making process concerning programming and planning of disaster risk reduction interventions.

2.2 Risk Governance

2.2.1 Risk Governance Theory

According to Sociological Theories and Disaster Studies, states of disaster involve physical damage that disrupts the routines of daily life (micro level) and simultaneously disrupts the social structures produced and reproduced by those routines (macro level) (Stallings 1997). The disruption must be impactful enough to prompt a response by state elements that leads to the development of a series of propositions about the process of routinization (Stallings 1997). Therefore, mitigation through risk reduction is the most rational option to minimize impact and accelerate recovery. Through analysis of risk, it is possible to predict the magnitude, potential occurrence, frequency, speed of onset, and spatial extent of harmful natural events or processes (called hazards) to minimize damage, loss, death, and injury (Wisner et al. 2015).

The urgency of disaster risk governance emerged at the beginning of the millennium. It then developed through the European Commission, which discussed the Science and Society Plan as a response to the macro-societal challenges. Subsequently, the International Risk Governance Council (IRGC)——formed in 2003——became a milestone in strengthening risk governance and influenced the increasing awareness of the aforementioned macro-economic urgencies; expanding to: managing external and environmental risks around waste, nuclear, climate change, and pandemics. IRGC, in its Introduction to the IRGC Risk Governance Framework, defines risk governance as the application of governance principles to identify, assess, manage, evaluate, and communicate risks in the context of multiple values and distributed authority. Furthermore, this governance emphasizes the stage of processing relevant risk information collected, analysis, and the results are communicated, as well as management in decision-making (IRGC 2017).

Risk governance is adapted from business models and described as the entire collaborative process of interaction and decision-making among stakeholders involved in collective issues related to risk control in the internal and external environment. This governance aims to adapt the business risk model to maintain and create sustainable value. Risk governance applies governance principles to identifying, assessing, managing, evaluating, and communicating risks in the context of plural values and distributed authority. It includes all essential actors involved, considering their rules, conventions, and processes. It is thus concerned with how relevant risk information is collected, analyzed, understood, and communicated and how management decisions are taken and communicated (Stein et al. 2019).

Van Asselt and Renn assess that governance will support highly contextual risk management practices. They will correlate with paradigm shift efforts that help professionals and policymakers get used to conducting more comprehensive risk assessments. Risk governance will be based on three main principles, namely (1) communication and inclusion; the first principle is communication, so risk governance must be reciprocal (two-way) between all stakeholders involved. Inclusion can be done in various ways, whether through deliberation, open forums, joint training, drafting negotiated regulations or policies, mediation, involvement of scientists, vulnerable groups and others; (2) Integration, in the context of integration, the most fundamental meaning is to support collaboration and synthesize all expertise, knowledge, experience and logical predictions; and (3) reflection, reflection relates to evaluation and review by institutions, organizations or communities regarding risk governance that has been carried out to continue making improvements (Van Asselt & Renn, 2011). Meanwhile, IRGC in 2005 suggested a risk governance process model based on four phases, namely (a) pre-assessment, (b) assessment, (c) characterization and evaluation, and (d) risk management (Renn et al. 2011).

Risk governance is also tied to the principles of openness and inclusiveness. In several of its provisions, the Sendai Framework emphasizes the urgency of involvement and partnership of the entire community. Disaster risk reduction requires empowerment, meaningful and inclusive participation, accessibility, non-discrimination, and exceptional attention to the most affected groups, especially low-income people. Good governance must also integrate into all policies and practices: gender, age, disability, and cultural perspectives, including promoting women's and youth leadership (UNISDR 2015). The principle of inclusiveness and gender insight has also become a standard for ASEAN countries since 2010, which is stated in point five of the AADMER 2010–2015 program, namely recognizing the unique needs and potential contributions of various stakeholder groups, especially children, the elderly, women, and people with disabilities. These subjects will contribute to disaster risk reduction, emergency response, and recovery, and there is a need to include gender, human and social security perspectives, equality issues, and transparency and accountability in the implementation and monitoring of work programs (Sari 2022).

2.2.2 *Transforming Risk Governance: Collaboration and Cooperation and Working Across Silos*

The Indian Ocean Tsunami in 2004 revealed that disaster could be happening across national boundaries. Tsunami impacts 2004 shows that disaster is a global problem that requires global and regional effort, and collaboration and initiatives to address this issue. This cooperation is critical to saving lives given the transboundary nature of tsunamis and the need to support developing countries build their resilience. The 2030 Agenda for Sustainable Development recognizes and reaffirms the urgent need to reduce the risk of disasters. Disaster risk reduction (DRR) is an integral part of social and economic development, it is essential and pre-requisite for sustainable development. There are many documents and global initiatives in disaster risk reduction. The Yokohama Strategy and Plan of Action for a Safer World (1994), as the first major international framework for disaster risk reduction, recognized the interrelation between sustainable development and DRR. Other Important International documents on disaster risk reduction are the Hyogo Framework of Action and Sendai Framework. In addition, the latest international document Paris Agreement also recognizes risk reduction governance. Article 8 of the Paris Agreement Parties recognize "the importance of minimizing and addressing loss and damage associated with the adverse effects of climate change, including extreme weather events and slow onset events, and the role of sustainable development in lowering the risk of loss and damage."

In response to the increasing threat of future disasters innovative and new transformations of risk governance are needed. The characteristics of disasters which often change, especially in the future with the large amount of natural damage that occurs do not rule out the possibility that there will be natural disasters that are bigger and more massive than before (Djalante and Lassa 2019) which causes the government and even the world to continue to evaluate and update strategies towards preventing and handling natural disasters in the future, as an effort to build a culture of resilience and reduce disaster risk. This of course needs to be done with a broad and consistent approach to the factors supporting the occurrence of natural disasters. (Aitsi-Selmi et al. 2016) Many researchers are calling for more Innovative and integrated governance approaches in dealing with complex problems posed by disasters (Djalante et al. 2011; Renn 2008; Fung 2006; Paton et al. 2010). Concepts of risk governance have been developed at the global level to deliver more effective DRR. Multi institutional Arrangement and Polycentric Framework and adaptive governance are the concepts used toward improving tsunami disaster risk reduction. An adaptive governance (AG) approach can be developed as an alternative method of managing complex social-environmental problems including disasters (Djalante et al. 2011). However, Denis Chang Seng suggests that "the polycentric and multi-layered institutions and structures synchronized with the decentralized political-administrative system are an ideal governance architecture for improved performance and for building national resilience to local and transboundary multi-hazard risks and disaster" (Chang-Seng 2012). In addition, polycentric multi-layered architecture and top-down technocratic Tsunami early warning system (TEWS) is not completely adequate if applied directly to local field due to different characteristics, problems of fit, adaptability, institutional diversity, and norms at that local level. Therefore, an ideal approach is a mixture model of the Early Warning System (EWS) process consisting of the local people-centered-adaptive approach and the national technocratic system approach to address the challenging issues of tsunami resilience (Chang-Seng 2012).

The Global Platform for Disaster Risk Reduction in Bali 2022 has called for transformation of risk governance mechanisms which emphasized that management of risk is a shared responsibility across sectors, systems, scales, and borders (UNDRR 2022b). Examples show that working closely across ministries and departments at all levels can help governments break institutional silos. To achieve the 2030 Agenda, disaster risk reduction must be integrated at the core of development and finance policies, legislation, and plans. Multi-stakeholder platforms are a way to implement adaptive governance (Djalante et al. 2011). Many governance approaches in developing countries are usually working in silos. For example, in coastal areas there are so many interests from many stakeholders, including government, industries, and local communities on land use planning, that it is difficult to address urbanization and people who live in coastal areas. Coastal areas are very susceptible to disasters including tsunamis. Urbanization has caused illegal settlement, especially in coastal areas in some urban or mega cities such as Jakarta. Informal settlements are particularly vulnerable to the impacts of climate-related risks including tsunamis (Bouwer et al. 2021). In addition, it is difficult for the relocation of local communities in the areas prone to disaster as they do not have alternative places to live, and their livelihood depends on these coastal areas. It is very challenging to consider a holistic and integrated approach in building disaster risk resilience in the face of rapid urban expansion. Land zoning and building and construction codes should also consider climate and disaster risk to ensure that new developments do not increase exposure to natural hazards Learning from Japan, there is a Japanese national law (2012) that regulated zoning to prevent tsunami disasters. It regulated that the areas expected to flood at a depth of 2 m or less are suitable for residential use, but areas expected to flood at a depth of more than 2 m are designated as non-habitable areas in the application of disaster hazard zone regulations. Land use in the areas of expected flood depth of more than 2 m is limited to industrial uses or parks; residential use is not permitted ("Tsunami Disaster Reduction Levels"). In such cases, the residential functions of villages and towns must be moved to higher ground or to more inland locations.

At the global level, UNDRR plays a crucial role in global and regional advocacy and communication initiatives, including bringing case studies, lessons learned, and expertise from the ground level to the attention of global and regional policy processes (UNDRR 2024). At the focal point of Disaster Risk Reduction in the United Nations systems, UNDRR supports the implementation of Sendai Framework for Disaster Reduction through systematic and structured engagement with partnerships and whole-of-society approaches. This partnership is the heart of effective disaster risk reduction (UNDRR 2024). Collaboration between governments, the United Nations, and other international organizations and stakeholder groups at large are necessary as collective action to build resilience and promote risk-informed decision-making and investment (UNDRR 2024). Howes et al. suggest that five key reforms are proposed to improve integrated disaster risk management include: developing a shared policy vision; adopting multi-level planning; integrating legislation; networking organizations; and establishing cooperative funding (Howes et al. 2015). As disaster risk landscape is becoming more complex and challenging. There is urgent needs of collaboration and holistic approach. Susana Djuangco, Director of the Philippines Office of Civil Defense, stated that disaster management should be everyone's business," Instead of working in silos, inter-cooperativeness and coordination should be the essence of all our DRR initiatives" (UNDRR 2022a).

At international level, besides, UNDRR which provides multi-hazards approach to understanding risk (Kruczkiewicz et al. 2021), other risk assessment framework exist across UN System to bolster coordination and cooperation across multiple intergovernmental organization, non-governmental organization, national, local governments (Kruczkiewicz et al. 2021). One of those assessments is Multi-sectoral Initial Rapid Assessment (MIRA). A MIRA is an inter-agency needs assessment and analysis process that is carried out within the first three days of a disaster under the guidance of the Humanitarian Coordinator. It forms the basis for the humanitarian country team's (HCT) joint strategic plan for emergency response. At ASEAN regional level, there are ASEAN risk and vulnerability Assessment Guidelines. RVA is recognized by the United Nations International Strategy for Disaster Reduction (UNISDR) as one of the most important elements of long-term Disaster Risk Reduction (DRR) and Disaster Risk Management (DRM). In addition, ASEAN also introduces the ASEAN Risk Monitor and Disaster Management Review (ARMOR) a multifaceted insight that encompasses culture, education, the use of data and technology, as well as government policies to highlight the steps needed to achieve sustainable disaster resilience in the region.

2.3 Inclusive Tsunami Disaster Management

A lot has been learned since the Indian Ocean tsunami. Natural disaster can exacerbate the vulnerabilities of pre-existing of population already at risk including poor, women, elderly, youth, and marginalized groups (The ASEAN Secretariat 2021). Therefore, resilience to disaster initiatives increases towards a more gender-responsive and inclusive approach. Disaster governance is often seen as a technocratic domain, that brings resources and knowledge together to work in the prevention, relief, and recovery sectors (Hilhorst et al. 2020). There are a global consensus toward a shift to inclusive disaster governance (Hilhorst et al. 2020).

An Inclusive Disaster Risk Management (DRM) includes a combination of top-down and bottom-up approach which combines disaster relief, emergency response, and recovery with Disaster Risk Reduction (DRR) strategies such as prevention, preparation, and mitigation to reduce vulnerabilities of those who are 'most at risk' during a disaster, as well as to increase their capacity. Political Declaration of High Level Review of the implementation of Sendai Framework also highlights the promotion of "the full, equal, meaningful and inclusive participation and contribution of women, older persons, persons with disabilities, migrants, Indigenous Peoples, and local communities in processes related to disaster risk reduction and in the design and implementation of disaster risk reduction policies, plans, and programs" (UN Resolution 2023). It is evident that women, children, elders, and poor people suffered the most because of the tsunami and need to be supported further during the reconstruction phase. In addition, older people, often neglected in normal situations are even more vulnerable in disasters. Despite there are international protocols such as the Madrid International Plan of Action on Ageing and the Sphere guidelines which foster society for all ages based on research from HelpAge confirms that in the tsunami they were not specifically targeted and, in some cases, the relief effort discriminated against them. In addition, in many cases, older men and women were unable to access health care, food, and cash due to discrimination, and a lack of information or support mechanisms.

There are many initiatives in mainstreaming gender and inclusion in disaster management at global and regional levels. Investing in gender-responsive and inclusive disaster risk reduction and resilience building is essential to achieve the Sustainable Development Goals, especially SDG 5 empower women and girls. In disaster management, the Sendai Framework emphasizes the importance of inclusivity and accessibility in disaster risk reduction (DRR) in order to increase the effectiveness of DRR. Furthermore, International donor such as the Global Facility for Disaster Reduction and Recovery (GFDRR) promote gender inclusive approach in disaster management. GFDRR is a multi-donor partnership that supports low and middle-income countries to understand, manage, and reduce their risks from natural hazards and climate change. In addition, almost all South Asian Countries have integrated social inclusion considerations in their budget cycles. The Sendai Framework for Disaster Risk Reduction (SFDRR) and the ASEAN strategy involve specific actions and principles designed to ensure comprehensive and equitable disaster preparedness and response and inclusive management of natural disasters. In addition, ASEAN has established a regional framework on Protection, Gender, in Disaster Management 2015–2025 (ARF-PGI) which is a significant improvement to the recognition of Gender Equality, Disability, and Social Inclusion (GEDSI) principles. ARF-PGI is in line with One ASEAN One Response which aims to share a common vision for promoting gender inclusion in disaster management in the ASEAN region (ARF-PGI 2021). The Framework also aims to support the implementation of the ASEAN Agreement on Disaster Management and Emergency Response (AADMER) work program 2021–2025 and other regional declarations and plans. The main activities that support gender protection and inclusion in the ASEAN region include: (i) Consolidating regional commitments on target actions for achieving inclusive disaster management across sectors; (ii) Identifying entry points for cooperation between National Disaster Management Organizations (NDMOs) and other organizations working similar issues; and (iii) Supporting ASEAN Member States (AMS) in setting priorities, indicators, and targets for measuring progress and baseline data and indicator in protection, gender, and inclusion (PGI) at the national level and regional level (The ASEAN Secretariat 2021). In addition, reaffirms the commitments of individuals AMS to CEDAW, the CRC, and the CRPD through The ASEAN Declaration on Gender-Responsive Implementation of the ASEAN Community Vision 2025 and Sustainable Development Goals, adopted in 2015. ASEAN also put importance on disabilities with the adoption of the ASEAN Enabling Masterplan 2021–2025: Mainstreaming the Rights of Persons with Disabilities (ASEAN Secretariat 2019).

The recognition of Gender Based Violence (GBV) in disasters as an issue that needs to be tackled by ASEAN has been shown in recent research led by the ASEAN sectoral bodies which highlighted the urgency of AMS acting to prevent and respond to both existing efforts and key gaps in not only addressing women's vulnerabilities, but also promoting women roles as leaders and agents of change in disaster management at both national and regional levels (ARF-PGI 2021). It is evident that working at ground level including engaging with community-based women's organizations, identifying women leaders, allowing open participation in capacity building, and engaging with local governments has had more immediate and huge impacts in empowering communities particularly women and the most vulnerable for disaster risk management (Gupta and Leung 2011). Incorporating climate and disaster resilience in special social protection programs will help strengthen the capacity of women, poor households, and vulnerable populations in responding to, and recovering from, disasters (ADB 2021).

Inclusive disaster management also recognizes knowledge of traditional communities (Traditional Knowledge) in recognizing symptoms of natural disasters (Nugroho 2018). UNDRR has also reported the emerging progress and challenges in integrating local, indigenous, traditional knowledge and practices in regions (UNDRR 2023a, b). Traditional people who have lived in the area for hundreds of years, certainly have understanding the geographical location of the region and the area which can be used as a reference in seeing existing symptoms of disasters. Inclusive Tsunami Disaster Management has also begun to be implemented by the Indonesian government, in this case, the regional government, by involving the community and using local wisdom, adopting traditional residential building models (Banuambaso/Sapo Oge/Banua Magau, Kataba, Fecjai Kinjai), in rebuilding houses. residents affected by the tsunami (Suntoro et al. 2023). The housing development in the coastal area was built with its unique characteristics in collaboration with the Ministry of Public Works and Public Housing (PUPR) to rebuild housing in Tsunami-prone areas along the coast and river estuaries (Fig. 2.1).

One of the permanent residential housing developments built by the government at the site of the 2018 tsunami on the edge of Taman Ria beach, Palu City, Central Sulawesi, Indonesia (Source: Ismeti)

One of the community houses that still has a local feel (Rumah Panggung) in Palu City, Central Sulawesi, Indonesia (Source: Ismeti)

Fig. 2.1 Post-tsunami permanent housing with local wisdom

Sima sami Bahous stated that one of the greatest strengths of Sendai Framework is the importance of local contexts and of communities shaping their own responses (UNDRR 2023a, b). UN Secretary General stated that adopting "a whole-of-society approach" that utilizes local knowledge and capacity to agree on such measures will dramatically scale up the effectiveness of measures to reduce disaster risk (UNDRR 2023a, b).

2.4 Toward Strengthening Tsunami Disaster Risk Governance Initiative

Disaster risk Reduction is a never ending task. Every time disasters like tsunamis strike, there are lessons to be learned and measures to be strengthened. Key lessons learned at global and regional levels are: a greater need for fostering and transforming risk governance towards more cooperation and collaboration working across silo, especially in ASEAN region; national and local approaches to foster more efficient planning; address climate change; biodiversity; poverty eradication; improving food, nutrition, health, education; implementing ecosystem-based approach in disaster risk reduction, and linking technology to people. Supporting a multi-stakeholder approach, including the voices of citizens, community groups, non-governmental organizations, academia, and the private sector are important policy in effectively building cooperation and collaboration. Kazuko Kohri stated that Sendai has adopted best practices and appropriate technologies to reduce risks in the face of future hazards (UNDRR 2023a, b).

There is a broader shift of risk governance at global level, a shift from monolithic structures of global governance to a wide range of organizations or polycentric governance and transforming risk governance toward working across silo. The importance of getting the global–local interface to work in tandem has been highlighted by recent hazard events, such as the 2011 Tohoku Earthquake and Tsunami, and the Indian Ocean Tsunami 2004. From a broad view of global and local shifts, a strategic role is becoming clearer at the national level for enhancing the relationships between the global and local levels (Jensen et al. 2015). In addition, the approach of risk governance is shifting to an inclusion and gender sensitive approach as many funding and donor of research projects such as Koneksi Research grant now emphasizes Gendered Disabilities and Social Inclusion (GEDSI) principle in running the projects. However, there are still gaps in the implementation of this global and regional risk governance at local level. It is stated that weak governance is linked to many other risk drivers such as poverty and inequality, poorly planned urban development, and globalized economic development (UNDRR 2006). To be most effective in reducing disaster risk, an integrated systems approach to governance, a strong coordination across sectors, and a delegation of responsibilities to the local level are needed (UNDRR 2024). In developing countries, sectoral egos, inadequate laws and regulations on disaster management, and inadequate funding at local level hinder the effective and fast response to disasters. Therefore, there is an urgent need to make a reform and to propose comprehensive laws and policies on disaster including addressing disaster as a result of climate change. Priority for Action 1 of the HFA called for the integration of disaster risk reduction into poverty reduction and other development strategies based on strong political determination by governments (UNDRR 2006). Unfortunately, many of these commitments in law and policy have not been translated into real priorities and investments (UNDRR 2006). Many local governments do not have adequate funding to respond to huge disaster events such as tsunami. Therefore, building the appropriate level of networked global-regional–local capacity will be critical for DRM in the twenty-first century.

Inclusive disaster risk governance is a way to better understand disaster risk as not all of the members of the population experience the impacts of disasters in the same way. Different groups of people have different vulnerabilities and exposure to disasters and states need to pay special attention to this. In addition, the evidence shows us that women's voice, agency, and leadership are still undervalued (UNDRR 2023a, b). The establishment of a Gender Action Plan for the Sendai Framework would be a very powerful step for elevating and strengthening the voices of women leaders. In developing countries, women's voices are undervalued. Significant evidence demonstrates that women are powerful agents of change during and after disasters. Therefore, there is an urgent need to empower women in making decisions and being involved in disaster risk governance at every level of government and local level. Governments should develop gender-responsive information-sharing mechanisms. One common approach is the establishment and support of organizational bodies that facilitate the flow of information between communities and assistance providers and that advocate on behalf of excluded populations.

2.5 Conclusion

The Hyogo Framework of Action (HFA) 2005–2015 and the Sendai Framework for Disaster Risk Reduction 2015–2030 are two global blueprints for disaster risk reduction. One of the priority actions in the Sendai Framework is strengthening disaster risk governance at global and regional, national and local levels. There are many initiatives and approaches in transforming risk governance, particularly regarding the calls for multi-sectoral and collaboration in addressing disaster risk reduction and strengthening laws, Policies and Plans for Comprehensive Disaster Risk Management and its effective implementation. Strengthening risk governance at local level requires cooperation from global and regional initiatives and efforts. Capacity building at local level in DRR is crucial and needs to be strengthened in the future. Strengthening risk governance at local level is imperative, as local actors are also first responders to disasters. In addition, many of the roadblocks to achieving the SDGs are at the local level——such as non-income poverty and inequality—— and must be addressed through SDG localization and by improving local and urban governance.

Risk governances are shifting toward more emphasis on strengthening prevention and preparedness approaches by fostering more efficient planning, addressing climate change issues, biodiversity, poverty eradication, improving food, nutrition, health, and education, implementing ecosystem-based approach in disaster risk reduction, and linking technology to people. The uncertainties of climate change is one factor in limiting the strategic disaster risk reduction planning at local level (Prabhakar et al. 2009). Despite this fact, many initiatives in integrating mainstreaming climate change into the development agenda. Planning for prevention and mitigation needs to be multi-hazard and multi-sectoral, clearly identifying roles, responsibilities, and coordination mechanisms for a wide range of sectoral actors.

In addition, global and regional initiatives are also urgently needing to work towards more inclusive disaster risk management policies for tsunami preparedness and response. Women, poor people, elderly, and youth should become a focus for safeguarding in case of disasters, with social safety nets specifically designed to be gender-responsive and targeting vulnerable groups as a priority. Inclusive disaster risk management——including recognizing and promoting traditional knowledge on disaster risk reduction, and empowering vulnerable groups women and children, and elder groups——needs to be strengthened in the future.

References

ADB (2021) Six ways Southeast Asia strengthened disaster risk management, ADB. https://www.adb.org/news/features/six-ways-southeast-asia-strengthened-disaster-risk-management

Agreement P (2015) United Nations framework convention on climate change, Paris Agreement. ed

Aitsi-Selmi A, Murray V, Heymann D, McCloskey B, Azhar EI, Petersen E, Zumla A, Dar O (2016) Reducing risks to health and wellbeing at mass gatherings: The role of the Sendai framework for disaster risk reduction. Int J Infect Dis 47:101–104. https://doi.org/10.1016/j.ijid.2016.04.006

ARF-PGI (2021) ASEAN regional framework on protection, gender, and inclusion in disaster management 2021–2025 Jakarta, ASEAN Secretariat, October 2021, https://asean.org/wpcontent/uploads/2021/11/PGI_Framework_2021-2025.pdf

ASEAN (2022) ASEAN Framework on Anticipatory Action in Disaster Management Jakarta, ASEAN Secretariat, June 2022, https://asean.org/wp-content/uploads/2022/06/ASEAN-Framework-on-Anticipatory-Action-in-Disaster-Management.pdf

ASEAN Secretariat (2019) ASEAN enabling masterplan 2025 mainstreaming the rights of persons with disabilities. https://asean.org/wp-content/uploads/2019/03/Publication-ASEAN-Enabling-Masterplan-2025-1.pdf

ASEAN Regional Framework on Protection, Gender, and Inclusion in Disaster Management 2021–2025 Jakarta. ASEAN Secretariat. https://asean.org/wp-content/uploads/2021/11/PGI_Framework_2021-2025.pdf

Bouwer R, Pasquini L, Baudoin MA (2021) Breaking down the Silos: Building resilience through cohesive and collaborative social networks. In: Environmental development. https://doi.org/10.1016/j.envdev.2021.100646

Chang-seng D (2012) Disaster risk preparedness. Graduated Research Series, UN University 5:246

Djalante R, Holley C, Thomalla F (2011) Adaptive governance and managing resilience to natural hazards. Int J Disaster Risk Sci 2(4):1–14. https://doi.org/10.1007/s13753-011-0015-6

Djalante R, Holley C, Thomalla F, Carnegie M (2013) Pathways for adaptive and integrated disaster resilience. Nat Hazards 69. Springer Netherlands. https://doi.org/10.1007/s11069-013-0797-5
Djalante R, Lassa S (2019) Governing complexities and its implication on the Sendai framework for disaster risk reduction priority 2 on Governance. Prog Disaster Sci 2:100010. https://doi.org/10.1016/j.pdisas.2019.100010
Fung A (2006) Varieties of participation in complex governance. Public Adm Rev, 66(s1): 66–75. https://doi.org/10.1111/j.1540-6210.2006.00667.x
Gupta S, Leung IS (2011) Turning good practice into institutional mechanisms: Investing in grassroots women's leadership to scale up local implementation of the Hyogo Framework for Action An in-depth study for the HFA Mid-Term Review.https://www.preventionweb.net/files/18197_201guptaandleung.theroleofwomenasaf.pdf
HelpAge International (2005) The impact of the Indian Ocean tsunami on older people Issues and recommendations
Hilhorst D, Boersma K, Raju E (2020) Research on politics of disaster risk governance: Where are we headed? Polit GovAnce 8(4):214–219. https://doi.org/10.17645/pag.v8i4.3843
Howes M, Tangney P, Reis K, Grant-Smith D, Heazle M, Bosomworth K, Burton P (2015) Towards networked governance: improving interagency communication and collaboration for disaster risk management and climate change adaptation in Australia. J Environ Plan Manag 58(5). https://doi.org/10.1080/09640568.2014.891974
IRGC (2017) Introduction to the IRGC Risk Governance Framework, Revised Version. IRGC. Lausanne and IRGC. https://doi.org/10.5075/epfl-irgc-233739
Jensen SJ, Feldmann-Jensen S, Johnston DM, Brown NA (2015) The emergence of a globalized system for disaster risk management and challenges for appropriate governance. Int J Disaster Risk Sci 6(1):87–93. https://doi.org/10.1007/s13753-015-0043-8
Jones M, Shanahan E, McBeth M (2014) The science of stories: Applications of the narrative policy framework in public policy analysis. Palgrave Macmillan
Kruczkiewicz A, Klopp J, Fisher J, Mason S, McClain S, Sheekh NM, Moss R, Parks RM, Braneon C (2021) Compound risks and complex emergencies require new approaches to preparedness. https://doi.org/10.1073/pnas.2106795118
Ministry of the Environment (2016) Eco-system based disaster risk reduction in Japan: A handbook for practitioners. Edited by Ltd Mitsubishi UFJ Research and Consulting Co. Chiyoda-ku, Ministry of the Environment, Tokyo
Nugroho SP (2018) Inclusive disaster management: Indonesia's message in AMCDRR 2018. Bnpb.Go.Id
Paton D, Bajek R, Okada N, McIvor D (2010) Predicting community earthquake preparedness: A cross-cultural comparison of Japan and New Zealand. Nat Hazards 54. Springer Netherlands
Prabhakar SVRK, Ancha S, Rajib S (2009) Climate change and local level disaster risk reduction planning: Need, opportunities and challenges. Mitig Adapt Strateg Glob Chang 14(1):7–33
Renn O, Klinke A, Van Asselt M (2011) Coping with complexity, uncertainty and ambiguity in risk governance: A synthesis. Ambio 40(2):231–246. https://doi.org/10.1007/s13280-010-0134-0
Renn O (2008) Chapter 1 white paper on risk governance : Toward an integrative framework 1 purpose and objectives of this white paper. Gov Int J Policy Adm (1):3–73
Sari S (2022) ASEAN mechanism in disaster management: A case study of disaster management in Indonesia 2010–2020. J Glob & Strat 16(1), 75–108. https://doi.org/10.20473/jgs.16.1.2022.75-108
Shi P, Li N, Ye Q, Dong W, Han G, Fang W (2010) Research on integrated disaster risk governance in the context of global environmental change. Int J Disaster Risk Sci 1(1):17–23. https://doi.org/10.3974/j.issn.2095-0055.2010.01.004
Srinivas H, Nakagawa Y (2008) Environmental implications for disaster preparedness: Lessons learnt from the Indian Ocean Tsunami. J Environ Manag 89(Issue 1):4–13
Stallings RA (1997) Sociological theories and disaster studies. University of Delware
Stein V, Wiedemann A, Bouten C (2019) Framing risk governance. Manag Res Rev 42(11):1224–1242. https://doi.org/10.1108/MRR-01-2019-0042

Suntoro A, Wahyuni T, Ismeti, Utomo NA (2023) Toponym Dan Kayori : Potential benefits of local wisdom to Kaili in disaster mitigation strategies in Central Sulawesi, Indonesia. Geography, vol 97. https://doi.org/10.22439/cjas.v37i1.5908

Tanaka N (2009) Vegetation bioshields for tsunami mitigation: review of effectiveness, limitations, construction, and sustainable management. Landscape Ecol Eng 5(1):71–79. https://doi.org/10.1007/s11355-008-0058-z

The ASEAN Secretariat (2021) Asean regional framework on protection, gender and inclusion in disaster management. Jakarta, Indonesia. https://asean.org/wp-content/uploads/2021/11/PGI_Framework_2021-2025.pdf

UNDRR (2006) Weak governance. https://www.preventionweb.net/understanding-disaster-risk/risk-drivers/weak-governance

UNDRR (2022a) Breaking silos to build resilience—Multi-hazard, multi-sectoral approaches to managing disaster risks. Undrr Org

UNDRR (2022b) From risk to resilience: towards sustainable development for all in a COVID-19 transformed word. Co-chairs' summary: Bali Agenda for Resilience

UNDRR (2023a) Inclusive disaster risk reduction means resilience for everyone. https://sendaiframework-mtr.undrr.org/news/inclusive-disaster-risk-reduction-means-resilience-everyone

UNDRR (2023b) Thematic report on local, Indigenous and traditional knowledge for disaster risk reduction in the Pacific. https://www.undrr.org/publication/thematic-report-local-indigenous-and-traditional-knowledge-disaster-risk-reduction

UNISDR (2005) Hyogo framework for action 2005–2015. Unisdr Org. https://doi.org/10.1007/978-1-4020-4399-4_180

UNISDR (2015) Sendai framework for disaster risk reduction 2015–2030. Geneva, Switzerland

UNDRR (2024) Engaging with stakeholders. Undrr Org

United Nation Office for Disaster Risk Reduction (UNISDR) (2023) Implementation of the Sendai framework for disaster risk reduction 2015–2030, pp 1–23

UN Resolution (2023) Political declaration of the high-level meeting on the midterm review of the Sendai Framework for Disaster Risk Reduction 2015–2030, Agenda item 18 (c) Sustainable development: Disaster risk reduction. https://unfccc.int/sites/default/files/GST/2023-06/A-RES-77-289%20-%20Political%20Declaration%20-%20Midterm%20Review%20Sendai%20Framework.pdf

Van Asselt MB, Renn O (2011) Risk governance. J Risk Res 14(4):431–449. https://doi.org/10.1080/13669877.2011.553730

Wisner B, Gaillard JC, Kelman I (2015) Theories and stories seeking to understand hazards, vulnerability and risk. In: Handbook of hazards and disaster risk reduction. The Routledge, pp 18–33. https://doi.org/10.4324/9780203844236.ch3

Yokohama Strategy and Plan of Action for a Safer World (May 1994) Guidelines for natural disaster prevention, preparedness and mitigation, world conference on natural disaster reduction Yokohama, Japan, 23–27

Chapter 3
All Hazards Approach and Its Importance in the Risk Reduction

Divya Suresh, Sujata Saunik, Takako Izumi, and Rajib Shaw

Abstract The All Hazards Approach (AHA) offers a paradigm shift in disaster management. It presents an alternative way of managing disasters, moving away from siloed planning for specific threats. And it prioritizes building comprehensive capabilities to address various emergencies. This chapter explores the concepts of systemic, transboundary risk and the trajectory of early warning systems. It discusses different aspects of AHA risk assessment, AHA risk perception, and AHA risk mitigation. Incorporating these key elements presents a conceptual framework of the All hazard approach, which is critical for preparedness for a full spectrum of hazards. Understanding and implementing this approach is of utmost importance in the present world, where the frequency and intensity of disasters are increasing. This chapter lists examples of the All Hazards Approach, drawing on successful implementations in Japan, the Philippines, California, and other countries. By highlighting these best practices, the chapter illuminates key challenges and how innovative technologies and approaches are leveraged in disaster response and management.

Keywords Disaster risk management · All Hazards · Risk assessment · Perception · Mitigation

3.1 Introduction: What is the All Hazards Approach?

The All Hazards Approach (AHA) is a comprehensive strategy for emergency preparedness, response, and recovery. It emphasizes planning and building capacities to address a wide spectrum of emergencies and disasters, both natural and human-made.

D. Suresh (✉) · S. Saunik · R. Shaw
Graduate School of Media and Governance, Keio University, Fujisawa, Japan
e-mail: divyas@keio.jp

T. Izumi
International Research Institute of Disaster Science (IRIDeS), Tohoku University, Tohoku, Japan

R. Shaw et al. (eds.), *Two Decades from the Indian Ocean Tsunami*, Disaster Risk Reduction, https://doi.org/10.1007/978-981-96-2669-4_3

The concept of AHA in disaster management revolves around being prepared for any type of disaster, regardless of its cause, size, or complexity. It's a proactive strategy that emphasizes planning, building capacity, and fostering collaboration to effectively manage emergencies (FEMA 2007).

3.1.1 Key Principles of AHA

- *Risk Assessment*: Identify potential hazards, their likelihood and impact, and prioritize them based on severity.
- *Preparedness*: Developing and testing emergency plans, training responders and communities, building communication systems, and securing resources.
- *Response*: Mobilizing resources and responders, coordinating efforts, and providing necessary services to the affected population.
- *Recovery*: Restoring normalcy and helping communities rebuild after a disaster (Mike and Gutierrez 2020; Donahue et al. 2012).

This chapter will provide valuable insights to the various stakeholders as discussed.

A. Policymakers

  - National and local government officials are responsible for disaster preparedness, risk management, and emergency response planning.
  - Legislators have a role in allocating resources and enacting policies that support disaster management initiatives.

B. Stakeholders in Disaster Management

  - Emergency management professionals working within government agencies or NGOs.
  - Representatives from communities vulnerable to diverse hazards.
  - Public health officials are crucial in ensuring community preparedness and response during emergencies.

3.2 Background of the All Hazard Approach Concept

The concept of AHA has been evolving over the last few years. Although AHA has been discussed by agencies like the Federal Emergency Management Agency, Centers for Disease Control and Prevention, etc., it has not been much discussed in academic work. It was first highlighted in the all hazard emergency operation planning guidebook (FEMA 1996) target to local authorities. Since then, the concept has evolved drastically to include new topics in it. For instance, in 2001, terrorism was included, and in 2011, NATECH, and in 2020 biological hazard, was included. Shaw (2024, 03 15). All hazards approach (AHA) for a resilient, sustainable, and inclusive society [conference presentation]. APRU 2024 Multi-hazards workshop, Tohoku University.

As outlined in Fig. 3.1, in the 2000s, when Information technology became an integral part of people's everyday lives, AHA's focus also turned toward IT and knowledge management. The OODA (observe, orient, decide, and act) framework, similar to PDCA (plan, do, check, and action), was used in an iterative decision-making process. The applicability of AHA for critical infrastructure, i.e., physical and cyber security of utility services, was discussed. At the beginning of the last decade, i.e., in 2012, WHO specified AHA could be linked to public health, i.e., in H-EDRM (health, emergency disaster risk management). During the pandemic period (2020–2021), the top hazard approach (THA) for resource constraint environment and analyzing the impact of COVID-19 with THA evolved.

The All Hazards Approach offers a paradigm shift in disaster management. It moves away from siloed planning for specific threats and instead prioritizes building comprehensive capabilities to address a wide spectrum of emergencies. This proactive strategy emphasizes risk assessment to identify potential hazards, analyze their likelihood and impact, and prioritize them for mitigation efforts. Through collaborative planning, training exercises, and resource allocation across various stakeholders (government agencies, NGOs, and communities), AHA fosters a unified approach to disaster preparedness, response, and recovery. This comprehensive approach can significantly enhance a community's resilience in the face of unforeseen disasters, minimizing damage and expediting recovery efforts.

Key Idea of AHA:

- *Shifting Focus*: AHA moves away from focusing on specific disaster types and instead prioritizes building capabilities to handle a broad spectrum of emergencies.

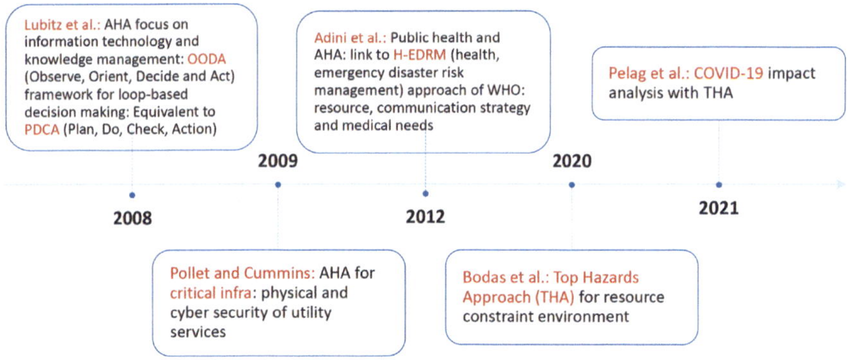

Fig. 3.1 Evolution of AHA over last 20 years. *Source* Rajib Shaw (2024, 03 15). All hazards approach (AHA) for a resilient, sustainable, and inclusive society [conference presentation]. APRU 2024 Multi-hazards workshop, Tohoku University

- *Proactive Preparation*: The emphasis is on preparedness through risk assessment, planning, training, and resource allocation. The goal is to anticipate potential disasters and be ready to respond swiftly and effectively.
- *Flexibility and Adaptability*: The AHA recognizes that disasters can be unpredictable. By having a comprehensive framework, communities and response teams can adapt to new or unforeseen hazards.
- *Collaboration and Coordination*: The AHA encourages collaboration between various stakeholders, such as government agencies, NGOs, and communities. This fosters better communication, resource sharing, and a unified response effort (FEMA 2007).

Outline of this study:

The following discussions of this chapter explore new risks, the significance of AHA risk assessment, perception, and mitigation. Finally, it presents a few global best practices, drawing a conceptual framework for AHA. The conclusion shows future directions.

3.3 Understanding New Risks

3.3.1 Link to Systemic Risk

In the current times, when the world is becoming increasingly interconnected and globalized, disasters are occurring frequently and unexpectedly because of unfavorable socio-economic and environmental challenges. We have created the potential for unintentionally transcending risks across geographic boundaries in the occurrence of a disaster at a large scale. The idea of systemic risk is that in the event of a disaster, one system or multiple systems could collapse, not just one part of a large system. Examples of systems that are prone to be affected are food and water supplies, health care, energy, transportation, and information and communications.

Some theoretical viewpoints from which systemic risk can be investigated are systems theory, network theory, and Complex adaptive systems theory. Systemic risks can be broadly categorized into four types: complexity, uncertainty, ambiguity, and ripple effects. To summarize the definition of systemic risks, it is global in scope, has cross-sector or transboundary implications, and usually has a tipping point while linked via non-linear cause-effect relationships. Also, the time of onset of the event and its effect are differential in nature, as well as the source of risk, i.e., endogenous versus exogenous, and signal strength, which are differentiators.

The Global Assessment Report on Disaster Risk Reduction 2019 (UNDRR 2019) provided the progress of different governments on the Sendai Framework, examined the risks from a systemic perspective, and noted the need for understanding from the global community on the dynamic nature of systemic hazards, novel risk governance frameworks, and novel risk-informed decision-making tools. The report refers to the earth as a single system containing systems of systems within it and describes why a compartmentalized approach to assessing and mitigating risk approach will be inadequate considering the current times. Therefore, the Global Risk Assessment framework that is proposed in the same report recommends the need for cross-disciplinary research to better comprehend risks in recent times and the interdependence of global economies where risk cascades and compounds, resulting in economic, social, and environmental implications.

The Global Assessment Report on Disaster Risk Reduction 2022 (UNDRR 2022a) has also highlighted the need to focus on systemic risk management to reduce disaster risk and build resilience, building global systems and resilient communities. The United Nations 'Common Agenda' document prescribes the use of an 'Emergency Platform,' 'Strategic Foresight and Global Risk Report' to holistically assess the existential and extreme risks that exist in today's world.

Systemic risks do not fall under the scope of traditional risk management practices, leading to the need to challenge new kinds of policy and governance issues. New approaches are to be devised based on systems thinking and the 'whole systems approach.' Also, the current approach of focusing on specific hazards and risk assessment should shift to the analysis of a transdisciplinary system, multi-stakeholder engagement, and action-oriented research on disaster risk reduction. To handle systemic risks, a new institutional architecture must be set up, and the governance system must be participative, responsive, and adaptive, which will help devise a multi-hazard, cross-disciplinary framework for the disaster research field (Mitra and Shaw 2023).

3.3.2 Link to Transboundary Risk

Some examples of transboundary risk that made its impact across nations, communities, and people of all socio-economic conditions are COVID-19, the SARS outbreak in East Asia, the food price crisis globally, the ebola outbreak in West Africa, the Bangkok floods, etc. These disasters exposed the need for mobilization of capacities and resources extending beyond the scope of the traditional risk management framework. Globally, most of the national-level risk management systems and frameworks are intended to respond and provide mitigation measures for risks that arise and affect people within the countries' borders. They lack the approach to look beyond the national boundary or lack adequate measures to handle the cascading effects of disasters occurring in some other parts of the globe.

The cross-border risk assessment poses additional challenges in disaster risk management exercises. Compared to the global models, the transboundary assessment of countries that share their borders, whose scale of analysis could be much smaller. Different countries use different models to assess various risks.

There could be various factors that affect international and transboundary cooperation toward disaster risk reduction. Regarding the example of seismic risk, hazard models could operate on the basis of different conditions for building classification, different intensity measures and return periods, different methods to work on fragility curves may be adopted, and varied levels of granularity for arriving at exposure assessment may be deployed. Because of these variabilities, harmonizing the tools and methodologies utilized for risk assessment is required. It is also required to channel risk assessment from the evaluation of a single risk to the analysis of multiple risks. The increased exposure to transboundary risks in recent times mandates the need for an integrated risk assessment framework.

A (Prabhakar et al. 2021) study recommends transboundary risk management techniques, such as adapting the ISO 31000:2018 and the ISO 22301:2019 Business Continuity Management System to strengthen already existing measures derived from international frameworks such as the Sendai framework. The risks are interlinked at three levels: global, national, and regional. This translates to the need to have disaster preparedness and response frameworks at global, regional, and national levels. It has been identified that transboundary risks can be managed through four steps: (1) Identify and recognize linkages, (2) shared risk analysis considering hidden vulnerabilities, (3) risk information sharing, and (4) devise globally coordinated solutions.

3.3.3 Link to EWS

Early warning is crucial in today's world to effectively reduce mortality arising from natural hazards. A World Bank study (Hallegatte 2012) noted that when developing countries can upgrade their early warning systems and hydro-meteorological information to that of developed countries, it could help save 23,000 lives per year. Enabling the capacity of early warning tools would be critical to addressing the climate risk scenarios highlighted in the Paris Agreement and the Sendai Framework for Disaster Risk Reduction. The UNDER estimated that countries whose multi-hazard early warning system is limited or moderate capacity resulted in eight times the mortality rate of countries with sufficient coverage.

In 2022, an assessment (UNDRR 2022b) was conducted to check the availability and accessibility of disaster risk reduction information and multi-hazard early warning systems. It revealed that the top 10 Asian Pacific countries with emerging risks and intensifying risks lacked enough forecasting capabilities and disaster risk knowledge, which they had to enhance to address sufficient coverage.

The Early Warning System for All initiative was devised in the same year to ensure every person on earth could be protected. It is based on four pillars of the multi-hazard early warning (UNESCAP 2024).

1. Disaster risk knowledge—data collection and risk assessment systematically.
2. Detection, observation, monitoring, analysis, and forecasting of hazards—create hazard monitoring and early warning services.

3. Preparedness and response capabilities—develop national and community response capabilities.
4. Warning dissemination and communication—risk information communication and early warning.

A study (Coughlan de Perez et al. 2022) noted that the lack of early warning protection arose not from the inadequate capacity of forecasting but rather from insufficient communication and response capacity. This can be addressed by bridging the gap in meeting the needs of the especially vulnerable populations such as women, the elderly, people with disabilities, etc. Building on an integrated people-centered early warning system approach helps strengthen the multi-hazard early warning systems by focusing on addressing the needs of vulnerable populations.

The future of early warning systems should leverage the capacity of emerging and innovative technologies, such as Artificial Intelligence, to improve the availability, accessibility, efficiency, and accuracy of the concerned systems. The equitable and sustainable provision of such advanced technologies to technologically poor countries could be ensured through the collaboration of the private sector and regional cooperation. It is anticipated that incorporating such technology could help synergize hazards and early warnings. It also provides the possibility of shifting to multi-hazard early warning systems from single-hazard early warning systems, resulting in economies of scale. It has not yet been fully possible to reap the benefits of the application of technology to support adapting to climate change, which is happening in the global north and needs to be developed in the global south as it is the place where climate change impacts are more pronounced.

Significance of AHA Risk Assessment, Perception, and Mitigation

3.4 AHA Risk Assessment

The cornerstone of effective disaster prevention is comprehending the interaction of hazards, vulnerability, and exposure. The risk assessment process outlines determining the risk's nature and extent, hazards analysis, and vulnerability evaluation in the existing conditions that hold the potential to harm the people, livelihood, services, and property that are exposed. Risk assessment is not limited to evaluating the magnitude of the likelihood of potential losses. It also involves identifying the causes and impacts of the losses. Therefore, Risk assessment is basic to disaster risk reduction and recovery. It should be intertwined with the policy and decision-making process and demands effective stakeholder collaboration.

3.4.1 Foster Risk Assessment at All Levels, from the International to the Community Level

Risk assessment needs to happen systemically at all levels: local, prefectural, and national. Assess with the local government and make risk information accessible and reachable to local people. We need to ensure local level personnel can conduct assessments through training. A risk assessment council or councils is/are needed (interdisciplinary team). The international community and universities can help.

3.4.2 Ensure that Capacity Building and Training is Practical

Strengthen capacity with technology. The way we support technology transfer needs to be focused on. Training is imperative. Contextual to local capacity building. Make it operational and practical. Local capacity building

3.4.3 Development of Integrated Risk Assessment Model

Comprehensive/Dynamic. All hazards. The council or teams should look at types of natural hazards (seismic, meteorology, hydrology), technological (haz mat releases, industrial fires, power outages), and anthropogenic (accidents and intentional events like terrorist attacks involving CBRN). Of course, it should also include rapid onset and slow onset disasters.

3.4.4 Data Sharing Mechanisms—Promote Open Data Practices

Experience from the past emphasizes the growing need to promote the creation and usage of open data. Because of the unavailability of some fundamental data, accurate assessments of climate and disaster risk are being hindered. We need a policy on data sharing. The information needs to be shared with leaders and other stakeholders. (Have a policy). Data sharing with updation. Strengthen the data-sharing mechanism. Gov. private sector international organization. Collaboration open data. Debriefing post-disaster event.

3.4.5 *Enable Creation of Risk Profile*

Attempt to obtain quantitative versus qualitative data to prioritize what is most important.

- Understanding the historical record of events, magnitude, scope, and impacts.
- Scenarios and expert judgment can also be used.
- Public Safety Canada has an All Hazards Risk Assessment tool to guide efforts.
- Consider Risk: hazard + exposure + vulnerability.
- We need to consider probability and impact.
- We must consider complex interactions (natech).
- Debriefing after disasters is essential.

This will result in a risk profile that will inform planning. If this is not developed, everything else that follows will be based on false assumptions, which could limit grant funding. Perhaps have the risk profile vetted to ensure it is complete and correct. It should be easily understood and actionable. Risk identification, evaluation, and prioritization.

3.4.6 *Promote Better Communication of Risk Information*

Clear communication is crucial throughout the risk assessment process. Important lessons can be drawn from the past on the need to close the gap while transitioning from risk awareness to action. Communicating risk information to each and every audience may vary and should be customized. It is also crucial to ensure the information-sharing models that center local actors can be adopted and implemented, as this could help all hazards preparedness and HA/DR missions be more effective.

Translating Disaster Risk information into macroeconomic risk is essential. Developing a tool that provides a graphical visualization of content that different stakeholders, specifically decision-makers, could meaningfully comprehend could be useful.

Also, people hold an aversion toward Risk communication. They focus on the negative aspects. There should be a positive narrative for risk communication, which could lead to a transition to a sustainable, environment-friendly society.

3.5 AHA Risk Perception

Risk perception is defined as an individual or group's valuation or judgment of consequences. Risk perception plays an influential role because the way people perceive risk impacts the way they visualize the outcome. The important ideas that shape risk perception are:

1. Enumerate the social goals and objectives surrounding the special concerns to specific risk sources by people or social groups.
2. What are the different ways in which people process information about risk sources? Regarding the acceptability of perceived risk, what kind of logical structure supports each judgment?
3. Exploring the motivational or cognitive biases behind the process by which people select information from varying risk sources.

To summarize, risk perception is a function of perceived average losses over time, trust in risk handling agencies, risk-related information & social amplification by media coverage, familiarity with risk due to personal experience, etc. The other influential factors for risk perception are the possibility of being killed rather than just injured, catastrophic potential, perception and exclusiveness of benefits, threats perceived by future generations, etc. (Ortwin Renn 1990; Gough 1990).

The four categories in which risk perception studies are viewed are psychological decision theory, socio-psychological judgment and attitude theory, classical decision analysis, and sociological systems theory and policy analysis.

It is important to improve our knowledge about risk perception, which forms the basis for risk communication and risk management. There are innumerable benefits to risk perception studies, which are as follows: identification of public concerns, improvising on the conflict resolution scenario, and valuable insights for devising effective risk communication programs.

3.5.1 Clearly Define the Methods, Audience, and Inform

Use risk assessment to shape risk perception. Shaping the risk perception to take appropriate risk response. People may be uninformed about these risks. Keep in mind that risk is a signifier of threat and a signal of opportunity. It can be a focusing event or issue. If we approach it correctly, we can open windows for change. (90 vs. 10% is not effective). Identify multiple target audiences (government, media, public, vulnerable groups, school, etc.). Interrelation of perception between the individual and the community. Determine the methods by which we are sharing information (government documents, media reports, social media, websites, newsletters, education in schools).

3.5.2 Periodically Review Perception

Recognize that there are or can be high emotions surrounding these assessment results. Constantly monitor risk perception and challenge existing views if and as needed. Ensure perception leads to actions.

3.5.3 *Consider the Broader Context of the Impact of Education*

Emphasis on Formal and Informal Education. School curriculum. Risk communication and public education are imperative to convey findings from the risk assessment. Ensure science is represented thoroughly and consistently.

3.5.4 *Promote Simulation Exercises*

Promote Experiential learning of potential impact. Broadening the scope of the simulation.

3.5.5 *Understand the Pros and Cons of Media/Social Media*

Increasing need to enhance and sustain awareness and understanding of climate change and risk reduction through all media and educational platforms. Make sure you have a plan in place to overcome misperception and mis/disinformation. Control the interplay of social media on risk perception through regulation, control, and education. Mis/disinformation management.

3.5.6 *Promote Dialogue with Political Leadership and Science-Based Policy Making*

Endeavor to fast-track the capacity building of local communities on DRR, adaptation and mitigation, on green and sustainable recovery, including ensuring that local development plans are science-based and risk-informed. Briefing/education of leadership. Science-based policy.

3.6 AHA Risk Mitigation

Risk mitigation can be defined as a series of measures that help prevent loss of life, property, etc. The overarching principle of risk mitigation in relation to disasters can be divided into two types: disaster risk reduction and disaster risk management (Gougelet 2016).

The key to renovating risk governance and management strategies is redefining development according to revealed and foreseen risks, innovating and integrating our risk information and knowledge management systems, prompting common grounds for cooperation for the common good, and strengthening the linkage between science, policy, and practice. Some other key aspects are as follows:

- Integrated risk information and knowledge management system
- Probabilistic risk assessment
- Impact-based forecasting and multi-hazard early warning system
- Loss-and-damage registry and fund
- MRV, including GHG inventory
- Sustainability reporting and transparency mechanism
- Climate finance mechanism, supported by climate rationale
- Carbon market mechanism
- Transitioning from risk frame to transformation frame
- Climate justice (Paris Agreement).

3.6.1 Promote Synergistic Measures for All Hazards

Develop a Hazard Mitigation Action Plan. However, there are other types of mitigation measures (monitoring chemical facility access, ensuring cybersecurity, etc.). Risk insurance for multi-hazard. Often related to early warning systems, land use, construction, and insurance.

3.6.2 Promote and Enable Cost-Effective Processes and Measures to Mitigate Risk

Short, medium, and long-term measures. Keep in mind the goals of prevention, reduction, minimizing impact, or sharing loss. Look for cost-effective measures for protection and risk-informed investments. We may need to build capacity to undertake this step (along with the others). Keep in mind scientific as well as traditional options. Understand countries may need assistance in this area due to capacity.

3.6.3 Application of Science and Technology in Disaster Preparedness

AHA Preparedness needs to be enhanced by incorporating technology for sending alerts and broadcasts to stakeholders. For instance, an earthquake early warning system is a relatively recent technology and can potentially improve response to dangerous levels of earthquake shaking. Given that mobile phones will continue to be the principal means of receiving alerts, educational efforts must center on the fact that when alerts are received, earthquake shaking is imminent, and protective action must be implemented immediately. Mobile alerts must incorporate more information to the extent possible, including earthquake magnitude, location (map), seconds remaining before arrival, and protective action.

3.6.4 Consider the Importance of Local and Indigenous Knowledge

Importance of local knowledge and citizen voice for urban planning. Promote sustainable urban planning/land use planning and development, given the continuing trend of rapid urbanization, increasing urban population density, and rural-to-urban migration. Also, facilitate efficient access of local communities to international and domestic climate finance and the transfer of knowledge and technology.

3.6.5 Foster Multi-Stakeholder Collaboration

There is a rising need to develop a platform for various stakeholders, such as governments, international organizations, businesses, civil societies, and investors, to foster collaboration toward scaling up risk mitigation efforts. The significance of multi-stakeholder collaboration lies in the strength of the stakeholders to jointly respond in a complex risk environment rather than being alone and well-planned preventive and response measures resulting in efficiencies of scale. The institutionalization of a multi-stakeholder approach will result in good governance. It enables the possibility of incorporating multiple views and reducing confirmation bias and other behaviors that usually cause divergence in risk response.

3.6.6 Cultivate and Promote Transboundary Measures

Since transboundary risks are risks that cannot be confined to any national or international boundary, a collective approach to a country-centric approach would be more just and equitable in the mitigation of the far-reaching consequences. Promoting transboundary measures such as developing global early warning systems and transboundary warning systems will support in preparing, responding, and enhancing resilience toward climate and other risks.

3.6.7 *Strategic Risk Mitigation and Response Planning*

There is a growing need to strengthen response planning measures by anticipating the events beforehand and ensuring capacities are in place at all levels through efficient preparedness of response training. The key to response planning is steering gender-equitable and universally accessible response training involving women, persons with disabilities, etc.

3.7 Disaster Response at the Local Level: Case Study Examples

3.7.1 *Operational Continuity Plan of Municipalities—Case of Japan*

Every municipality in Japan has a Business Continuity Plan or operational continuity plan. This is very similar to what corporations have. The government (cabinet office) has enabled the policy framework on business continuity plans to conduct disaster response at the local level, for example, the BCM guideline in 2005. Pre-decided agreements with private partners are made to conduct risk assessments. The business continuity planning adoption ratio has increased significantly for large-scale companies from 18.9% in 2007 to 100% and for medium-scale companies from 12.4% in 2007 to 50%. Yet another risk management process to maximize capacity and minimize losses enabled through legal measures is the ISO22301/22313 Business Continuity Management system (2012) (2019). Takahiro Ono (2024, 03 14). AHA Risk Assessment [conference presentation]. APRU 2024 Multi-hazards workshop, Tohoku University.

3.7.2 *Risk Assessment and Innovative Systems for Understanding Risk and Renovating Risk Governance and Management—Case of the Philippines*

The Philippines had devised climate change legislation exactly one month after the flooding. The following year, a new law, the Disaster Risk Reduction and Management Act, was implemented. National and local climate action planning are incorporated into their legal system. With the enactment of laws and policies related to climate change and disaster management, such as the People Survival Fund Law, the Renewable Energy Act of 2008, the Climate Change Act of 2009, and the Green Jobs Act of 2016, the Philippines became the pioneer in climate change and disaster management laws.

Each Local government unit has its own unique set of climate and disaster risks, and therefore Philippines employs bottom-up planning whereby local governments are involved in risk-informed development and investment planning. Yet another interesting initiative is Academe 2019: Accelerated Climate Action for Transformation (ACT Local), in which academia trained local governments to accelerate climate action. Emmanuel M.de Guzman (2024, 03 15). Risk assessment and innovative systems for understanding risk and renovating risk governance and management [conference presentation]. APRU 2024 Multi-hazards workshop, Tohoku University.

3.7.3 Assessing Earthquake Warning Behavioral Response and System Evaluation: Earthquake Reporting System in California

The webpage 'Did you feel it?' is a citizen science tool that collects data in the form of a questionnaire asking volunteers about their experience of earthquake shaking and its effects and asking them to mark their location. By collecting data from large groups of people, a real-time online map is created that helps the public and emergency responders better visualize the effects of earthquakes. A color-coded intensity map is created based on the intensity values computed at various earthquake-reported locations. People are becoming data providers by sharing valuable observations. Also, the platform stands to educate people on the widely misunderstood concept of the difference between macroseismic intensity and earthquake magnitude.

The web portal is a place where citizens can participate and share their traumatic experiences, thereby providing some sort of emotional support to them. The important aspect of this whole initiative is a human perspective that contributes to an important sociological dimension in comprehending how people behave, respond, and perceive risk in times of emergency (Quitoriano and Wald, 2020) James D. Goltz (2024,03 15). Hazard warnings for earthquakes and assessment of behavioral response [conference presentation]. APRU 2024 Multi-hazards workshop, Tohoku University.

3.7.4 Integration of S&T in Disaster Management—Case of South Korea

South Korea has been top-ranked in the UN e-government survey for over a decade. Disaster management is one of the core projects in the digital government strategy. The key aspects are public value (high responsibility in the public sector), effective participation and coordination, and rapid and accurate information gathering and dissemination. Digital transformation has been observed as a trend not just in disaster management but also in various policy areas. Knowledge-based planning and decision-making with Information Technology have been promoted.

Efforts have been taken to integrate S&T in disaster management and put more into practice, not just stand as the application of technology. The role of e-governance and e-office should be the basis of the mitigation plan. Data should be ensured that it is not siloed/compartmentalized but shared across departments. The best-case example of this is Korea safe-net, a unified network that connects and provides information simultaneously to disaster-related institutions. It is a single, national-scale communication network that supports one channel of command and control & integrated response powered by PS-LTE (public safety—long-term evolution). Another example would be public and private partnered emergency drone operations for disaster risk response, which the South Korean government has implemented. Private drone operators participate in disaster response and support investigating damages in the affected areas. Yoon-kyung (2024, 03 15). Science and Technology policy for effective disaster management [conference presentation]. APRU 2024 Multi-hazards workshop, Tohoku University.

3.8 A Conceptual Framework of AHA

See (Fig. 3.2)

3.9 Way Forward

To assess the complexities of disaster risk reduction, it's important to consider the call for systemic risk management outlined in the Global Assessment Report on Disaster Risk Reduction 2022 (UNDRR 2022a). This demonstrates a move toward new ways of thinking about systems and adopting a 'whole systems approach.' Due to increasing transboundary risks, it is important to align risk assessment tools and methods. By integrating risk assessment frameworks like ISO 31000:2018 and ISO 22301:2019, the measures offered by international frameworks like the Sendai Framework can be improved.

Bridging gaps in meeting the needs of vulnerable populations is crucial to addressing climate risk scenarios and improving early warning systems. Leveraging emerging technologies like Artificial Intelligence can enhance the efficiency and accuracy of early warning systems, promoting equitable access across technologically poor countries.

Risk assessment is essential at all levels. Promoting multi-stakeholder collaboration and cultivating transboundary measures will be required to achieve this. By implementing cost-effective processes, utilizing local knowledge, and engaging in dialogue with political leadership, risk mitigation can be enhanced, aligning with the Sendai Framework for Disaster Risk Reduction goals.

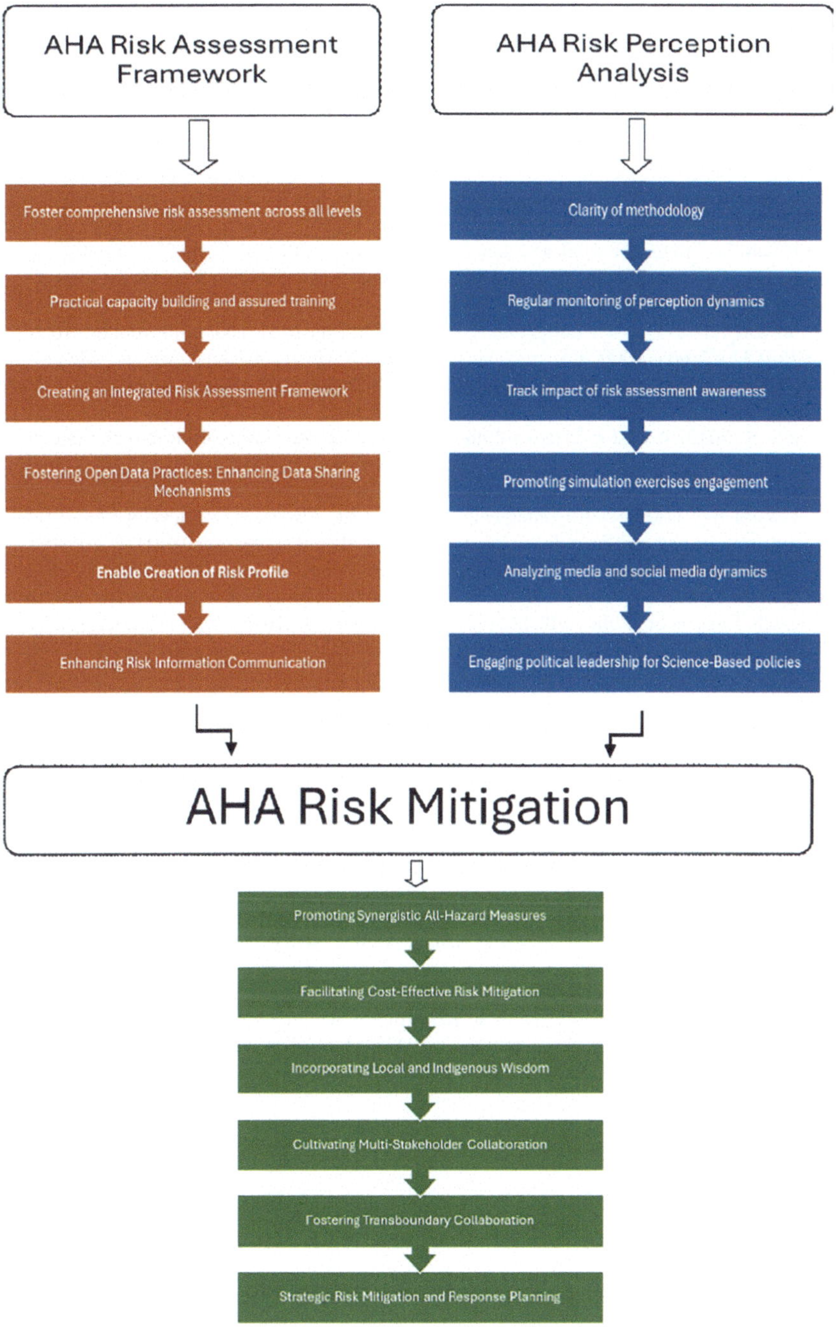

Fig. 3.2 Conceptual framework of AHA. *Source* Authors own

References

Asia-Pacific Disaster Report ESCAP (2023) Retrieved March 28, 2024, from https://www.unescap.org/knowledge-products-series/asia-pacific-disaster-report

Coughlan de Perez E, Berse KB, Angelico Depante LC, Easton-Calabria E, Pierre Evidente ER, Ezike T, Heinrich D, Jack C, Mahar Francisco Lagmay AA, Lendelvo S, Marunye J, Maxwell DG, Binte Murshed S, Garimoi Orach C, Pinto M, Poole LB, Rathod K, Van Sant C (2022) Learning from the past in moving to the future: Invest in communication and response to weather early warnings to reduce death and damage. https://doi.org/10.1016/j.crm.2022.100461

Donahue DA, Cunnion S, Balaban CD (2012) The all needs approach to emergency response

FEMA (2007) National preparedness guidelines

Gougelet RM (2016) Disaster mitigation. Ciottone's Disaster Med, 160. https://doi.org/10.1016/B978-0-323-28665-7.00027-3

Gough JD (1990) A review of the literature pertaining to "perceived" risk and "acceptable" risk and the methods used to estimate them

Hallegatte S (2012) A cost effective solution to reduce disaster losses in developing countries: Hydro-meteorological services. Early Warn, Evacuation. https://doi.org/10.1596/1813-9450-6058

Mike A, Gutierrez DAM (2020) Emergency management: All-Hazards. Encyclopedia of public administration and public policy, Third Edition, 1–4. https://doi.org/10.1081/E-EPAP3-120053208

Mitra A, Shaw R (2023) Systemic risk from a disaster management perspective: A review of current research. Environ Sci Policy 140:122–133. https://doi.org/10.1016/J.ENVSCI.2022.11.022

(PDF) Risk perception and risk management: a review. Pt. 1. Risk perception (1990). Retrieved March 28, 2024, from https://www.researchgate.net/publication/277734287_Risk_perception_and_risk_management_a_review_Pt_1_Risk_perception

Prabhakar SVRK, Issar R, Bakar ABA, Yokoo M (2021) Mitigating transboundary risks by integrating risk reduction frameworks of health and DRR: A perspective from COVID-19 pandemic. Environ Resil Transform Times COVID-19, 63. https://doi.org/10.1016/B978-0-323-85512-9.00014-0

Quitoriano V, Wald DJ (2020) USGS "Did you feel it?"—Science and lessons from 20 years of citizen science-based macroseismology. Front Earth Sci 8:530048. https://doi.org/10.3389/FEART.2020.00120/BIBTEX

UNDRR. (2019). United Nations Office for Disaster Risk Reduction. https://gar.unisdr.org

UNDRR (2022a) GAR2022: Our World at Risk (GAR) | UNDRR. https://www.undrr.org/gar/gar2022-our-world-riskgar

UNDRR (2022b) Global status of multi-hazard early warning systems 2022 | UNDRR. https://www.undrr.org/publication/global-status-multi-hazard-early-warning-systems-2022

UNESCAP (2024) Asia-Pacific Disaster Report | ESCAP. https://www.unescap.org/knowledge-productsseries/asia-pacific-disaster-report

Chapter 4
The Evolution of Global Tsunami Warning and Mitigation System in Indian Ocean Basin in Twenty Years

Harkunti Pertiwi Rahayu

Abstract The achievement in twenty years is not a minor one. A huge leap of achievement has been enhanced, from zero to 4.0 early warning and mitigation system. With nonexistence of a warning system and lack of people's awareness and preparedness for tsunami caused over 230 thousand people killed in the region during the 2004 devastated Indian Ocean tsunami. To respond to this event, a warning system was established in 2005 under the mandate of the Resolution of UNESCO IOC-XXIII-12, named Indian Ocean Tsunami Warning and Mitigation System. The principles of people centre and end-to-end approaches were used in the development of the early warning system, showing the responsibility sharing among regional, national and local stakeholders for each warning function. Currently, the 4.0 hazard detection and forecast are led by Australia, India and Indonesia as tsunami service provider for the 24 member countries in Indian Ocean, as well as using 4.0 technology for threat evaluation and alert formulation to warn their people at risk. However, biggest challenges remain at the downstream component to respond to the global commitment that by 2030 all the people living at tsunami prone area should be prepared and be resilient and all the tsunami prone area should be protected by early warning system.

Keywords Tsunami warning and mitigation system · People centre and end-to-end warning system · Response capacity · Resilience

H. P. Rahayu (✉)
Chair of WG1, Indian Ocean Tsunami Warning and Mitigation System (2015–2024), Hyderabad, India
e-mail: harkunti@itb.ac.id

The School of Architecture, Planning and Policy Development, ITB, Bandung, Indonesia

R. Shaw et al. (eds.), *Two Decades from the Indian Ocean Tsunami*, Disaster Risk Reduction, https://doi.org/10.1007/978-981-96-2669-4_4

4.1 The Landscape of 2004 and Initiation of the Global Tsunami Warning and Mitigation System in Indian Ocean Basin

The devastating impact of the 2004 Indian Ocean Tsunami, with over 230 thousand people killed in Indian Ocean countries, was mainly due to the nonexistence of a tsunami warning system in the region (Rahayu 2012). By 2004, the landscape of the region was characterized by several factors, i.e. lack of understanding of tsunami risk, inadequate seismic observation to detect undersea earthquakes, no real-time sea level monitoring to verify tsunami waves, no system to forecast tsunami threats, no national focal institution for disseminating warning to the people at risk, lack of community's tsunami awareness and preparedness and no response planning for mega disaster (Rahayu et al. 2020). Meanwhile, mitigation toward tsunami threat hardly existed in the region, as shown by over one million coastal communities displaced in the Indian Ocean region during the most destructive disaster.

However, months after the 2004 event as indicated by 2005 baseline information obtained from regional capacity assessment for tsunami preparedness among 16 Indian Ocean countries, the region's capacity to respond to tsunami was still low (UNESCO-IOC 2020). The existing warning mechanism in the region was mainly for receiving tele-tsunami bulletins from the Pacific Tsunami Warning Center and the Japan Meteorological Agency, with very low capacity for a national or regional warning centre to monitor and warn the regional and locally generated tsunami. There were lack of 24/7 warning centre, low capacity for receiving real-time seismic data, and low availability of real-time or near real-time sea level data.

Therefore, this 2004 mega-tsunami event has become a *wake-up call* for the establishment of a regional and national *tsunami early warning and mitigation system* to save all people in tsunami-prone region.

To reflect at the above issues, the *Intergovernmental Oceanographic Commission of UNESCO (IOC-UNESCO)* has been requested by international communities for the development of an entirely new *tsunami warning and mitigation system* for the Indian Ocean region through several international and regional meetings. Such global meetings were Special ASEAN Leaders' meeting in Jakarta in January 2005, UN Conference on Small Island Developing States in Port Louise——Mauritius in January 2005, Special Session on Indian Ocean Disaster during UN World Conference on Disaster Reduction in Kobe——Japan in January 2005, UNGA Resolution 59/279 in New York January 2005), Phuket Ministerial Meeting on Regional Cooperation for Tsunami Early Warning Arrangements in Phuket-Thailand January 2005, GEO Communique in Brussels January 2005, and several other technical meetings held in India, China and Indonesia during the year of 2005. Then, during the XXIII Session IOC Assembly in Paris 21–30 June 2005, the *Indian Ocean Tsunami Warning and Mitigation System* was formally established through the Resolution of IOC-XXIII-12 (UNESCO-IOC 2005). The system is then known as *UNESCO Intergovernmental Oceanographic Commission/Intergovernmental Coordination Group Indian Ocean Tsunami Warning and Mitigation System* called *UNESCO IOC/ICG*

IOTWMS with 24 member-states. The system was formed as a coordinated network of national systems, known as *system of system* in the region. The system is then also part of four *global early-warning system network for all ocean-related hazards.* They are *Caribe-Early Warning System* with 32 member-states/16 territories, *North East Atlantic and Mediterranean Tsunami Warning System* with 40 member-states, and *Pacific Tsunami Warning System* with 46 member-states. Most of these global tsunami warning systems were established after the 2004 tsunami event, except the Pacific which was first convened in 1968, to respond to 5 major destructive Pacific tsunamis in the previous 19 years, to reduce the loss of life.

The two basic concepts of early warning systems, i.e. people centre and end-to-end, were adopted in the development of IOTWMS to build a sound and solid system. The people centre early warning system, endorsed by UNISDR Third Early Warning Conference in Bonn 2006, consisted of the four pillars of early warning system, i.e. risk knowledge, monitoring and warning service, dissemination and communication, and response capability. As shown in Fig. 4.1, Risk Knowledge as the first pillar is a systematically process to collect data and undertake risk assessments, characterized by recognizing tsunami hazard and vulnerability, patterns and trends of hazard threat, and availability of risk maps and data. Monitoring and Warning Service as the second pillar aims to develop hazard monitoring and early warning services, characterized by using right parameters, scientific-based forecast, and accurate and timely warnings. Dissemination and Communication as the third pillar focuses on communicating risk information and early warnings, characterized by the ability to reach all, and well-understood, clear and useable information. Then, Response Capability as the fourth pillar aims to build national and community response capabilities, characterized by the existence of response plans, local capacities, people prepared and ready to response.

During the development of IOTWMS, the four pillars of people centre early warning system was embedded in the end-to-end system. The end-to-end system addressed the role of regional, national, and local responsibilities sharing in developing functional warning system for tsunami and other hazards that threatened communities in the Indian Ocean region (U.S. Indian Ocean Tsunami Warning System (U.S. IOTWS 2007), see Fig. 4.2. For example, risk knowledge is embedded in all stages of end-to-end warning system. While monitoring and warning service is integrated in hazard detection and forecast as well as in threat evaluation and alert formulation. Meanwhile, dissemination and communication are integrated into alert dissemination and public safety messages. The response capability is embedded in preparedness and response.

The end-to-end system covers the whole chain of warning system. It starts from hazard detection and forecast, which is under responsibility at regional level, for example by IOTWMS. Followed by the threat evaluation and alert formulation as well as the alert dissemination and public safety messages, which are under responsibility of at national level of Indian Ocean member countries. Finally, the preparedness and response, which is responsibility of at local level of each 24 Indian Ocean member countries.

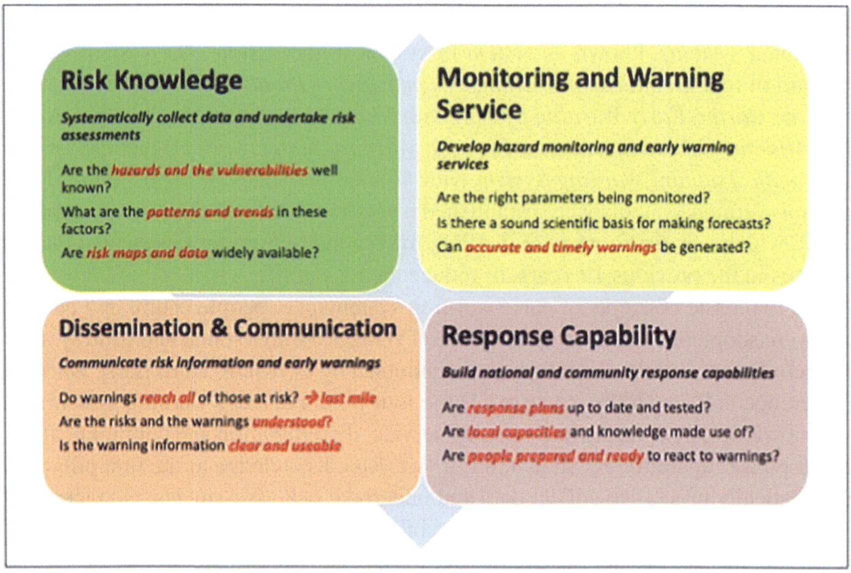

Fig. 4.1 People centre early warning system (adapted from UNISDR 2006)

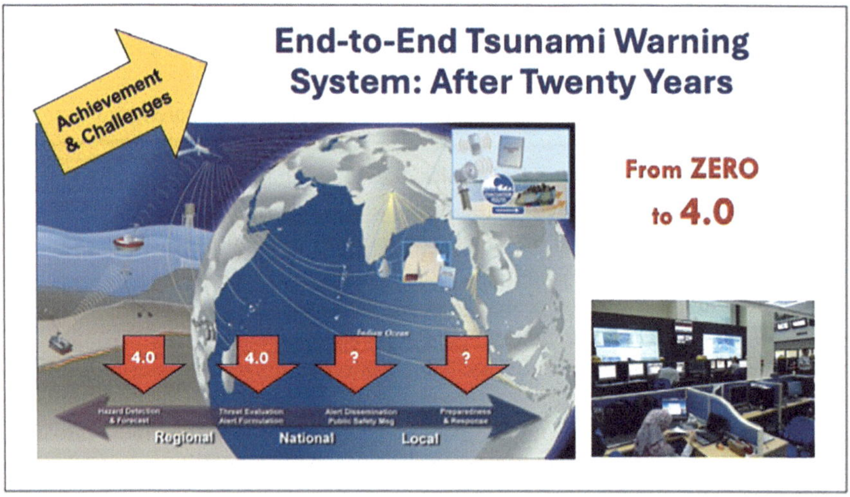

Fig. 4.2 Current State of end-to-end tsunami warning system (Adapted from U.S. IOTWS 2007)

This end-to-end concept applies to all Indian Ocean member countries, except for Australia, India and Indonesia. Currently, these three countries function as TSP—tsunami service provider to the regions due to their capacity and ability to conduct hazard detection and forecast. The role of these three countries besides providing tsunami warnings for their own people but also for other people in the region, through such IOTWMS mechanism. The progression of TSPs is discussed more in the next section as the evolution of IOTWMS.

4.2 Evolution of Tsunami Warning and Mitigation System in Indian Ocean Since 2005

As described in Sect. 4.1, at the initial stage of the development in 2005, the IOTWMS was designed as a "system of systems" concept, which is a network system with multiple *Tsunami Service Providers (TSPs)* providing information of warning to 24 Member States' *National Tsunami Warning Centres (NTWCs)*. Based on the ICG meeting in Mombassa Kenya in March 2007, the National Tsunami Warning Centre (NTWC) of Indonesia along with India, Australia, Malaysia, Thailand and Iran have been mandated as TSPs for the region. The TSPs are expected to generate real-time alert services, warning information/bulletin products for all NTWCs within the Indian Ocean region. However, since 2011 with progressive capacity, performance and high commitment of the government of Australia, India and Indonesia, then the 6 TSPs reduced into 3 TSPs representing these three countries, i.e. TSP India with the INCOIS—Indian National Centre for Ocean Information Services, TSP Indonesia with the BMKG—Meteorological, Climatological, and Geophysical Agency, and TSP Australia with the BOM—Bureau of Meteorology have taken the lead for the Indian Ocean region. During the process development of TSPs 2005–2011, the tsunami information/bulletin for Indian Ocean countries was provided voluntarily by *Pacific Tsunami Warning Center (PTWC)* and *Japan Meteorological Agency (JMA)* as Interim Advisory Service (IAS) until the completion of the Tsunami Service Providers (TSP) function.

From the perspective of upstream component, i.e. hazard monitoring and detection, there were very limited real-time seismometers and almost no real-time earthquake monitoring networks operating in the Indian Ocean. However, after 20 years after the 2004 event, several countries have been able to operate real-time seismic networks, especially these 3 TSPs which are capable of estimating earthquake parameters in near real-time, less than 5 min after each earthquake event. Further, from sea level monitoring perspective, prior to 2004 there were only 4 tide gauges in the Indian Ocean available, but now they operate real-time tsunameter, such as bottom pressure recorders and tide gauge networks and monitoring real-time sea-level data (tsunami confirmation within 30–60 min) (UNESCO-IOC 2019). To have a better understanding of the three TSPs mechanism, Fig. 4.3 describes the data and information flow. The threat information produced from both seismic detection and sea monitoring combined with tsunami model results is then disseminated to 24 NTWC's member countries using multi-mode tools, such as GTS, SMS, fax, email and web. Then each NTWC analyses the threat information then issues the warning, which is disseminated to National Disaster Management Office (NDMO), Local Disaster

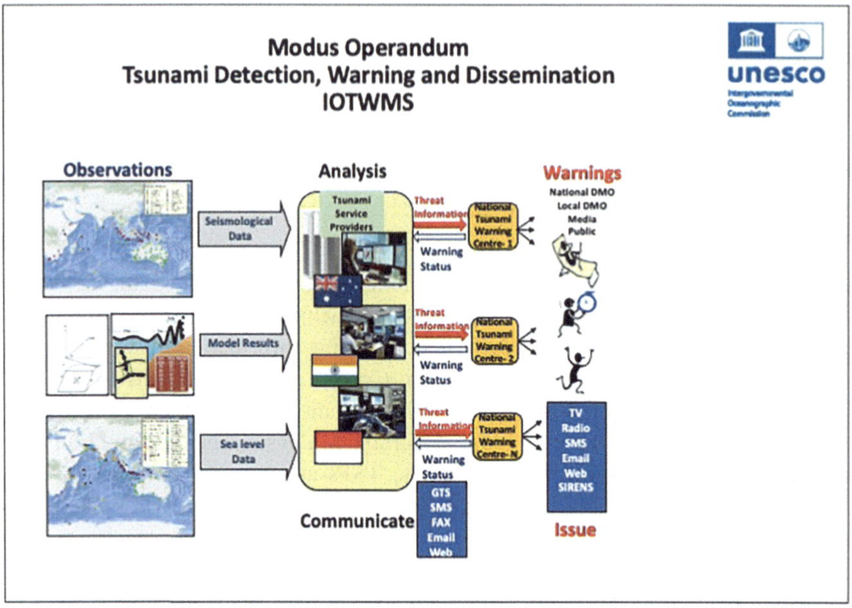

Fig. 4.3 IOTWMS modus operation tsunami detection, warning and dissemination (*Source* Karyono et al. 2024)

Management Office (LDMO), media, public as well as TV, radio, SMS, Web, sirens (Fig. 4.4).

Currently, the TSPs produce threat information and warning status in the forms of service level 1 and 2 to all 24 NTWCs of member countries. Service level 1 consists of earthquake source information to determine potential tsunami threat and travel times. Service Level 2 states more specific tsunami watch information comprising of offshore/coastline estimated wave amplitude, tsunami estimate time of arrival at the coastline, as well as the potential threat zones, sea level verification information and confidence levels as applicable (UNESCO-IOC 2022). However, intensive development has been carried out by the TSPs using advanced technology and modelling tools to enhance the service level of 3, where each NTWCs under bilateral agreements with TSPs will develop and implement an inundation mapping-based tsunami warning information system.

The products of TSPs consist of 4 bulletins. Bulletin 1 is the earthquake information. Bulletin 2 is the tsunami forecast. Bulletin 3 is the tsunami forecast and tsunami observation. Bulletin 4 is the tsunami service finalization. Based on these four bulletins, the 24 NTWCs use their own alert formulation and analyses and are solely responsible to disseminate the warning and issue the alert to their National Disaster Management Offices (NDMO) and Local Disaster Management Offices (LDMO) using their own national warning chain mechanism. To achieve the assurance of the TSPs' products and performances, regular communication tests are needed to be conducted with all the 24 NTWCs as recipients.

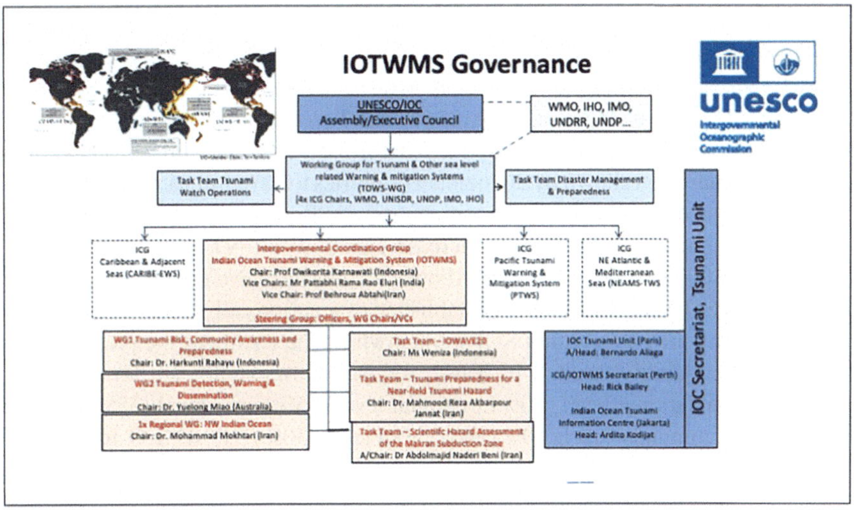

Fig. 4.4 Tsunami warning and mitigation system governance (*Source* Rahayu 2021)

To strengthen the capacity of IOTWMS, *Standard Operating Procedures (SOP)* training workshops combined with *Indian Ocean Wave Exercise* (IOWave) activities have been conducted on a regular basis besides communication test. The IOWave exercise is conducted every two years, the SOP training workshop is conducted annually, and then the communication tests are performed every six months. The SOP training workshops aim to enable the 24 NTWCs to develop their own national SOPs in compliance with SOPs of the three TSP products. The communication test aims to test the communication infrastructures for warning dissemination and to validate the TSPs' performance for dissemination, the NTWCs' reception of the warnings, and their feedback to the TSPs. Meanwhile, the IOWave exercises are intended to test the readiness of the end-to-end system and the resilience and responsiveness of all at-risk coastal community.

From the perspective of preparedness and response, after the 2004 event, several IOWave exercises were conducted, i.e. IOWave09, IOWave14, IOWave16, IOWave18, IOwave20 and IOWave 23 (Weniza and Kumar 2024). These exercises showed escalation of participated countries, i.e. from 18 countries in 2009 up to 24 countries in 2018, 20 countries in 2020 and 19 countries in 2023. The level of exercise intensity has also increased. For example, the number of tsunami source scenario increased from 1 to 4. Meanwhile, a number of participating countries started having communities' evacuation, as well as the involvement of media and number of tsunami-ready community. The high commitment to test and improve the warning and mitigation system was shown during pandemic, the IOWave20 exercise was still being conducted to test the readiness of the communication system only, without any movement of people for evacuation.

In twenty years, the evolution of global tsunami warning and mitigation system in the Indian Ocean basin has shown a very significant increase. The early warning for all EW4A global commitment has been shown in the four pillars of IOTWMS, Risk Knowledge; Detection, Monitoring, Analysis and Forecasting; Warning Dissemination and Communication; and Preparedness and Response Capabilities. Risk knowledge capacity has been shown by 20 member countries to have full national capability to assess and/or receive threat information from TSP and advice/warn the coastal communities. There is a significant increase capability of NTWC with 24/7 staff from 31% in 2005 and 90% in 2018. About 19 countries have established warning dissemination SOP, and 18 countries established evacuation call SOP, while 12 countries have community evacuation SOPs. Meanwhile, the testing of the preparedness and response capability through IOWave exercises shows significant level of exercises, i.e. at the national level (70%), regional level (55%), city level (35%), village level (50%), community level (50%) and school level (30%) ((UNESCO-IOC 2020).

4.3 The Governance of Indian Ocean Tsunami Warning and Mitigation System

During the establishment of the IOTWMS in 2005, the governance of IOTWMS consisted of six working groups, covering the expansion of the three principal pillars of a warning system, i.e. (1) Risk Assessment and Reduction (collect data and undertake tsunami hazard and risk assessment); (2) Detection, Warning and Dissemination (develop hazard detection, monitoring and early warning services and communicate threat information and early warnings; and (3) Awareness and Response (build national and community response capabilities) with the support of the Secretariat based in Perth, Australia to coordinate all activities. Later in 2012, these six working groups were merged into three, representing the pillars of warning system, as stated above.

However, for effectiveness in governance of IOTWMS, at the 2015 ICG meeting in Oman, these three working groups were merged into two, i.e. Working Group 1 on Tsunami Risk, Community Awareness and Preparedness and Working Group 2 on Tsunami Warning, Detection and Dissemination and supported by North-West Indian Ocean Sub-regional Working Group, several Task Teams such as capacity assessment of tsunami preparedness, IOWaves, tsunami preparedness for a near-field tsunami hazard and scientific tsunami hazard assessment. WG1 covers activities related to increasing capacity for hazard assessment, risk assessment, use of hazard/risk assessment on DRR policies, plans and guides, as well as emergency response policies, evacuation planning, tsunami exercises, education and community awareness. While, WG2 has responsibility for increasing capacity for detecting, monitoring and analysis of tsunami threats; organizational performance to function 24/7; TEWS thresholds; and TEWS downstream warning chain. However, during the 2022 ICG IOTWMS meeting, a new working group was added to the governance, i.e. WG3 Tsunami Ready Implementation.

The governance of ICG/IOTWMS is shown in Fig. 4.3. It consists of Working Groups and Task Teams. All Working Groups will work closely with the Task Teams established for certain purposes for certain periods, i.e. Task Team for IOWave, Task Team for CAPT (Capacity Assessment for Tsunami Preparedness), Task Team for Tsunami Preparedness for Near Field Tsunami Hazards, Task Team for Scientific Hazard Assessment of the Makran Subduction Zone. The governance of IOTWMS is dynamic and compliant with current and future needs. The endorsement is taken during ICG meeting which is conducted every two years. For example, to achieve target of the Ocean Decade Tsunami Program, a new Working Group and several Task Teams are needed, i.e. Tsunami Ready Working Group, Critical Infrastructure Task Team and Atypical Tsunami Task Team.

Further from a governance perspective, there has been a significant increase in tsunami policies, plans and guidelines during the two decades. The existence of tsunami policy at national level (19 countries) and at local level (12 countries). Meanwhile, the disaster risk reduction plan existed at the national level (15 countries) and at the local level (11 countries). While there were 14 countries that have national tsunami guidelines and 12 countries have local tsunami guidelines.

4.4 Gaps and Achievements

As mentioned in the previous section prior to the 2004 event the tsunami detection, warning and mitigation had not been significantly in place in the Indian Ocean region. However, there have been two capacity assessments conducted in the region, i.e. 2005 and 2018. The 2005 survey aimed to measure the baseline data on capacity of the three pillars to be used for the development plan of IOTWMS. They were information on hazard/risk assessment; detection, warning and dissemination; and awareness and response. While the 2018 assessment aimed to measure the updated baseline for tsunami preparedness capacity in the region for further improvement. The focus of the 2018 assessment consisted of five sections, which were elaborated from the aforementioned three pillars of warning system. These were: (1) country profile; (2) capacity of tsunami-related policies, plans and guidelines; (3) capacity of risk assessment and reduction; (4) capacity of detection, warning and dissemination; (5) standard operating procedures; (6) capacity of tsunami exercise; and (7) capability of public awareness, preparedness and response.

A greater number of IOTWMS member-states responded to the 2018 assessment, i.e. 21 out of 24, compared to the 2005 assessment, i.e. 16 out of 24. However, the 2018 in-depth assessment results indicated a considerable improvement across the three pillars of the IOTWMS since 2005.

Based on the two-assessment survey, Fig. 4.1 was developed to map further the gaps and achievements from the perspective of the warning system's three pillars.

The gaps are very useful to recognize the needs and priorities for future IOTWMS capacity development and improvement at both regional and national levels.

4.5 To Address Capacity Gaps and Support Requirements

As also shown in Table 4.1, overall significant improvement was recognized at the second pillar of TEWMS, i.e. detection, warning and dissemination, followed by the first and third pillars. They are hazard and risk assessment then awareness, preparedness and response.

The capacity gaps and support requirements identified from the 2005 and 2018 assessments are intended to be used for development strategy for future capacity development in the Indian Ocean region, in order to achieve the two objectives of IOTWMS Medium Strategy 2019–2024. According to the UNESCO IOC TS 144, the two overall objectives are first to achieve an interoperable system that is able to provide timely and effective advice to National Tsunami Warning Centers (NTWCs), based on coordinated Member State contributions. Second is all Indian Ocean communities at risk are aware and prepared for tsunami threat, able to reduce risk, save lives and protect property and infrastructure.

The pillar of *detection, warning and dissemination* consisted of 3 components, i.e. detection, warning and dissemination capacity, supported by the capacity of SOP and tsunami exercise, have shown the most significant improvement. Most Member States are well understood that the priority of development should be on the detection, warning and dissemination component, otherwise no warning system exists. However, there is still strong and consistent improvement needed to improve this pillar. They are such as: (1) providing real-time seismic and sea-level data for tsunami threat analyses; (2) advancement of tsunami modelling for threat forecasts; and (3) updating software tools for tsunami threat forecast or tsunami modelling. In addition, increasing the frequency and quality of tabletop and other related tsunami warning exercises is important to test the sound SOP as well as to reduce the potential complacency among member countries which had no experience with recent tsunami. Based on the principles of early waning for all, the involvement of community evacuation in each member country for an end-to-end tsunami exercise will not only provide a bigger picture of TEWMS performance in the regions but also strengthen the last miles, i.e. community at risk. The community at risk may include village level, community themselves and school level. Further, there is a need to conduct the inclusive tsunami exercise, to response SDG's *"no one left behind"*. This year, there are initiatives conducted by WG1 collaborated with Bandung Institute of Technology and University of Huddersfield to develop an inclusive SOP for tsunami evacuation to accommodate the need for a special school for disabilities to evacuate students to safe place in the safe way.

From the perspective of people centre early warning system, the component of knowing the risk is the most important part. This *Risk Assessment and Reduction* pillar consisted of the risk assessment, reduction and legal framework developed

Table 4.1 Gap and achievements of the capacity of the three TEWS pillars

TEWS pillars	Component	2005 baseline (%)		2018 baseline (%)		Gaps
1. Hazard and risk assessment	Risk assessment and reduction	Tsunami hazard evaluation Use of bathymetry and topography in the model	44 25	Tsunami hazard assessment Use of bathymetry Use of topography	100 85 80	Significant increased
		Tsunami vulnerability assessment	22	Tsunami risk assessment conducted	75	Significant increased
		Numerical modelling for tsunami inundation	22	Numerical modelling for hazard assessment (PTHA)	35	Sufficiently increased
	Policies, plans and guidelines	Policy for disaster warning formulation, dissemination and response	59	Tsunami Policy: At national level At local level	90 60	Local level policy needed to be improved
		Plans for tsunami emergency and evacuation	19	Tsunami disaster risk reduction plan: at national level At local level At community level	75 55 40	DRR plan should be improved at local and community level
				Guidelines for Tsunami at national and local level	70 60	Further improvement needed
2. Detection, warning, dissemination	Detection, warning, dissemination	Tsunami warnings received from AIS (PTWC and/or JMA)	94	National capability to assess and/or receive potential tsunami threat information and advice and/or warn coastal communities	100	Constant improvement to aligned with the trend and demand TEWS technology

(continued)

Table 4.1 (continued)

TEWS pillars	Component	2005 baseline (%)		2018 baseline (%)		Gaps
		24/7 warning centre	94	24/7 warning centre	90	Slightly decrease become major concerned
		Access to seismic networks (national/ international)	41	Access to seismic networks (national/ international)	90	Significant increase, however intensive improvement needed to cope with technology advancement
		Access to sea level networks (national/ international)	41	Access to sea level networks (national/ international)	85	Significant increase, however intensive improvement needed to cope with technology advancement
	Standard operating procedures	–		SOP warning dissemination SOP evacuation call SOP community evacuation	90/80/60	Focus on improvement of the SOP of evacuation call and community evacuation are needed
3. Awareness, preparedness and response	Tsunami exercises	Public awareness towards tsunami threat	37	Tsunami exercise: at national, regional, city, village, community and school level	70/55/35/50/ 50/30	Need more effort for tsunami exercise at all levels
	Awareness, preparedness and response	–		Tsunami education material	65	Need more development on education material

(continued)

Table 4.1 (continued)

TEWS pillars	Component	2005 baseline (%)		2018 baseline (%)		Gaps
		Community education and preparedness for tsunami	47	Tsunami signage	25	Tsunami signage and other supporting infrastructure for evacuation need improvement
		Public education for tsunami	6	Public evacuation maps	25	Still far from sufficient
		School curricula	12	School curricula	45	Still far from sufficient
		Media training	22	Media arrangement SOPs	80	Need consistent improvement

based on the assessed and reduced risk. Though there is a significant effort among the member countries to conduct tsunami hazard and risk assessment, however the effort for the development of numerical modelling for probabilistic tsunami hazard assessments is still low (35%). Therefore several actions are required to: (1) develop pool of experts in the region in carrying out tsunami hazard and risk assessments; (2) provide publicly accessible data for tsunami hazard and risk assessments; (3) increase capacity of hazard mapping and inundation mapping to be used in evacuation mapping; (4) part of pool of expert approach, capitalize existing capacity of Member States to deliver training on hazard and inundation mapping; (5) capacity building for city, village and community level for tsunami risk assessments; and (6) provide guidelines for developing risk maps, evacuation maps and action plans. Meanwhile, for the component of *Policies, Plans and Guidelines*, there are strong need to support member countries to provide accessible best practice national tsunami policies, plans and guidelines, especially for the prevention and mitigation, preparedness. Also, to encourage all tsunami-prone countries in the region to have local level capacity building to produce stand-alone tsunami policies, plans and guidelines, or as part of a multi-hazard document.

There tsunami exercise is basically needed to measure the capacity of two pillars, i.e. *tsunami detection, warning dissemination*; and *Awareness, Preparedness and Response*. Among the TEWMS three pillars, *Awareness, Preparedness and Response* have shown the lowest capacity achievement. In fact, this component is the most important, without the ability and readiness of community at risk to response to the warning, the warning system will not be effective. To improve the response capability, to support member countries to improve their SOPs, especially at the interface between upstream and downstream notification of warnings, including a 24/7 emergency operation centre which is able to receive and convey threat and warning information from the NTWC to the people at risk, following response and decision-making criteria. Besides the existence of associated resources for evacuation planning, i.e. evacuation route as well as supporting infrastructure such as signage and shelter.

Further, it is essential to support: (1) the SOP development of community evacuation; (2) capitalizing good practice of SOPs across Member States; (3) capacity building through sharing good practices regarding many types of evacuation infrastructure; (4) mainstreaming multi-level tsunami exercises into cities, villages, communities and schools; (5) capacity building through sharing good practices of public engagement materials; (6) access to educational materials, such as teaching kits; (7) integrating tsunami *awareness and preparedness* into the school curricula; and (8) participate in the global Disaster Risk Reduction Day (13 October) and World Tsunami Awareness Day (5 November).

4.6 Way Forward

The latest global commitment, the UN Decade of Ocean Science for Sustainable Development, aims to build a safe ocean where the people live in harmony with the sea and oceans for their livelihood and lifestyle without fear of coastal hazards such as tsunamis. To actualize this idea, the UNESCO IOC has developed a framework of action known as UN Ocean Decade Tsunami Program, with a target that by 2030: (1) 100% coast at risk have access to actionable and timely tsunami warnings of all identified sources; and (2) 100% communities at risk be prepared and resilient to tsunamis, see also (UNESCO-IOC 2023).

Considering this tsunami program framework for action to protect communities from the world's most dangerous wave, the Indian Ocean Tsunami Warning and Mitigation System has developed a strategic pathway in the context of UN Ocean Decade through strengthening the 3 pillars which compliance to the interpretation of Tsunami Dedicated Decade (Table 4.2).

The activities for strengthening Pillar #1 Risk Assessment and Reduction focus on enhancing technology for tsunami risk assessments, so the countries know their expected threat and vulnerability, and are able to identify and prioritize the at-risk communities. Meanwhile, the activities to reinforce Pillar #3 Awareness and Response focus on enhancing 100% of communities at risk of tsunami to prepare for and become resilient to, tsunamis by 2030 through the implementation of the Tsunami Ready-like program using penta-helix participatory engagement, i.e. national government, research institutions, universities, other scientific organizations, commercial and non-government organizations. However, to enhance tsunami risk assessment, reduction and community preparedness, several challenges need to be considered, such as the lack of active and consistent engagement of all relevant organizations nationally; crossing the silos between upstream and downstream tsunami warning

Table 4.2 Strategic pathway in context of the UN Ocean decade

The 3 pillars of IOTWMS	Ocean decade tsunami program
Risk assessment and reduction: *hazard and risk identification and risk reduction*	• Promote advance tsunami risk assessments model to increase the capacity of the member countries to identify and prioritize the at-risk communities
Detection, warning and dissemination: *rapid detection and warning dissemination down to the last mile*	• Promote advance detection and monitoring systems, including seismographs, coastal tide gauges and deep ocean tsunameters • Promote new technologies to address observational gaps • Enhance timely and effectively warn the coastal and maritime communities
Awareness and response: *public education, emergency planning and response*	• Achieve 100% of communities at risk to be prepared for and resilient to tsunamis by 2030

components; integrating preparedness for single hazards into a multi-hazard framework; multi-hazard community education; and maintaining the commitment of government, community, business entities, academia, broadcast media.

The activities for continuous improvement for the *Pillar #2 Tsunami Warning, Detection and Dissemination* are needed to focus on three targets, namely: (1) deployment of advanced technologies that address observational gaps; (2) accessible real and near-real-time data; and (3) accessible data, tools and communication platforms, protocols and training to warn coastal and maritime communities timely and effectively of tsunami risk.

References

Karyono K, Jijjavarapu P, Greenwood R (2024) UNESCO IOC ICG/IOTWMS Working group2: Tsunami detection, Warning and Dissemination Report

Rahayu HP, Comfort LK, Haigh R, Amaratunga D, Khoirunnisa D (2020) A study of people-centered early warning system in the face of near-field tsunami risk for Indonesian coastal cities. Int J Disaster Resil Built Environ 11(2):241–262. https://doi.org/10.1108/IJDRBE-10-2019-0068

Rahayu HP (2012) Integrated logic model of effective tsunami early warning system [Kochi University of Technology]. http://hdl.handle.net/10173/897

Rahayu HP (2021) Tsunami risk, awareness, and preparedness report, chair, Working group 1. Report to Intersessional Meeting of ICG/IOTWMS. https://oceanexpert.org/document/29335

U.S. IOTWS (2007) Tsunami warning center reference guide supported by the United States Agency for International Development and partners. https://nctr.pmel.noaa.gov/education/IOTWS/program_reports/TsunamiWarningCenterGuide.pdf

UNESCO-IOC (2005) Twenty-third session of the assembly, resolution XXIII-12—Intergovernmental coordination group for the Indian Ocean Tsunami Warning and Mitigation System. https://oceanexpert.org/document/17951

UNESCO-IOC (2019) Indian Ocean Tsunami Warning and Mitigation System (IOTWMS): Medium term strategy,2019–2024. Paris, UNESCO. Technical Series No. 144. Report prepared by: ICG/IOTWMS Steering Group. https://oceanexpert.org/document/24772

UNESCO-IOC (2020) Intergovernmental oceanographic commission technical series 143—Capacity assessment of tsunami preparedness in the Indian Ocean Status Report 2018. https://oceanexpert.org/document/27303

UNESCO-IOC (2022) Fact sheet: ICG/IOTWMS Working Group 2 on Tsunami detection, warning and dissemination. https://oceanexpert.org/document/32006

UNESCO-IOC (2023) Research, development and implementation plan for the ocean decade tsunami programme. IOC Technical Series No 180. https://unesdoc.unesco.org/ark:/48223/pf0000386603

UNISDR (2006) Early Warning—from concept to action: the conclusions of the third international conference on early warning. In: Wiltshire A, Amlang S (eds), pp 1–33. International Strategy for Disaster Reduction. https://www.unisdr.org/2006/ppew/info-resources/ewc3/checklist/English.pdf

Weniza W, Kumar A (2024) Exercise IOWave23—report task team. https://oceanexpert.org/document/33778

Chapter 5
Systemic, Complex, and Compound Hazards and Its Consideration in Risk Reduction Framework

Yafang Wen, Yanwu Zhang, Yinan Gao, Arunabh Mitra, and Rajib Shaw

Abstract The management of disasters has given increasing attention to systemic risks over the past two decades. This term refers to the potential chain reactions triggered by one or more disasters across social systems, impacting sectors such as the economy, communication, transportation, and healthcare. The catastrophic 2004 Indian Ocean tsunami, which resulted in massive casualties and economic losses across South Asia, serves as a prime example of systemic risks. This chapter delves into the theory and historical development of systemic risks, highlighting their main manifestations during the Indian Ocean tsunami. It also explores the evolution and current shortcomings of systemic risk management since the event. The chapter's objective is to enhance the resilience of societies against complex disasters from a systemic risk perspective and to provide a reference framework for future developments in systemic risk management

Keywords Disaster risk reduction · Indian Tsunami · Systemic risk

---

Y. Wen (✉) · Y. Zhang · Y. Gao · A. Mitra · R. Shaw
Graduate School of Media and Governance, Keio University, Fujisawa, Japan
e-mail: wen.yafang.study@keio.jp

Y. Zhang
e-mail: yanwu@sfc.keio.ac.jp

Y. Gao
e-mail: sxyxgyn@qq.com

A. Mitra
e-mail: arunabhmitra@gmail.com

R. Shaw
e-mail: shaw@sfc.keio.ac.jp

R. Shaw et al. (eds.), *Two Decades from the Indian Ocean Tsunami*, Disaster Risk Reduction, https://doi.org/10.1007/978-981-96-2669-4_5

5.1 Introduction

The term "systemic risk" refers to the possibility that a localized event may trigger a chain reaction, resulting in widespread disruptions across a variety of industries and sectors. Originally conceptualized within the financial sector, systemic risk has broadened to encompass disaster management, reflecting the interconnectedness of modern social, economic, and environmental systems (Helbing 2013). Intricacy, lack of certainty, ambiguity, and cross-border implications are characteristics of this risk (Mitra and Shaw 2022). Understanding systemic risk is crucial for disaster resilience. Traditional risk management approaches often focus on single hazards in isolation, neglecting the cascading effects of disasters. In contrast, systemic risk assessment helps identify these interconnections, allowing for more comprehensive risk reduction strategies that address the root causes of vulnerability and enhance adaptive capacities across multiple sectors (UNDRR 2019).

Furthermore, systemic risk in disaster management provides a holistic view of potential hazards. By considering the interconnected nature of modern societies, systemic risk analysis facilitates the identification of critical infrastructure and services that, if disrupted, could lead to widespread societal impacts. This approach promotes integrated risk management strategies that take into account compound and cascading hazards (OECD 2014). Effective systemic risk management involves cross-sector collaboration, continuous monitoring, and adaptive planning. It requires a shift from reactive to proactive measures, emphasizing prevention, preparedness, and resilience-building. By doing so, integrating systemic risk into disaster management frameworks enables policymakers to better anticipate the indirect and long-term consequences of disasters, thereby reducing their overall impact on communities (Linkov and Trump 2019).

A compelling illustration of systemic risk in action is the 2004 Indian Ocean tsunami, which serves as a poignant case study for understanding systemic risk in disaster management. On December 26, 2004, a massive undersea earthquake off Sumatra triggered devastating tsunamis, impacting 14 countries, and causing approximately 230,000 deaths (National Geographic 2005). Besides causing immediate destruction and loss of life, the tsunami triggered cascading effects across a wide range of sectors. Among the effects of the tsunami were disruptions of transportation networks, interruptions of communication systems, and overcrowding in healthcare facilities. The economic impact was profound, with significant losses in tourism, fisheries, and agriculture. The disaster highlighted the vulnerabilities of interconnected systems and underscored the need for comprehensive risk management approaches that account for systemic interdependencies (Kahn 2005). The response to the tsunami also underscored the importance of international cooperation and early warning systems in mitigating systemic risk. Efforts to rebuild and enhance resilience were multifaceted, focusing on strengthening infrastructure, improving disaster preparedness, and fostering community engagement. These initiatives highlighted the critical role of systemic risk management in disaster recovery and long-term resilience (UNESCO-IOC 2007).

Taking these insights into account, this chapter examines the 2004 Indian Ocean tsunami from the perspective of systemic risk, with the goal of shedding light on the evolution of risk management practices. It explores the theoretical underpinnings, historical developments, and current challenges in managing systemic risks, providing a reference framework for enhancing societal resilience against complex and compound hazards.

5.2 Evolution of Systemic Risk Management Since 2004

5.2.1 Major Developments in Systemic Risk Theory

As a result of the interconnectedness and interdependence of markets and financial systems, the concept of systemic risk first originated in the financial sector. This early understanding was influenced by network theory and the study of complex systems, emphasizing how small perturbations in one part of the system could lead to significant disruptions elsewhere (Mitra and Shaw 2022). In disaster management, systemic risk theory gained traction as researchers recognized parallels between financial systems and social systems impacted by disasters. The complex adaptive systems (CAS) theory, an extension of systems theory, highlights how systems self-organize, adapt, and evolve, which is crucial for understanding how disasters can trigger cascading effects across interconnected systems (Helbing 2013).

From this initial understanding to its practical implications, the historical development of systemic risk can be traced back to several key events that highlighted the importance of risk management and assessment integrated. It was the 2004 Indian Ocean tsunami that demonstrated the catastrophic potential of interconnected risks at its earliest and most significant. The tsunami not only caused immediate devastation through flooding and destruction of infrastructure but also had far-reaching secondary effects on healthcare, the economy, and communication systems across multiple countries (National Geographic 2005). The disaster sparked a global interest in systemic risk, driving both theoretical research and practical application. According to the International Risk Research Group (IRGC), systemic risk is a risk that arises from complex interactions and may have a significant impact on the system as a whole. Over the past two decades, systemic risk research has evolved from traditional single-risk analyses to frameworks that integrate multiple risks and their interactions. More typical is the integrated risk management framework and the emergence of complex systems theory, a system that emphasizes cross-disciplinary collaboration and systems thinking and proposes that risk management should take into account the interdependencies between different risks. A framework that proposes to recognize and manage risks under different risk conditions from a systemic perspective. This theory is reflected in many countries' disaster management policies, such as the comprehensive disaster management strategies of Japan and the European Union (Andersson and Herzog 2016) (Fig. 5.1).

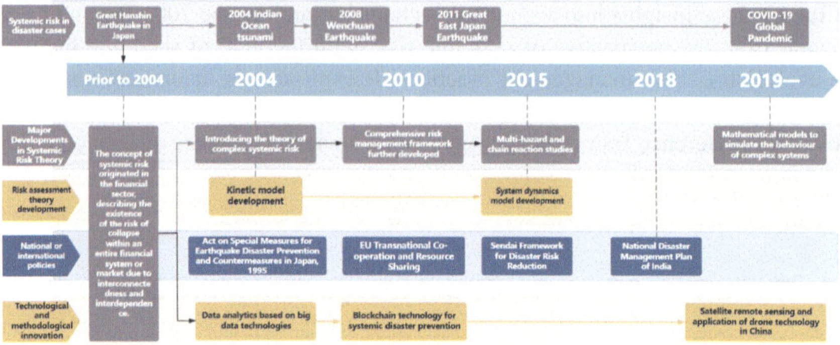

Fig. 5.1 The evolution of systemic risk management since 2004

Since the 2004 Indian Ocean tsunami, systemic risk theory has evolved through a number of significant stages of development, from traditional single-risk analyses to a framework that integrates multiple risks and their interactions (UNDRR 2015). The dramatic impact of the 2004 Indian Ocean tsunami not only triggered a global focus on systemic risk, but also prompted researchers to realize that natural disasters are not just a single event, but a complex systemic problem that involves multiple interconnected factors. To address this complexity, complex systems theory was introduced to study interactions between different elements and their impact on overall system stability. After the Indian Ocean tsunami, researchers realized the intricate linkages between natural disasters and social systems, such as the impact of infrastructure collapse on relief efforts, and thus complex systems theory was used to model the chain reaction after a disaster to develop more effective contingency plans.

From 2010, the Comprehensive Risk Management Framework (CRMF) was further developed. CRMF emphasizes cross-disciplinary collaboration and systems thinking, proposing that risk management should take into account the interdependencies between different risks (European Commission 2013). This framework was proposed when the 2004 Indian Ocean tsunami demonstrated the limitations of single-risk management. This prompted researchers to consider multiple risks and their interactions. CRMF has been reflected in many countries' disaster management policies, such as the comprehensive disaster management strategies of Japan and the EU (NDMA 2014). In the aftermath of the Great East Japan Earthquake in 2011 and the Fukushima Nuclear Accident in 2011, multiple hazards and their knock-on effects were demonstrated to be insufficiently responded to. Prompted researchers to apply scenario analysis to disaster early warning systems to improve early warning and emergency response capabilities.

During 2015–2019, multi-hazard and chain reaction research became a significant area of systemic risk management. Scenario analysis and system dynamics modelling were widely used to analyze the dynamic interactions of factors within a system by constructing different future scenarios and simulating the behavior of complex systems through mathematical models (NDMA 2014). Additionally, complex network analysis has also been introduced to study different nodes and their connectivity in a system to reveal the propagation paths and key nodes of systemic risks. This theory is widely used in financial systems, supply chain management and infrastructure networks to identify and manage critical nodes and weak links.

5.2.2 *Advances in Risk Assessment and Mitigation Strategies*

Risk assessment uses advanced models and tools to identify and assess systemic risks. For example, methods such as scenario analysis, system dynamics modelling, and complex network analysis are used to understand and predict disaster chain reactions. A substantial amount of progress has also been made in mitigation strategies, including improvements to early warning systems, optimization of contingency plans, and enhancement of social infrastructure.

Scenario analysis assesses potential risks and impacts under each scenario by constructing different future scenarios. For example, in Japan and Indonesia's disaster early warning systems, early warning and emergency response capacities have been improved by modelling different scenarios. (Valdés and Coombs 2018). System dynamics models, on the other hand, use mathematical models to simulate complex systems and analyze the dynamic interactions of factors within the system. The Indian Ocean tsunami demonstrated the multiple impacts of disasters on various systems, prompting the development of more comprehensive dynamic models. In the Netherlands water management system, system dynamics models are used to simulate sea level rise and flooding, improving disaster preparedness. Our ability to use such assessment models in order to predict and explore future developments is certainly crucial in the face of increasingly dramatic changes in climate patterns.

5.2.3 *Integration of Systemic Risk in National and International Policies*

Many countries and international organizations have integrated systemic risk management into their policy frameworks. For example, the Sendai Framework for Disaster Risk Reduction (2015–2030) published by the United Nations International Strategy for Disaster Reduction (UNISDR) emphasizes systemic risk management. In addition, several countries have formulated relevant laws, regulations, and policies aimed at enhancing their capacity to respond to systemic risks. More representatively, the Japanese government recognized the need to strengthen systemic risk management and initiated a series of measures to enhance its response capacity (Australian Government 2020). The Act on Special Measures for Earthquake Disaster Prevention and Countermeasures was adopted, which specifically addresses earthquake hazards, emphasizing comprehensive precautionary measures for earthquake disasters and the construction of earthquake early warning systems (Hatamura et al. 2014).

Through international cooperation and policy coordination, the United Nations International Strategy for Disaster Reduction (UNISDR) aims to reduce disaster risk. The global impact of the 2004 tsunami prompted United Nations Member States to recognize international cooperation. Governments are encouraged to develop and enact systemic risk management policies under the Sendai Framework, such as India's National Disaster Management Plan (BNPB 2018). The European Union (EU) strengthens systemic risk management through transnational cooperation and resource sharing. After the 2004 tsunami, EU member states recognized the need to work together in response to transnational disasters and formulated the EU Strategy on Disaster Reduction (EUSDR), which promotes collaboration and resource sharing among member states to improve prevention and mitigation capacity.

5.2.4 Technological and Methodological Innovation

Many countries and international organizations have integrated systemic risk management into their policy frameworks. For example, the Sendai Framework for Disaster Risk Reduction (2015–2030) published by the United Nations International Strategy for Technology-based Disaster Reduction contributes significantly to systemic risk management. Geospatial information systems (GIS), remote sensing, and other technologies are being developed, big data analysis and artificial intelligence have made disaster monitoring and risk assessment more accurate and efficient. For instance, satellite remote sensing and unmanned aerial vehicle (UAV) technologies enable rapid access to real-time images of disaster areas, providing critical data support for post-disaster reconstruction. Several emerging technologies, especially those related to autonomous control and unmanned detection, have shown strong functionality and development potential for systemic risk detection and management. How to use these technologies correctly and better for systemic risk detection will undoubtedly become a crucial discovery for systemic risk management in the future.

Remote sensing technology improves disaster monitoring and assessment accuracy by acquiring real-time data from disaster sites through satellites and drones. In China, for example, monitoring earthquakes and floods through satellite remote sensing technology has improved emergency response efficiency (Li and Wang 2017). Big data analytics and AI technologies leverage massive amounts of data to predict disaster risks and trends. The increased demand for data processing and analytics capabilities following the 2004 tsunami has fueled the development of these technologies. In the United States, early warning systems were improved by analyzing hurricane data through big data and AI technologies (NDRRMC 2014). Blockchain technology, on the other hand, improves disaster data credibility and transparency through its transparency and non-tamper ability. This ensures the authenticity and transparency of information in disaster emergency response and resource allocation.

5.3 Systemic Risk Manifested in Previous Disaster Cases Since 2004

5.3.1 Systemic Risks in the 2004 Indian Ocean Tsunami

The 2004 Indian Ocean tsunami, triggered by a 9.0–9.3 magnitude undersea earthquake off Sumatra, resulted in immediate catastrophic impacts and triggered chain reactions across various sectors. The tsunami affected 14 countries, causing approximately 230,000 deaths, and displacing over 1.7 million people (National Geographic 2005). There were extensive casualties and significant economic losses, especially in the hard-hit countries of Indonesia, Thailand, Sri Lanka, and India.

Specifically, the tsunami resulted in extensive fatalities and injuries, with entire communities destroyed within minutes. Indonesia suffered the highest number of casualties, with over 170,000 deaths reported in Aceh province alone. In Sri Lanka, the tsunami killed approximately 35,000 people, while India and Thailand reported death tolls of 16,000 and 8,000, respectively. Economic losses were equally devastating. In Indonesia, the Aceh province's economy, heavily reliant on agriculture and fishing, was decimated. In Thailand, the tourism sector, particularly Phuket and the surrounding islands, was severely impacted. Hotels, resorts, and other infrastructure were destroyed, leading to significant job losses and economic instability. Sri Lanka's agricultural sector also suffered, with vast tracts of farmland rendered unusable due to saltwater intrusion (Mitra and Shaw 2022).

Furthermore, the tsunami severely disrupted communication networks, making rescue and relief efforts difficult. In many affected areas, telecommunications infrastructure was completely destroyed, isolating communities, and hampering information flow. Transport infrastructure, including roads, bridges, and ports, was extensively damaged, further complicating aid delivery and the movement of people and goods. In Aceh, the destruction of the main airport and seaport delayed international aid. Healthcare facilities were overwhelmed by the sudden influx of injured individuals. Many hospitals and clinics were damaged or destroyed, leading to a critical shortage of medical supplies and personnel. The lack of clean water and sanitation facilities in temporary shelters also contributed to the spread of disease, exacerbating the public health crisis (UNESCO-IOC 2007).

5.3.1.1 The Impact of Economy Systems

Turning to Thailand, the tourism sector, particularly Phuket and the surrounding islands, suffered extensive damage. The destruction of hotels, resorts, and other tourism infrastructure resulted in significant job losses and economic instability. The tourism industry, which accounted for a significant portion of Thailand's GDP, experienced a severe downturn as tourists avoided the region in the aftermath of the disaster. Local businesses dependent on tourism, such as restaurants, shops, and tour operators, also suffered, leading to widespread economic hardship. The Thai government and international organizations launched efforts to rebuild the tourism infrastructure and promote the region as a safe travel destination, but recovery was slow and challenging (Rodriguez et al. 2006).

Similarly, in Indonesia, particularly in the Aceh province, the tsunami had a devastating impact on the fishing industry. The powerful waves destroyed fishing boats, gear, and coastal infrastructure, leading to significant livelihood losses for local communities. Fisheries disruption not only impacted local food security but also had broader economic implications as fish exports declined, affecting national revenue. The Indonesian government, with support from international donors, implemented programs to provide refurbished boats and fishing equipment to affected communities. However, the recovery process was slow due to the scale of the devastation and the need for comprehensive infrastructure rebuilding (Rodriguez et al. 2006).

Likewise, in Sri Lanka, the tsunami inundated vast tracts of agricultural land with seawater, causing long-term soil salinity issues and affecting crop yields and food production. The eastern and southern coasts, where rice paddies and other crops were prevalent, were particularly hard hit. Farmers faced significant challenges in restoring their land to a productive state, and many were forced to rely on food aid as soon as the disaster occurred. The Sri Lankan government and international aid organizations launched initiatives to rehabilitate the agricultural sector, including soil restoration projects and the distribution of salt-tolerant crop varieties, but full recovery took several years (Rodriguez et al. 2006).

5.3.1.2 The Impact of Social Systems

In addition to these economic and infrastructural impacts, the 2004 Indian Ocean tsunami's impact extended beyond immediate physical and economic damages, deeply affecting social systems within the affected regions. The disaster exposed and exacerbated existing vulnerabilities and inequalities within societies. One of the most profound social impacts was the displacement of over 1.7 million people, leading to significant social disruption. Communities were forced to relocate to temporary shelters, often overcrowded and unsanitary. This displacement disrupted social networks and community cohesion, making recovery and rebuilding efforts more challenging. The loss of homes and livelihoods created a sense of insecurity and instability among survivors, many of whom struggled with the psychological trauma of losing loved ones and their way of life (Mitra and Shaw 2022).

Psychologically, survivors experienced immense trauma. The sudden loss of loved ones, homes, and livelihoods resulted in widespread grief, anxiety, and depression. Survivors often reported nightmares, flashbacks, and severe anxiety symptoms related to post-traumatic stress disorder (PTSD). The long-term psychosocial impact included increased incidences of mental health issues, which required sustained support and intervention. In Aceh, community-based mental health programs were established to provide counseling and support services, but the stigma associated with mental health issues often hindered access to these services (Rodriguez et al. 2006).

Healthcare systems were also severely affected. The healthcare system was overwhelmed by sudden medical demand. The destruction of healthcare facilities and the loss of healthcare workers exacerbated the strain on the system. In many affected areas, there was a critical shortage of medical supplies, clean water, and sanitation facilities, leading to the spread of diseases such as cholera and dysentery. The long-term impact of the tsunami highlighted the need for robust healthcare infrastructure and emergency preparedness plans to manage large-scale crises. Efforts to rebuild and strengthen healthcare systems included the construction of refurbished hospitals and clinics, the training of healthcare workers, and the establishment of early warning systems for future disasters (UNESCO-IOC 2007).

Moreover, many schools were destroyed or repurposed as shelters, disrupting thousands of children's education. The loss of educational facilities and the psychological impact on children had long-term implications for their development and well-being. In the immediate aftermath of the disaster, many children could not attend school, leading to significant interruptions in their education. School rebuilding and the provision of psychosocial support for children were essential components of the recovery process. Initiatives to restore education included the construction of temporary learning spaces, the distribution of school supplies, and the implementation of trauma-informed teaching practices to help children cope with their experiences (Rodriguez et al. 2006).

5.3.2 Systemic Risk Manifested After the 2004 Indian Ocean Tsunami

As a result of this awareness, the Japanese Tohoku earthquake and tsunami and the Fukushima nuclear disaster in 2011 were also critical cases. These events underscored the interplay between natural and technological hazards, showing how natural disasters can trigger technological failures with widespread and long-term consequences. The cascading effects of this disaster prompted a reevaluation of existing risk management frameworks to incorporate systemic risk perspectives (Mitra and Shaw 2022).

As an example of systemic risk in contemporary times, one can look at the COVID-19 pandemic, where a public health crisis rapidly escalated into a global economic and social disaster. The pandemic highlighted vulnerabilities within healthcare systems, supply chains, and economies, reinforcing the necessity for a holistic approach to managing systemic risks.

5.4 Current Challenges of the Systemic Risk

Despite many advances in systemic risk management, some notable shortcomings remain. Existing risk management approaches often lack a comprehensive understanding of complex system interactions. Additionally, risk assessment models often struggle to provide accurate predictions under conditions of uncertainty and extreme events. For example, in the 2017 Mexico City earthquake, ground subsidence triggered by the over-pumping of groundwater resources led to large errors in disaster risk assessment models, and as a result, many buildings were damaged in the earthquake (LACSC 2020). In the Christchurch earthquake in New Zealand, the failure of the model to adequately take into account the chain reaction of the earthquake led to an underestimation of building damage. The following section analyzes the challenges of current system risk development.

5.4.1 Insufficient Collaboration Capabilities Among Stakeholders

Political factors also largely influence systemic risk management implementation. Inconsistency in policy formulation and implementation is a major problem. Lack of collaboration and information sharing among different government departments leads to the inefficient implementation of disaster management policies. The Philippines' response to typhoons, for instance, the lack of coordination between different departments led to uneven distribution of relief materials, which affected relief efforts (Katada et al. 2007). In addition, unclear authority and responsibility and poor collaboration between local and central governments led to confusion in disaster response.

5.4.2 Lack of Public Awareness of Systemic Risks

Cultural factors can also have a significant impact on systemic risk management. Firstly, differences in risk perceptions and coping attitudes across cultures, affect the acceptance and implementation of disaster management policies (Qiao and Cheng 2012). For example, underestimation of risk and distrust of disaster responses may exist in some cultures, leading to difficulties with policy implementation. Secondly, community and public participation is also a key factor. In many countries, public awareness and capacity to participate in disaster management are low, and there is inadequate effective public education and training, leaving communities and individuals ill-prepared to assist each other. For example, in Japan, although the government has taken several measures to strengthen disaster management, there are still deficiencies in public awareness of and response to nuclear power plant risks, resulting in an ineffective community and public response to the Fukushima nuclear accident aftermath.

5.4.3 Data and Predictive Technology in Need of Improvement

Accurate data and efficient predictive models are the foundation of systemic risk management. However, current data collection and sharing mechanisms remain inadequate, leading to limitations in risk assessment and decision support. For example, many developing countries lack advanced technical means to obtain and process disaster data. This results in incomplete and inaccurate data, making it difficult to assess risk effectively. Therefore, establishing a global data-sharing platform and improving data quality are critical tasks for the future. Moreover, the accuracy of prediction models is limited: the accuracy and reliability of existing prediction models still need to be improved when dealing with complex and changeable disasters. Models often rely on historical data, but their predictive power may be insufficient when dealing with new or extreme events. In 2011, an earthquake occurred in northeastern Japan, causing a huge tsunami. Although Japan has a highly developed earthquake and tsunami warning system, the size of this earthquake and the height of the tsunami exceeded prediction models. The Fukushima nuclear power plant design, in particular, failed to fully take into account such a large-scale tsunami risk, leading to serious consequences (Pletcher and Rafferty 2024).

5.5 Recommendations for Enhancing Resilience Against Systemic Risks

It is mentioned in the Asia–Pacific Disaster Report 2021 that the impact of climate change is chain-linked. Firstly, it will have a serious impact on vulnerable groups and slow down the process of poverty reduction (ESCAP 2021). Secondly, climate change threatens the physical and mental health and legal rights of women, the elderly, and children. At the same time, disasters caused by climate change will directly affect the lives of ordinary people, the economy, society, and homeland security as well as their health, and living conditions (Hui-Min et al. 2021). Based on the literature on systemic risk, this section will provide how to better respond to systemic risks brought about by disasters due to global warming.

5.5.1 Raise Public Awareness of Systemic Risks

As compared to catastrophic events, systemic risks are typically underestimated (Schweizer and Renn 2019). This phenomenon occurs because systemic risks are very complex, and it is difficult to identify and quantify the cause and effect between multiple influencing factors and specific effects, and therefore systemic risk is often considered nonlinear (Schweizer and Renn 2019). In addition, systemic risks are related to the well-known common pool problem, that is, the impact of each actor on systemic risks is insignificant, and therefore there is a lack of motivation for change. Human society is seriously underestimating the impact of systemic risks because the public lacks a basic understanding of them (Renn et al. 2022). Based on this, this article believes that improving the country's comprehensive understanding of systemic risks, paying attention to systemic risk management, eliminating inconsistencies in supervision, and controlling free-riding behavior can alleviate social conflicts that arise after disasters. In addition, from the perspective of community disaster reduction, we should focus on cultivating public awareness of disaster risks and implementing systemic risk management policies in grassroots communities.

5.5.2 Promote Interdisciplinary Theoretical Research

Based on the systemic risk theory, disasters will impact infrastructure, economy, social culture, and other aspects, seriously affecting the normal life of human beings. To date, comprehensive research into disaster management is still lacking in most regions around the world (Schweizer and Renn 2019). Therefore, there is a need to promote the integrated development of science and technology (technical electricity, water conservancy engineering, and construction science, etc.), natural science (environment, disaster), social science (policy, social culture), and other multi-disciplinary approaches. Efforts should be made to integrate basic disciplines with the frequency and scope of the impact of floods, hurricanes, earthquakes, and other disasters. This will promote systemic risk theory improvement and development. In addition, attempts to establish a comprehensive theoretical measurement framework for systemic risk are also based on interdisciplinary research. A complete system risk measurement framework can better predict the impact a region may suffer after a disaster occurs, and preventive measures can be taken in advance for disasters. After a disaster occurs, a complete system risk management framework can reduce losses as much as possible, narrow the scope of impact, speed up post-disaster recovery, and avoid more social problems.

5.5.3 *Build an Information Sharing Platform Among Stakeholders*

Events' impacts on politics, economics, social, culture, and ecology need to be considered when assessing systemic risks (Hui-Min et al. 2021). From a national perspective, it is necessary to promote collaborative cooperation among various government departments, establish efficient information sharing and communication mechanisms, and jointly formulate and implement risk management strategies, such as cooperation between meteorological departments and transportation departments to guarantee the normal operation of the transportation system under extreme weather; Communications and disaster management departments working together to convey disaster warnings in a timely manner. In addition, the government should promote public–private cooperation, including social organizations' role in disaster relief, to ensure that social rescue forces can function in a timely manner after disasters and minimize other problems caused by untimely rescue, cooperation between medical resources and the government, and complete information resources. Ensure that the affected people receive immediate treatment to avoid other health diseases after the disaster. There is a close relationship between social media and the government, which has a profound impact on the media (Mavrodieva et al. 2019). In order to avoid social panic after a disaster, residents should be instructed on how to view the disaster in the right way before and after it occurs. From a global perspective, it is imperative to strengthen cooperation between countries, as well as exchanges among countries and academic staff from non-governmental organizations, and to improve disaster management. To jointly respond to regional or global systemic risks, it is imperative to share information and resources.

5.5.4 *Strengthen Data and Technological Support*

Advanced science and technology can help predict disasters more accurately, and governments can formulate targeted measures to mitigate systemic risks. In addition, international cooperation and collaboration are essential. To create standardized protocols for data collection, sharing, and analysis, governments, international organizations, NGOs, and the private sector are required to collaborate. A global data-sharing platform could leverage advances in technology such as cloud computing, blockchain, and artificial intelligence to ensure data integrity, accessibility, and security. For example, the African Union is promoting the Pan-African Disaster Management System, which aims to enhance African countries' disaster management capacity through data-sharing and technical cooperation (UNDP 2023).

5.6 Discussion

This chapter discusses and analyzes systemic risk theory. Since the Indian Ocean tsunami in 2004, systemic risk theory has been recognized and paid attention to by the international community. It plays a crucial role in disaster risk reduction. Since 2004, systemic risk management theory has undergone significant development, moving away from traditional single-risk analyses to a framework that integrates multiple risks and their interactions. Complex systems theory, integrated risk management frameworks, research on multiple hazards and chain reactions, and the introduction of emerging technologies are driving systemic risk management in a more comprehensive and effective direction. Despite the many challenges, systemic risk management will continue to play a significant role in mitigating disaster impacts and enhancing the resilience of societies as theories and technologies continue to advance.

Although systemic risks are evolving rapidly, challenges remain. There are still many barriers to the implementation of systemic risk management, including political, technological, and cultural factors. Uncoordinated policy formulation and implementation, lack of cross-sectoral cooperation, and differences in risk perceptions in cultural contexts limit systemic risk management effectiveness. To overcome these barriers, countries need to enhance overall disaster response capacity by raising public awareness of systemic risks, facilitating the development of data sharing and prediction technology, promoting dialogue among stakeholders, and promoting the development of systemic risks through interdisciplinary theoretical research.

5.7 Conclusion

In the face of the increasing intensification of climate change and rising disaster risks, humanity must fully recognize the complex impacts of disasters and their significant effects on the economy, society, and culture. Systemic risk management requires close cooperation and exchange between different disciplines and places higher demands on interdisciplinary disaster management research. In today's highly globalized world, systemic risks can have regional and even global impacts. Therefore, while improving their own disaster risk management policies, governments must also emphasize regional and global disaster management collaboration, promote disaster data sharing, increase the development of relevant scientific technologies, and enhance disaster prediction capabilities.

Furthermore, systemic risk management should not remain at the theoretical level but should extend to grassroots communities, increasing public awareness of systemic risks and strengthening community resilience to disasters, thereby minimizing the impact of disasters on human society.

References

Andersson C, Herzog S (2016) Systemic risk and comprehensive risk management. J Risk Res 19(6):721–739. https://doi.org/10.1080/13669877.2015.1042500
Australian Government (2020) Royal commission into National Natural Disaster arrangements. https://naturaldisaster.royalcommission.gov.au/
African Union (2019) Pan-African disaster risk reduction framework. https://au.int/en/articles/pan-african-disaster-risk-reduction-framework
BNPB (2018) Indonesia disaster management reference handbook. https://www.bnpb.go.id/
European Commission (2013) EU strategy for disaster risk reduction. Retrieved from https://ec.europa.eu/echo/what/civil-protection/disaster-risk-reduction_en
Emergency Response Law of the People's Republic of China (2007) Source: emergency response law of the People's Republic of China (Order of the President No. 69). https://www.gov.cn/
ESCAP U (2021) Resilience in a riskier world: managing systemic risks from biological and other natural hazards. Available at: https://repository.unescap.org/bitstream/handle/20.500.12870/3811/ESCAP-2021-RP-Asia-Pacific-Disaster-Report-2021.pdf?sequence=1
Goh XP, Radhakrishnan M, Zevenbergen C, Pathirana A (2017) Effectiveness of runoff control legislation and Active, Beautiful, Clean (ABC) Waters design features in Singapore. Water 9(8):627. https://doi.org/10.3390/w9080627
Hatamura Y, Abe S, Fuchigami M, Kasahara N, Iino K (2014) The 2011 Fukushima nuclear power plant accident: how and why it happened, vol 73. Woodhead publishing. https://doi.org/10.1016/C2014-0-00845-3
He X, Zhang XY (2021) Research on climate change adaptation: international negotiation issues and domestic action strategies. Science Press, Beijing
Helbing D (2013) Globally networked risks and how to respond. Nature 497(7447):51–59. https://doi.org/10.1038/nature12047
Hui-Min LI, Xue-Chun WANG, Xiao-Fan ZHAO, Ye QI (2021) Understanding systemic risk induced by climate change. Adv Clim Chang Res 12(3):384–394. https://doi.org/10.1016/j.accre.2021.05.006
Kahn ME (2005) The death toll from natural disasters: The role of income, geography, and institutions. Rev Econ Stat 87(2):271–284. https://doi.org/10.1162/0034653053970339
Katada T, Kimura H, Kodama M (2007) A study on the desirable utilization of flood hazard maps for disaster risk communication. J Jpn Soc Civ Eng Ser D 63(4):498–508. https://doi.org/10.2208/jscejd.63.498
Kan R (2022) Knowledge-action network on emergent risks and extreme events. Environ Sci Policy, 140:122–133. https://www.risk-kan.org/knowledge-action-network-on-emergent-risks-and-extreme-events-risk-kan/risk-kan-motivation-and-more/
Latin American & Caribbean Seismological Commission (LACSC) (2020) Improving earthquake prediction models in Latin America. https://lacsc.org/en/
Li J, Wang S (2017) Advances in remote sensing applications for disaster management. J Remote Sens 41(4):771–785. https://doi.org/10.1080/01431161.2016.1275059
Linkov I, Trump BD (2019) The science and practice of resilience. Springer International Publishing, Cham. https://www.oecd.org/content/dam/oecd/en/topics/policy-sub-issues/water/council-recommendation-on-water.pdf
Mavrodieva AV, Rachman OK, Harahap VB, Shaw R (2019) Role of social media as a soft power tool in raising public awareness and engagement in addressing climate change. Climate 7(10):122. https://doi.org/10.3390/cli7100122
Mitra A, Shaw R (2022) Systemic risk from a disaster management perspective: A review of current research. Environ Sci Policy 140:122–133. https://doi.org/10.1016/j.envsci.2022.11.022
National Disaster Management Authority (NDMA), India (2014) National disaster management system. https://ndma.gov.in/en/
National Geographic (2005) The deadliest Tsunami in history? National Geographic News. https://www.nationalgeographic.com/science/article/deadliest-tsunami-in-history

National Disaster Risk Reduction and Management Council (NDRRMC), Philippines (2014) Typhoon Haiyan (Yolanda) situation report. https://ndrrmc.gov.ph/

Nuclear Regulation Authority, Japan (2012) The Fukushima Daiichi nuclear accident report. https://www.nsr.go.jp/data/000067226.pdf

OECD (2014) Boosting Resilience through Innovative Risk Governance. OECD Publishing. https://doi.org/10.1787/9789264209114-en

Pescaroli G, Alexander D (2018) Understanding compound, interconnected, interacting, and cascading risks: a holistic framework. Risk Anal 38(11):2245–2257. https://doi.org/10.1111/risa.13128

Pletcher K, Rafferty JP (2024) Japan earthquake and tsunami of 2011. Encyclopedia Britannica. https://www.britannica.com/event/Japan-earthquake-and-tsunami-of-2011

Qiao G, Cheng X (2012) The improvement of legal system and management mechanism for natural disaster prevention in China: Lessons from the United States experience. Wuhan University Institute of Environmental Law. http://riel.whu.edu.cn/view/2124.html

Renn O, Laubichler M, Lucas K, Kröger W, Schanze J, Scholz RW, Schweizer PJ (2022) Systemic risks from different perspectives. Risk Anal 42(9):1902–1920. https://doi.org/10.1111/risa.13657

Rodriguez H, Wachtendorf T, Kendra J, Trainor J (2006) A snapshot of the 2004 Indian Ocean tsunami: Societal impacts and consequences. Disaster Prev Manag: Int J 15(1):163–177. https://doi.org/10.1108/09653560610654310

Schweizer PJ, Renn O (2019) Governance of systemic risks for disaster prevention and mitigation. Disaster Prev Manag: Int J 28(6):862–874. https://doi.org/10.1108/DPM-09-2019-0282

UNDP (2023) Data-driven disaster risk reduction in Africa: Shaping a resilient future. https://www.undp.org/africa/press-releases/data-driven-disaster-risk-reduction-africa-shaping-resilient-future

UNDRR (2019) Global assessment report on disaster risk reduction 2019. United Nations Office for Disaster Risk Reduction. https://www.undrr.org/publication/global-assessment-report-disaster-risk-reduction-2019

UNESCO-IOC (2007) Indian ocean tsunami warning and mitigation system. UNESCO. https://jdoss1.jodc.go.jp/jodcweb/info/ioc_doc/Technical/IOC_Technical_Series_77.pdf

United Nations Office for Disaster Risk Reduction (UNDRR) (2015) Sendai framework for disaster risk reduction 2015–2030. https://www.preventionweb.net/sendai-framework/sendai-framework-for-disaster-risk-reduction

Valdés J, Coombs S (2018) Post-event analysis of Mexico City's 2017 earthquake: Groundwater extraction-induced subsidence. Environ Res Lett 13(6):064027. https://doi.org/10.1088/1748-9326/aac438

Chapter 6
Multicultural Nursing Approach on Primary Health Care to Enhancing Globalized Community Resilience in Japan

Sakiko Kanbara

Abstract This chapter explores the evolving health landscape in Southeast Asia, where communicable diseases have historically posed significant challenges. Primary Health Care (PHC) approaches have addressed local issues. However, with globalization, the Asia region has also seen an increase in people's mobility and digital transformation. The 2004 tsunami devastated entire communities, highlighting various social determinants of health. The reconstruction and recovery process necessitated the adoption of new lifestyles and coping mechanisms. This paper aims to discuss how people have managed their health and well-being amidst these evolving risks and what future collaborations are necessary to enhance community resilience and health outcomes in Japan and Asian countries from the 20th anniversary of the 2004 Indian Ocean tsunami. This study positioned mutual aid disaster risk reduction in a multicultural society as "primary health care focusing on social determinants of health in peacetime" and considered it as an interaction of multiculture from a nursing perspective. The study examined the principles of primary health care: person-centered strategies, equitable access, citizen participation, multisectoral collaboration, and appropriate technology utilization. The results underscored the need for comprehensive policies that strengthen networks with all people, especially vulnerable populations, regardless of the type or timing of the disaster.

Keywords Multicultural nursing · Primary health care · Community resilience · Disaster risk reduction · Social determinants of health

S. Kanbara (✉)
Kobe City College of Nursing, Kobe, Japan
e-mail: sakikokanbara@gmail.com

R. Shaw et al. (eds.), *Two Decades from the Indian Ocean Tsunami*, Disaster Risk Reduction, https://doi.org/10.1007/978-981-96-2669-4_6

6.1 Introduction

The Sendai Framework for Disaster Risk Reduction 2015–2030 (UNISDR 2015) has been critical and aligns with SDG 11. The increasing severity of natural disasters underscores the need to integrate health considerations into policies and disaster risk reduction (DRR) and promote health investments as integral to the well-being of society rather than isolated expenditures (Benson et al.). The framework aims to reduce disaster risk and losses and emphasizes the complex relationship between resilient cities and health. The physical environment includes clean water and air, healthy workplaces, adequate housing, transportation services, and safe communities.

Among the 17 Sustainable Development Goals (SDGs) set for 2030, SDG 3 stands out with its call to "ensure healthy lives and promote well-being for all at all ages" (UN 2015). This goal requires robust primary health care (PHC) systems that can adapt to local conditions while providing universal health coverage (UHC), reducing economic hardship, and ensuring quality health services for all. PHC interventions that address broader social and environmental determinants of health have promised more significant benefits than those focusing solely on treating individual patients (Rifkin et al. 1988). However, the interconnectedness of all SDGs highlights the complex linkages between health and other areas, such as hunger, gender equality, clean water, sanitation, and climate change action.

According to WHO's "Solid Fact" report, factors other than genetics and access to healthcare, such as socioeconomic status, play a crucial role in health outcomes. It also points to the need to develop partnerships with environmental services to collect appropriate climate, weather, and environmental data to facilitate managing and communicating climate-related health risks. Viewed through the lens of disaster risk reduction also reported that vulnerability to hazards such as earthquakes and floods could be reduced by strengthening coping capacities. Top contemporary vulnerabilities include (1) increasing population density and the number of people aged 60 and over and (2) environmental changes and declining human capacity to cope with natural disasters (OECD 2003).

Therefore, in 2018, WHO launched a comprehensive framework called Health Crisis and Disaster Risk Management (H-EDRM) as a standard for people's health and well-being in health crisis management, including disasters and infectious diseases, which has substantial social implications rather than being a stand-alone measure of vertical administration. It calls for strengthening health systems that enable communities to engage, plan, and take risk-reduction actions as a system-wide responsibility (WHO 2019).

Globally, vulnerable populations are often considered at risk and face inequality, lack of healthcare, poor economic conditions, discrimination, and violence (Marmot 2007; Koh et al. 2011). These challenges are further complicated by individual, structural, and situational factors such as age, gender, literacy, and differences in economic, political, social, and environmental conditions. These factors increase the risk of health disadvantages and diseases in the host country. Traditionally, vulnerable populations encounter significant barriers to accessing health care, influenced by cultural differences in medical practices and perceptions (Lebanon et al.). Access to health care remains a fundamental human right and an essential aspect of health policy and health services research (OHCHR 2008; WHO 2017). The COVID-19 pandemic significantly strained health services at the physical, mental, economic, and social levels. Strengthening PHC, the front line of all health systems, is critical for health systems to remain robust against such a significant health crisis (OECD 2021). A new World Bank report also urges countries to increase investment in PHC (Barış et al. 2022). The COVID-19 public health crisis highlighted the significant challenges vulnerable populations face and the importance of ensuring timely access to health care for multilingual and multicultural communities and promoting a concerted global response. It also highlighted the need for healthcare providers to build cultural competence and advocate for addressing the social determinants that affect vulnerable populations. This includes working with local healthcare providers, community health centers, and NGOs to enhance service delivery to diverse populations.

A cross-disciplinary framework is needed to rethink PHC in the future. Community health crisis management also requires knowledge and changes in people's values and behavior patterns to practice individual self-care. The role of nurses needs to be strengthened to support disaster risk reduction efforts and respond to local disaster prevention policy and community revitalization efforts. Policies and education must be reevaluated from the human security perspective to promote global health beyond national and regional inequalities.

Particularly in Japan, with its declining and super-aging population, natural disasters, and diverse health challenges, transitioning to prevention and risk-reduction measures in households and communities is essential to reduce the medical burden and spread of disease. The COVID-19 pandemic is a significant challenge to the health system because of the physical distance and social isolation contributing to information gaps. The COVID-19 pandemic highlights the importance of addressing the physical distance and lifestyle changes contributing to information gaps and social isolation. Communities today are increasingly mobile due to occupational and lifestyle choices, making self-care and primary health care at the individual, family, and community levels increasingly important.

6.2 Inclusive Primary Health Care for Multicultural Disaster Resilience

Losses due to crises have significant short-, medium-, and long-term impacts on the economy, society, health, culture, and environment, and human characteristics are distributed along a continuum that is not universally diagnosable as positive or negative, normal, or abnormal. The traditional family or community is no longer the standard unit of life. One of the most important cultural developments is that many people live alone or in nuclear families, a society where traditional mutual aid is inadequate. People are constantly on the move, and within a given space, there is a mix of cultural and social groups with different levels of racism, gender, generational inequality, and poverty. Focusing on human behavior, people today work in communities with knowledge about healthy living, safe societies, and local disasters. In an era characterized by uncertainty and long-term risks, individual lifestyle choices are increasingly influenced by health issues and behaviors at the local cultural level. Instead of helping each other within a community, these people are helping each other as "people" in global virtual societies and local mass gatherings. "People" and the normative assumptions under which they live are constantly changing over time. People also depend on the attitudes and culture of their local communities. We must continue to understand the nature of inclusive communities, considering the variety of relationships across borders. The most crucial factor is that many people live alone or in nuclear families, resulting in a society where mutual assistance and care are inadequate. This includes understanding the unique challenges that vulnerable populations face during disasters. Considering the various cross-border relationships and the nature of inclusive communities, we need to understand the unique challenges vulnerable populations face during disasters and help each other.

Therefore, the PHC approach advocated in this paper must also be person-centered, reflecting the nuances and resources of each society to ensure that no one is left behind (Sandier et al. 2004). This study adopts a PHC approach, recognizing that while individual control over some health determinants is limited, community health outcomes can be dramatically improved through collective action on multicultural coexistence. The necessary approach, as seen in the PHC principles that serve as a reference for mutual aid, can be viewed in contemporary Japanese society in BOX1(1–4). Furthermore, in line with the Astana Declaration used in the current strategy, future-oriented initiatives are also necessary to advance PHC for all residents, including vulnerable populations (5–8).

6.3 Nursing Ethics and Disasters

Next, we explore the clues to the ideal form of citizen participation and the communities that support it from the viewpoint of PHC. In examining disaster risk reduction in our multicultural daily lives, a comprehensive understanding of both health and disaster risk reduction will be discussed from the perspective of nursing science.

Nursing is an evolving practice that constantly changes and adapts to society. Unexpected experiences not only lead to lessons to be learned immediately afterward but also carry them with us as we continue our research on nursing practice. Figure 6.1 shows disaster events and important initiatives on disaster health as well as disaster nursing capacity development in Japan that collect unanticipated experiences and, through conceptual analysis and theory building, continue to think tirelessly about how to incorporate them into the next generation of nursing research and educational practice. Through this process, we must face the challenges of nursing researchers in

Japan and globally. Nursing does not situate the human person as a mere biological fact. The concept of humanity, its practical extension, and a significant part of the history of humanitarianism have been characterized by attempts to conceptualize the nature of desirable human attributes. Nursing, which, along with biological medicine, emphasizes the dignity and life of people, is theorized within the meta-paradigm of “human,” “environment,” “health,” and “nursing” (Donaldson and Crowley 1978). Three disciplinary themes can be described: The first is “an interest in the principles and laws governing human life processes, well-being, and optimal functioning, whether in illness or health.” The second is “an interest in human behavior patterns in interaction with the environment in critical life situations.” The third is “an interest in the processes by which positive changes in health status are affected.”

The International Council of Nurses has a Code of Ethics for all nurses worldwide. It affirms all persons’ human rights, integrity, and dignity, guarantees appropriate care for all persons without discrimination, and is to be practiced by nurses. Disasters, conflicts, and political divisions have increased the pressure on nurses, and the COVID-19 epidemic highlighted the difficulties nurses face. In 2021, a new code was published, reinforcing the content of the most recent code, the 2012 code. It also promotes equity and social justice, respect for the natural environment and climate, challenging unethical behavior, and the impact of technology in an ever-changing world, including digitized communication and artificial intelligence. In cross-cultural work, there is a new emphasis on the responsibility and accountability of nurses. It is necessary to assume personal responsibility for maintaining professional and practical knowledge while being informed about cultural values, needs, and respect.

Nurses live and work in their communities, and in the event of a disaster, they are often the frontline healthcare providers in and around the disaster area. They provide critical insight into the community’s immediate needs and collaborate with various organizations. This environment is markedly different from the typical workplace and requires constant contingency planning and problem-solving. Protecting individuals from the risk of illness and disasters is a complex issue, and conventional practices based on the idea of protecting individuals through manuals for one-off illnesses and disasters have not yet developed solutions for individuals who face multiple disabilities and risks throughout their lives. Social structural issues, such as age and gender disparities, have not been sufficiently addressed. Additionally, the rapid onset of new risks often surpasses the pace of improvements in vulnerability management, making it more difficult to identify affected individuals and distribute assistance effectively. When considering health and survival risks during disasters, the medical model typically provides a comprehensive view of diseases managed by healthcare professionals. On the other hand, the human security approach emphasizes development and social science perspectives, which are crucial in a more unpredictable society. Both approaches aim to safeguard the intrinsic value of human life. The human security model offers a more compassionate response to disasters by acknowledging potential freedoms. As a result, community development practitioners in disaster management and public health should adopt a human security

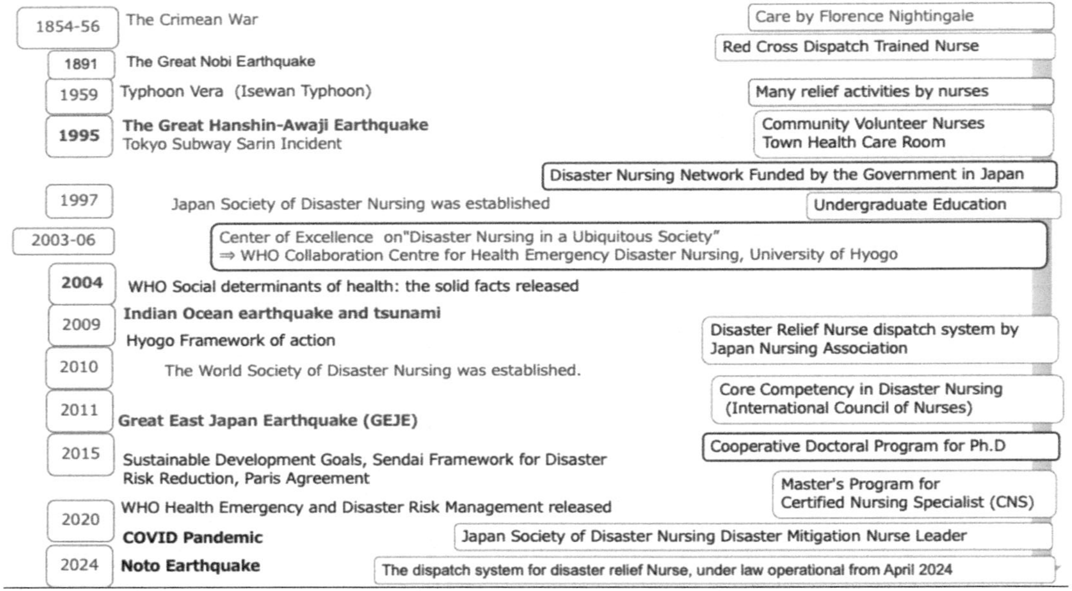

Fig. 6.1 Disaster event and important initiative (Left) on disaster health and changing disaster nursing capacity (Right box)

perspective in their efforts. This shift underscores the need for a new ethical framework that integrates the principles of human science and technology ethics within nursing. Additionally, collaboration with disaster scientists is essential to explore and address the fundamental question of "how should humans live?" with a strong emphasis on cultural context.

Figure 6.2 shows Leininger's nursing theory as "cultural care theory, on Social Determinant of Health by Marmot," and author modified the subject of universality and diversity of cultural care, which has been studied and utilized by nursing students worldwide. This is supplemented by the PHC, as mentioned in earlier principles. Cultural care, according to Leininger, refers to values, beliefs, and patterned lifestyles that are learned and passed on by assisting, supporting, and empowering individuals and groups to maintain health, cope with illness, disability, and death, and enhance their lifestyles (Leininger 2002, p. 51). Variety refers to the mutuality and variability regarding meanings, patterns, values, and lifestyles about care, which aid and support the expression of human care (Leininger 2002, p. 47). Universals encompass meanings, values, beliefs, lifestyles, and care symbols that are evident across many cultures and can offer support to individuals. These elements are common and share similarities across different cultural contexts. This highlights the reality that diverse cultures exhibit distinct nursing behaviors, values, and attitudes toward health and illness. By acknowledging these cultural variations, nurses can deliver care that is culturally tailored to meet the specific needs of individuals. The objective is to provide care that aligns with the cultural context, enhancing the well-being of clients, families, and cultural groups, thereby improving the quality and relevance of nursing care services for people from various cultural backgrounds.

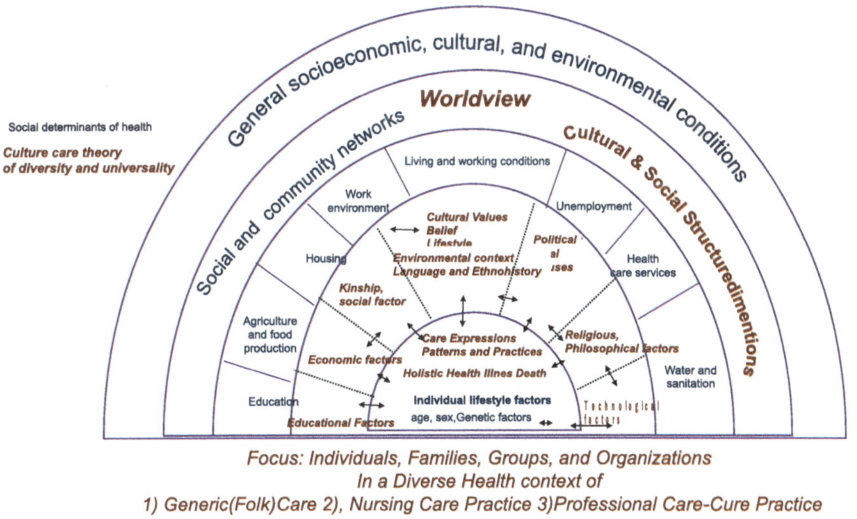

Fig. 6.2 Cultural care theory on social determinants of health

To achieve culturally aligned nursing care for individuals from diverse and similar cultures, it is essential to recognize the importance of culturally congruent practices. Surviving a disaster often involves enduring a recovery process that can extend over months or even years. Key nursing priorities in primary health care during such times include ensuring people's safety, protecting their health and well-being, and maintaining a stable living environment. For example, disasters often lead to sudden shortages of essential items and experiences from past events highlight significant health issues that emerge in the aftermath. A common scenario during emergencies is the inadequate provision of balanced nutrition in shelters, where meals rich in carbohydrates but lacking other essential nutrients can result in health complications over time.

Furthermore, the distribution of relief supplies often falls short of reaching the affected population in a geographically or socially equitable manner. The long-term and severe consequences of disasters include the sudden loss of primary livelihoods, prolonged poverty, hunger, and overall decline in health and well-being. As people lose loved ones, homes, and possessions, they are forced to live in temporary shelters or alternative facilities. The disruption of relationships with family, friends, and community members, along with the loss of critical support networks, exacerbates their vulnerability. Difficulty in securing new employment and loss of income sources create a vicious cycle that leaves survivors more susceptible to future shocks. Therefore, it is crucial to adopt a long-term perspective on risk reduction that goes beyond mere preparedness and response. The changes in people's lives due to disasters often reveal their pre-existing vulnerabilities, and cultural differences can pose significant challenges in preparing for and recovering from such events.

6.4 Discussion

We discuss the challenges in changing PHC in the community over the past twenty years and the necessary initiatives for the future based on its principles, from the perspective of disaster nursing experience.

6.4.1 Equitable Distribution and Access (Equity): Addressing Barriers to Access to Health Care for Vulnerable Populations

Access to health care is a fundamental human right and an essential concept in health policy and health services research (OHCHR 2008; WHO 2017). Historically, access to health care has been viewed primarily as a service utilization issue, with theories such as Shengelia et al. focusing on guarantees as "the probability of receiving necessary health interventions conditional on medical necessity" (Gulliford et al. 2002).

In Japan, vulnerable populations face significant barriers to accessing health care, including language barriers, cultural differences, and economic difficulties. Despite the high standard of medical care and universal health insurance, the Japanese healthcare system is often inaccessible to those who do not speak Japanese or are unfamiliar with local medical practices. Language barriers are especially severe, hindering effective communication regarding medical procedures and symptoms. Many vulnerable populations come from countries where the health care system is less robust and systematic than in Japan. They may need help with the complexity and sophistication of Japanese medical facilities, especially when they need a specific department or advanced treatment. In addition, access to such specialized medical care is expensive and scarce. Furthermore, many cultures rely on home remedies or wait for symptoms rather than seek professional advice, a tendency reinforced by the lack of access to appropriate medical care abroad.

Disaster management similarly requires greater cultural competency to improve access for this diverse population. This includes language support through interpreters and multilingual materials and a deeper understanding of the diverse disaster cultures and practices of different cultures. For vulnerable populations to live in better security, more comprehensive disaster management policies, improved education, and a whole new approach. Disaster management that addresses their practical and cultural needs must be considered.

6.4.2 Strengthening Citizen Participation: Involvement of Diverse Populations in Community Action Plans

Encouraging the participation of vulnerable populations in community activities is a way to strengthen Japan's healthcare system and a necessary step toward a more comprehensive and effective public health strategy. As Japan becomes more culturally diverse, including vulnerable populations in community planning activities is crucial to developing a comprehensive public health framework, and ensuring that vulnerable populations are recipients of public services and participants is essential for a democratic society to function effectively and promote an inclusive living environment. It is also essential to raise awareness of disaster risk reduction and provide opportunities to learn about it regularly. To promote this participation, improved language services and cultural competence are necessary. Cultural background significantly influences beliefs and practices, affecting risk reduction-seeking behaviors and perceptions of action. Training in cultural competence is essential to reduce misunderstandings and provide targeted assistance to people of diverse backgrounds. This includes involving vulnerable populations in needs assessments, program planning, and policy-making processes. Viewing established communities not merely as target groups but as platforms for voicing concerns and contributing insights can address specific health issues and foster a sense of local belonging among vulnerable populations.

Organizational barriers often prevent vulnerable populations from participating, but their involvement in local policymaking ensures that their cross-cultural specific needs and perspectives are considered and addressed. For example, regular promotion of multicultural community development can lead to mutual aid in a disaster. By providing appropriate information to vulnerable populations, they can take the same self-help and mutual-help actions in an emergency. Even healthy people may face unforeseen difficulties during emergencies, and awareness of vulnerable populations can lead to proactive safety measures.

During the public health crisis of COVID-19, collaboration among international students across various disciplines highlighted a coordinated approach to addressing the health and legal challenges vulnerable populations face. Unlike natural disasters, the global nature of the pandemic has allowed for consultation through online platforms to reduce potential health disparities. Mutual aid is crucial to minimizing the impact of disasters. These individuals can serve as essential volunteers and influencers and have the potential to effectively disseminate needed information across different age groups and cultural backgrounds.

6.4.3 Multisectoral Collaboration: Integrating Public Services for Diverse Populations

Effective integration and support of vulnerable populations requires a cohesive and comprehensive approach that transcends traditional sectoral boundaries. Multisectoral cooperation, including health care, education, communications, and local government, must address vulnerable populations' complex, comprehensive, and intertwined needs. A strategic administrative framework is needed to integrate services between different sectors.

This collaboration is essential to improve medical care and disaster preparedness and adequately address all aspects of vulnerable populations' lives, including legal assistance, employment, and social integration. These factors significantly impact their well-being and ability to thrive in their new environment. For example, unfamiliarity with the legal system can lead to exploitation and inadequate protection in the workplace. As a result, they cannot pay medical bills, and experience increased stress at work and in their daily lives. At the same time, difficulties with social integration can exacerbate feelings of isolation and stress, which can further affect mental health, especially during disasters. After the Great Hanshin-Awaji Earthquake, Setting up a familiar consultation service for daily life, such as health and legal services, is also necessary. The activities of the "Town HealthCare post," a volunteer group of the Nursing Association that began after the Great Hanshin-Awaji Earthquake, and consultations with NPOs have revealed that vulnerable populations face language barriers, cultural differences, and a lack of information on medical and welfare services. By sharing local issues and future visions with residents, including citizens, neighborhood associations, NPOs, and businesses, the bureau has promoted community development by taking advantage of their expertise and sharing roles and responsibilities. Kobe City government also started to create soil to accept internationalization, made many efforts to reduce disaster risk, and established the Medical Industry City and the Health Creation City. Such a holistic approach is a critical issue for vulnerable populations and the inclusiveness of social and economic health in the country.

6.4.4 Appropriate Technology: Improving Accessibility for Vulnerable Populations Through Digital Solutions

Teleconsultation has proven to be a valuable tool for improving access to healthcare, especially for those who have difficulty visiting healthcare facilities due to language barriers or lack of familiarity with the healthcare system. Introducing digital health services, with consultations in various languages, will allow vulnerable populations to receive timely medical advice and treatment without visiting a healthcare provider in person. It is essential to ensure that these technological solutions are well-designed and accessible to all population segments. Technology can play a pivotal role in ensuring that all residents, regardless of language or cultural background, have equitable access to the medical care they need. This will benefit the people and strengthen the country's disaster response capabilities. On the other hand, it should be remembered that implementing technology solutions faces several challenges, including privacy issues, the digital divide, and the costs associated with developing and maintaining quality digital health services.

6.4.5 Building Knowledge and Capacity: Strengthening Training Programs for Diverse and Inclusive Health and Disaster Management

The need for training programs to address the health needs of vulnerable populations is increasingly evident. These programs must include culturally sensitive care and services for diverse populations, considering factors such as age, gender, and literacy level. They must be comprehensive, addressing professional knowledge, cultural competence, legal rights, and communication skills. Such programs should be dynamic, incorporate participant feedback, and adapt to people's changing needs and emerging health issues. Training methods could include multicultural dimensions and partnerships with vulnerable communities.

6.4.6 Secure Resources for Integrated Health and Social Programs

The social determinants of health—the conditions under which people are born, grow, live, work, and age—are influenced by the distribution of money, power, and resources. Addressing these determinants requires a collaborative approach across various sectors.

In the case of healthcare, economic barriers often discourage access to healthcare services due to high costs. For vulnerable populations in Japan, economic constraints severely hamper access to health care, and the high cost of medical services delays needed treatment. Despite Japan's universal health insurance system, which ostensibly reduces healthcare costs, the survey found that lack of awareness of health insurance coverage remains a barrier. Another economic barrier is the lack of price transparency, which discourages people from seeking medical care because of the possibility of unexpected expenses. This problem was prominently noted in one study, where uncertainty about the cost of healthcare led many to forego needed healthcare services. During the COVID-19 crisis, frequent examinations became a financial challenge for vulnerable populations and the government, reflecting the broader economic impact of the health emergency. It is more important to empower communities to allocate funds according to local health and social needs and increase resource use's effectiveness and transparency. This will encourage community participation in budget decisions and increase the credibility and accountability of health activities. The same could be analogy for disaster response and disaster risk reduction.

6.4.7 PHC Required in Future Society

In today's digital age, the appropriate use of technology and data is the cornerstone of PHC, enabling communities to support each other toward self-reliance and optimal resource use (Kanbara and Shaw 2022). The development of digital health platforms exemplifies a versatile solution to the numerous challenges that vulnerable populations face in accessing health care services. For example, providing services and information in multiple languages on these platforms effectively bridges the language barrier that often prevents non-native Japanese speakers from accessing appropriate healthcare. Features such as online appointments, accessible multilingual electronic medical records, and telemedicine significantly reduce these populations' logistical and communication challenges. Language barriers were the most common obstacle for foreigners during the COVID-19 pandemic, affecting their ability to obtain critical information and receive needed treatment and vaccinations. Evidence shows that language barriers critically impair early access to health care. These challenges extend beyond health care to community resilience. Interpreting services are available, but their reliability and quality are often questioned, and many rely on informal interpreters. This digital divide is further exacerbated by reduced interaction in the community and increased reliance on the Internet for information, which may bias the information accessed by vulnerable populations. The lack of adequate risk reduction information and the inability to utilize such information highlights the need for informal networks of neighbors, colleagues, and family members to play a more critical role in information dissemination. However, challenges remain, as low literacy rates and limited access to technology hinder the effective dissemination and use of health-related information. Although English is commonly used to provide disaster information, accessibility remains problematic. Living conditions and reasons for migration significantly impact how people access and perceive disaster information. Integrating disaster information with everyday information sources and community activity networks is vital to promoting multicultural community development and raising residents' awareness of information barriers.

6.5 Conclusion

Global migration of people is increasing alarmingly. On the other hand, a large-scale disaster expected to affect the country will be severely damaging and challenging to cope with. Concerns for the health and safety of new migrants are expected to increase, especially during crises and disasters. Migrants also could be influential volunteers who save lives and understand different cultures. In the age of infodemics, it is necessary to effectively disseminate reliable and essential information that reaches different age groups who understand their own culture. Second, mutual aid in one's country may also be optimized globally. Similarly, the media is an influential stakeholder in disaster risk reduction.

Mutual aid that considers people's cultures is not just about providing supplies and temporary support but also about people working together and learning how to enrich their lives and stay safe by themselves, which will lead to the realization of actual disaster risk reduction. These things will lead to global citizens understanding each other's thoughts, values, and customs, and it is considered essential to resolve these issues as they think together about how to create a safer community. Japan's Ministry of Internal Affairs and Communications (MIC) has stated the importance of people of different nationalities and ethnicities living together as members of the local community while acknowledging each other's cultural differences and trying to build equal relationships. Local governments, faced with severe problems such as education and medical care for vulnerable populations in Japan, have responded earlier than the government and, in the process, have continued their efforts toward multicultural conviviality, and their philosophy is the same. Based on these nursing-driven experiences and considerations, it is essential to improve Japan's disaster risk reduction education and information provision methods from a mutual aid perspective and promote efforts so everyone can live safely and comfortably in Japan.

6.6 Future Directions: Rethinking Health and Well-Being: The Role of Dynamic Spiritual Well-Being in Multicultural Community

As we explore the integration of primary health care and disaster risk reduction within the context of enhancing multicultural community resilience, it becomes essential to reconsider our understanding of health and well-being, particularly the significance of spiritual well-being. The World Health Organization (WHO) initially defined health (1948) as "a state of complete physical, mental, and social well-being and not merely the absence of disease or infirmity," linking it to the fundamental human rights enshrined in the WHO Charter. While this definition has remained unchanged, debates about its adequacy have persisted, particularly its omission of spiritual well-being. In 1998, during the review of the WHO Charter at the 101st session of the Executive Board, a proposal was made to revise this definition by including the terms "spiritual" and "dynamic," reflecting growing concerns over the neglect of the human spirit in the face of material and economic advancements. This discourse highlighted critical questions about the nature of well-being: its definitions, the criteria by which it can be measured, and the interrelationship between its physical, mental, social, and spiritual dimensions.

While WHO has yet to incorporate spiritual well-being into its definition of health formally, revealed distinct from mental well-being, can be crucial in building resilience, offering individuals and communities a sense of purpose, meaning, and inner strength in the face of adversity. Some disaster survivors are healthy, financially well-off, have enough food, clothing, and shelter, and have no problems with family relationships or social status. In other words, they may feel "unhappy" even though their physical and social being is "fulfilled and good. On the other hand, some have a lifelong incurable disease and are severely affected by disaster, yet their faces never cease to show a gentle smile. In other words, some people feel "fortunate" even though their physical being is "unfulfilled. There are also those who, because of a disaster, find a kind of happiness that they had never noticed before. If the WHO definition of well-being is that all three aspects of being (physical, mental, and social) are "complete" or "fulfilled," then these examples would be considered examples of non-well-being. However, as we experience and observe these examples in the communities before and after the disaster, we realize that not only are "fulfilled, good state" and "well-being" different concepts but that both concepts are necessary to think about well-being and that it is essential to consider their interrelationships.

As we seek to enhance PHC and community resilience, we must incorporate spiritual well-being into our approach, ensuring that health care and disaster preparedness strategies are holistic and inclusive.

There is no doubt that the central point of WHO's appeal to the world through this definition of well-being is to help more people achieve well-being. It is only natural that the world should work together and support each other to create a system where everyone can be physically healthy, have shelter from the rain and dew, have food, and be respected and protected as human beings. There is an impending necessity to rethink the statement to explore the fourth dimension of health—the Spiritual well-being of individuals in local communities with global citizens.

Well-being is a state in which others can understand the data they can experience with their five senses and judge whether or not the subject person's state has reached well-being. The "objective data" is the data that can be used to understand the physical well-being of a person and to judge the correctness or incorrectness of his/her understanding. Others can judge a person's physical well-being. In addition, by taking data on whether a person has a certain level of income, food, clothing, shelter, family relationships, and stable social status and examining the data, a third party can judge whether the person is in a state of social well-being. To create such a data set and support system and to encourage supportive actions, it is essential to visualize the state of well-being in numerical form and to objectively measure the basis for such visualization and use it as primary data.

There is a growing recognition of the need to expand our focus beyond the traditional health triangle of physical, mental, and social dimensions when addressing the health of multicultural communities. Recent discussions on spiritual health, often referred to as the "fourth dimension" of health, highlight its significance. Spiritual health is understood as a state in which an individual navigates daily life challenges in a way that fosters the realization of their full potential, meaning, life purpose, and inner fulfillment. This state can be achieved through processes such as disaster response, relocation, recovery, and rebuilding a healthier, more self-actualized life. By cultivating spiritual health, individuals may enhance their sense of well-being, empowering them to take proactive roles in their lives. It also serves as a resilient coping mechanism, enabling individuals to address broader health challenges, including those related to planetary health, and contributing to a more meaningful and creative experience in both anticipated and unforeseen circumstances.

References

Bahar A, Koçak HS, Samancıoğlu Bağlama S, Çuhadar D (2020) Can psychological resilience protect the mental health of healthcare professionals during the COVID-19 pandemic period? Dubai Med J 3(4):133–139

Barış E, Silverman R, Wang H, Zhao F (2022) Walking the talk: reimagining primary health care after COVID-19. World Bank Publications

Benson C, Clay E (2004) Understanding the economic and financial impacts of natural disasters. The World Bank

Blomberg B, Mohn KGI, Brokstad KA, Zhou F, Linchausen DW, Hansen BA, Langeland N (2021) Long COVID in a prospective cohort of home-isolated patients. Nat Med 27(9):1607–1613

Dahlgren G, Whitehead M (2021) The Dahlgren-Whitehead model of health determinants: 30 years on and still chasing rainbows. Public Health 199:20–24

Donaldson SK, Crowley DM (1978) The discipline of nursing. Nurs Outlook 26(2):113–120

Edward J, Hines-Martin V (2015) Exploring the providers perspective of health and social service availability for immigrants and refugees in a southern urban community. J Immigr Minor Health 17:1185–1191

Fawcett J (1984) The metaparadigm of nursing: present status and future refinements. Image: J Nurs Sch 16(3):84–87
Germain S, Yong A (2020) COVID-19 highlighting inequalities in access to healthcare in England: a case study of ethnic minority and migrant women. Feminist Legal Stud 28(3):301–310
Guadagno L (2020) Migrants and the COVID-19 pandemic: An initial analysis
Gulliford M, Figueroa-Munoz J, Morgan M, Hughes D, Gibson B, Beech R, Hudson M (2002) What does 'access to health care' mean? J Health Serv Res Policy 7(3):186–188
https://www.who.int/publications/m/item/weekly-epidemiological-update-on-COVID-19-29-june-2021
Kanbara S, Shaw R (2022) Disaster risk reduction regime in Japan: an analysis in the perspective of open data, open governance. Sustainability, 14, 19. https://doi.org/10.3390/su14010019
Koh HK, Piotrowski JJ, Kumanyika S, Fielding JE (2011) Healthy people: a 2020 vision for the social determinants approach. Health Educ Behav 38(6):551–557
Leininger M (2002) Culture care theory: a major contribution to advance transcultural nursing knowledge and practices. J Transcult Nurs 13(3):189–192
Marmot M (2007) Achieving health equity: from root causes to fair outcomes. The Lancet 370(9593):1153–1163
OECD (2021) Realising the potential of primary health care. OECD Health Policy Studies, OECD Publishing, Paris,. https://doi.org/10.1787/a92adee4-en
OECD (2003) Large-scale disasters. https://doi.org/10.1787/9789264020207-3-en
OHCHR (2008) The right to health. https://www.ohchr.org/sites/default/files/Documents/Publications/Factsheet31.pdf
Rathore FA, Farooq F (2020) Information overload and infodemic in the COVID-19 pandemic. J Pak Med Assoc 70(5):S162–S165
Rifkin SB, Muller F, Bichmann W (1988) Primary health care: on measuring participation. Soc Sci Med 26(9):931–940
Sandier S, Pari V, Polton D, Thomson S, Mossialos E, World Health Organization (2004) Health care systems in transition, France
UNISDR (United Nations International Strategy for Disaster Reduction) (2015) Sendai framework for disaster risk reduction 2015–2030
United Nations (2015) Transforming our world: the 2030 agenda for sustainable development. NY, USA, New York
WHO (2017) Human rights and health. Retrieved June 29, 2021, from Who.int website: https://www.who.int/news-room/fact-sheets/detail/human-rights-and-health
WHO (2019) Health emergency and disaster risk management. Retrived https://www.who.int/publications/i/item/9789241516181
Wilkinson RG, Marmot M (eds) (2003) Social determinants of health: the solid facts. World Health Organization

Chapter 7
Understanding the Mechanism of the 2004 Indian Ocean Tsunami as a Complex Disaster

Fumihiko Imamura and Anawat Suppasri

Abstract The 2004 Indian Ocean tsunami caused human casualties as well as various other types of damage because it struck an area that did not have physical defense structures such as breakwaters and seawalls and lacked nonstructural measures for tsunami awareness, such as warning systems and evacuation procedures. The widespread tsunami flooding of coastal lowlands triggered a chain of destruction to buildings; damage to social infrastructure, such as coastal roads, bridges, railways, and ports; erosion of coastal land; and destruction of ships, oil tanks, and automobiles. Such a complex disaster occurs when multiple hazard/disaster events with different characteristics are closely related. This study explains the actual situation in each region regarding wide-area complex disasters caused by earthquakes, large tsunamis, and other events as well as (1) hazards (predisposing factors), (2) impacts and damage, and (3) points that particularly increase the impact to express the relationship between each event in a flow. The results can be compared with the complex disaster flow chart of the 2011 Great East Japan Earthquake to deepen our understanding of the actual state of complex disasters.

Keywords Giant tsunami · Oceanic propagation · Types of tsunami damage · Fragility functions · Complex disasters

7.1 Introduction

7.1.1 Overview of the Indian Ocean Tsunami

The Mw 9.1 earthquake that occurred in the Indian Ocean off the northwest coast of Sumatra, Indonesia, at 7:58 a.m. on December 26, 2004 (local time), caused strong shaking in various places, followed by a large tsunami. The tsunami caused extensive damage to Indonesia's Sumatra Island, which was close to the epicenter, as well as

F. Imamura (✉) · A. Suppasri
International Research Institute of Disaster Science (IRIDeS), Tohoku University, Sendai, Japan
e-mail: imamura@irides.tohoku.ac.jp

R. Shaw et al. (eds.), *Two Decades from the Indian Ocean Tsunami*, Disaster Risk Reduction, https://doi.org/10.1007/978-981-96-2669-4_7

Thailand, India, Sri Lanka, the Maldives Islands, and other parts of the Indian Ocean, and spread as far as the east coast of Africa and Antarctica (Suppasri et al. 2015). The twenty-first century has been an increased period of earthquake activity, and large tsunamis due to large earthquakes, similar to the Great East Japan Earthquake that occurred in 2011, have occurred (Mori et al. 2022).

In the northern region of Sumatra near the epicenter, the tsunami reached coastal areas such as Banda Aceh approximately 15 min after the earthquake. Although more than two hours passed between the earthquake and the tsunami in areas such as around Phuket Island in Thailand, the east coast of India, all the islands of Sri Lanka, and the Maldives Islands, there was no history of tsunamis, and no one experienced shaking caused by the earthquake. The area was hit by a tsunami in a defenseless situation because there was no advance response; there was no tsunami warning system in place. Furthermore, since the earthquake occurred on Sunday morning and during the Christmas holidays of 2004, many tourists, including foreigners, were killed. As a result, the number of victims of this tsunami is reported to have exceeded 220,000 people along the entire coast of the Indian Ocean (UN report). Similarly, on New Year's Day 2024, an earthquake occurred in Japan's Noto Peninsula, causing strong shaking, tsunamis, liquefaction, landslides, and fires (Suppasri et al. 2024). Similarly, the victims were families and tourists who had returned home after recovering from COVID-19 infection, and the government and other disaster response teams were also understaffed due to the holiday. This reflects the characteristics of disasters that occurred on different days. (Imamura et al. 1988).

Overviews of tsunami heights in various locations have been reported (Choi et al. 2005; Choi and Imamura 2005). The tsunami was 15–30 m high on the west coast of northern Sumatra and 6–12 m high on the coast of Banda Aceh in northern Sumatra. Banda Aceh city has low-lying areas that extend up to approximately 5 km from the coast, and even in areas approximately 2.5 km away from the coast where the ground height is approximately 1.5 m, the inundation depth is 4–5 m. In Thailand, the elevation is 5–20 m, mainly in Phuket; on the east coast of India, it is 5–10 m; and in Sri Lanka, it is 5–15 m.

This large-scale tsunami struck an area that did have physical structures such as breakwaters and seawalls and did not have prior tsunami awareness, warning systems, evacuation systems, or other nonstructural measures in place; thus, loss of life and various kinds of damage occurred, including widespread flooding of coastal lowlands; destruction of wooden and brick buildings; damage to coastal infrastructure, such as roads, bridges, railways, ports and airports; and erosion of coastal land, ships, oil tanks and motor vehicles. In addition, drifting water caused salt damage to farmland. A disaster that occurs when multiple disaster events of different natures are closely related is called a complex disaster. A complex disaster is defined as follows: (1) Any phenomenon related to natural hazards (hazard or predisposition) that causes another disaster phenomenon in a chain reaction. (2) Different types of complex disasters are recognized, such as complex disasters that cause and expand damage, e.g., damage caused by a tsunami after an earthquake, and complex disasters that occur when two or more natural disasters with completely different causes coincide coincidentally. It is suggested that definition 1) applies to the 2004 Indian Ocean tsunami.

Disasters are defined by exposure to and vulnerability to hazards. Tsunamis are hazards that increase in wave height and current speed as they approach shallow waters, causing great damage to coastal areas. Damage caused by tsunamis extends over a wide range of areas, both on land and in shallow waters. Damage to land includes human casualties, damage to houses, damage to facilities (seawalls, water gates), fire damage, economic damage (disruption of logistics, suspension of services), damage to lifelines (water supply, sewage, electricity, gas, and communications), traffic damage (roads, railways), and agricultural damage (saltwater intrusion into rice fields). Examples of disasters in the sea area include facility damage (coastal facilities such as breakwaters), ship damage, fishery damage, ground damage (scouring due to earth and sand movement), and oil and timber spills (causing fires and coastal environmental pollution). To reduce tsunami damage, it is essential to clarify the causes, impacts, and extent of damage and develop a modeling method that can predict the impact on each region on the basis of exposure and vulnerability. Furthermore, as land use along coasts has become more complex, large-scale tsunami disasters need to be viewed as chain disasters, as they can become complex and compound, similar to Indian Ocean tsunamis. This chapter explains the actual situation in each region regarding wide-area complex disasters caused by earthquakes and large tsunamis, as discussed by Suppasri et al. (2021), to describe (1) hazards (predisposing factors), (2) impacts and damage, and (3) points that particularly increase the impact and express the relationship between each event in a flow. The results are compared with the complex disaster flow chart of the subsequent 2011 East Japan earthquake to deepen our understanding of the actual state of complex disasters, which were caused mainly by giant tsunamis.

7.1.2 Outline of Earthquake/Tsunami Occurrence and Disasters

The occurrence of the M9.2 earthquake was measured by seismometers installed worldwide, and its magnitude and epicenter were immediately identified by the Japan Meteorological Agency, the US Geological Survey, and other agencies. The epicenter of the earthquake was off the coast of Sumatra, Indonesia, on the ocean floor 10 km northeast of Simulue Island near the northern tip of Sumatra (3 °N latitude, 96 °E longitude, 30 km depth), and the rupture in the crust caused the earthquake to occur in the Nicobar Islands to the north. It spread to the Andaman Islands and reached a total length of approximately 1000 km. This subduction zone earthquake generated a large tsunami, which occurred in an earthquake nest where the Indo-Australian plate was subducting toward the Eurasian plate. As shown in Fig. 7.1, the Indo-Australian plate is subducting below the Eurasian plate at a rate of approximately 3–5 cm per year. The estimated magnitude of the earthquake was 9.2, making it one of the largest

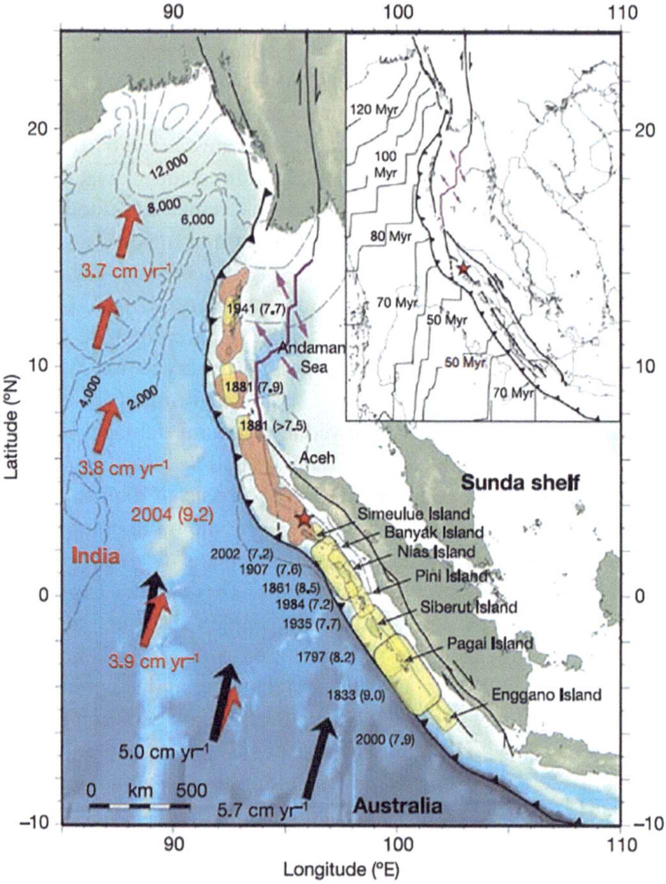

Fig. 7.1 Tectonic setting and ruptures of major interplate earthquakes, including the 2004 Sumatra–Andaman earthquake, along the Sunda megathrust (Subarya et al. 2006)

earthquakes in the world and the second largest since 1900 (the largest was the 1960 Chile earthquake at M9.5).

7.1.3 Tsunamis Propagating Throughout the Indian Ocean

Once the fault model of the earthquake is estimated, changes in the uplift and subsidence of the ocean floor are calculated, which changes the water mass above it. This is the main mechanism by which tsunamis occur due to submarine earthquakes. Afterward, tsunami propagation from the epicenter through the ocean and through the coasts of each country can be reconstructed via computer numerical calculations on the basis of bathymetric and topographic data and longwave theory, which is the principle of tsunami propagation (Imamura et al. 1988; IOC 1997; Lobkovsky and Mazova 2007). Figure 7.2 shows the results of the tsunami propagation calculations and shows how the tsunami propagated 1, 2, 3 and 4 h after occurrence. Here, to understand the propagation characteristics in the open ocean, rupture processes such as the earthquake rupture speed are not considered. Although this diagram is difficult to understand, the water level of a tsunami immediately after its occurrence is distributed as negative values (undertow) on the east side and positive values (overflow wave) on the west side.

A tsunami hit Banda Aceh in northern Indonesia, which was close to the epicenter, immediately after the earthquake occurred. One hour later, the tsunami was largely divided into two tsunamis heading east toward the Andaman Sea and another heading toward the Indian Ocean, but their tip positions were different. The average depth of the Andaman Sea from the epicenter to Thailand is approximately 800 m, which is relatively shallow considering the open ocean, and the tsunami propagation speed was approximately 280 km/h. On the other hand, because the sea area from the epicenter to Sri Lanka was deep, with an average depth of approximately 4000 m, the tsunami propagated at an extremely fast speed of approximately 70 km/h. This tsunami has a wave period of 40–50 min, and the wavelength varies depending on

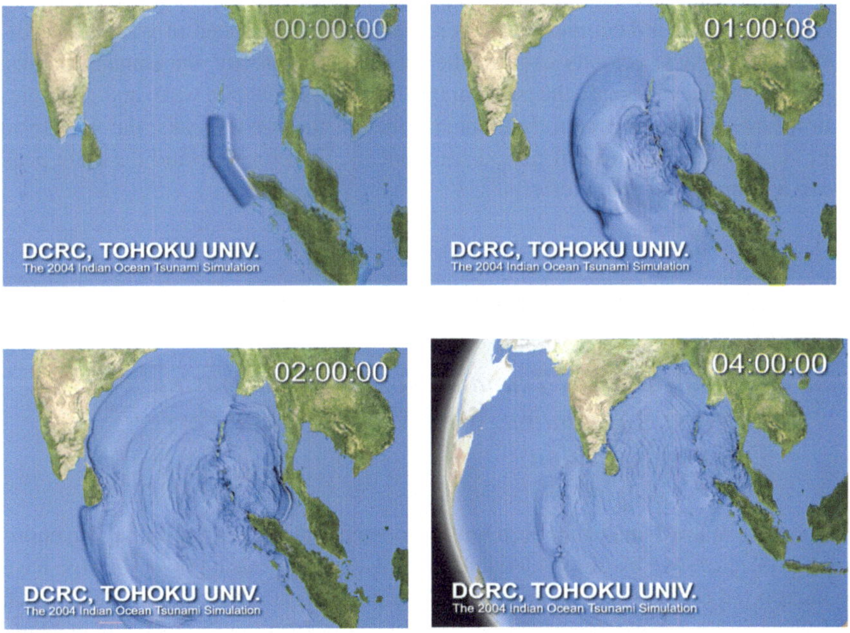

Fig. 7.2 Propagation of the Indian Ocean tsunami (showing the situation 1, 2, 3 and 4 h after occurrence). This CG was constructed from the tsunami simulation using a grid size of 2 arc-min in spherical coordinates with 2 s of simulation. The calculation methods and conditions at 2 arc-min (approximately 1.855 km at the equator) can be found in other studies (Imamura et al. 1988; IOC 1997; Suppasri et al. 2011a, b)

the propulsion. It tended to be shorter in shallow waters and longer in deeper waters. Two hours later, an undertow was expected to reach eastern Thailand near Phuket. The tip of the tsunami propagating to the west was a pushing wave, and it was about to reach the east coast of India; in Sri Lanka, it could be seen that it had started to wrap around behind the island. Furthermore, four hours later, the tsunami passed through the islands of the Maldives and continued to propagate far toward the east coast of Africa.

Generally, as a tsunami spreads across a wider ocean, its height decreases. Additionally, the direction and height of tsunami waves can vary depending on the topography of the ocean floor, and they are reflected and scattered by continents and islands, making them more complex. However, the tsunami height is usually smaller than 2 m in the open ocean, which is small compared with the length of hundreds of kilometers, making it difficult to recognize the tsunami offshore, and the impact it has on ships is small. When a tsunami that has propagated through the open ocean reaches coastal areas, the water depth becomes very shallow, and the height of the tsunami increases, causing damage. Owing to the effects of ocean floor topography (refraction, diffraction, energy gathering effects, etc.), wave height and wave direction can vary depending on location, resulting in different damage conditions.

Furthermore, when the tsunami water level rises above normal sea level, it becomes a push wave that can travel upland or upriver. The propagation speed of a tsunami slows as it becomes shallower, approaching the speed of water particles. If the water particle velocity exceeds the propagation velocity, the tsunami breaks. Even after the waves break, the tsunami becomes a bore-shaped wave and continues to propagate or run upstream. Immediately before the wave breaks, the waveform slope becomes steep, so short-period components often form behind the main peak due to nonlinear dispersion effects.

7.2 Examining Direct Damage Caused by the Tsunami

7.2.1 Classification by Tsunami Hazard (Predisposing Factors): Inundation, Flow Velocity, Fluid Force, Erosion/deposition

Historical tsunami damage has been limited to human casualties and damage to buildings, as described in historical documents. However, various types of damage have been reported due to changes in land use and production activities in coastal areas, and damage has been categorized on scales such as tsunami intensity, as in Shuto (1993) and Papadopoulos and Imamura (2001). In some cases, building damage ratios were calculated using flow depth and tsunami height as indices, as in Hatori (1964). Figure 7.3 schematically shows the relationship between tsunami intensity and damage. This is the result of adjusting the occurrence of each damage under six categories and is the first step in evaluating tsunami damage as a complex disaster.

However, there is no indication of their relationship or linkage. For Indian Ocean tsunamis, Suppasri et al. (2011a, b), Koshimura et al. (2009), and others have estimated fragility functions (relationships between tsunami characteristics and damage) for Thailand and Indonesia. Because a large amount of valuable data was obtained, the incidence of damage (fragility function) became continuous, and the reliability increased. Furthermore, in addition to the tsunami flow depth, the flow velocity and fluid force are added as external force factors, which are introduced in the following section.

However, recent tsunami research and field surveys have revealed that in addition to the tsunami scale and tsunami height, the main external force factors include the wave force, flowing water force, buoyancy force, and uplift force because tsunamis and floating debris can collide with buildings. This has also been shown to be a factor in leakage. The wave pressure (the wave force per unit area) of a tsunami that floods flat land after overthrowing a vertical seawall is thought to be distributed under hydrostatic pressure. On the other hand, the wave pressure when the tip of a

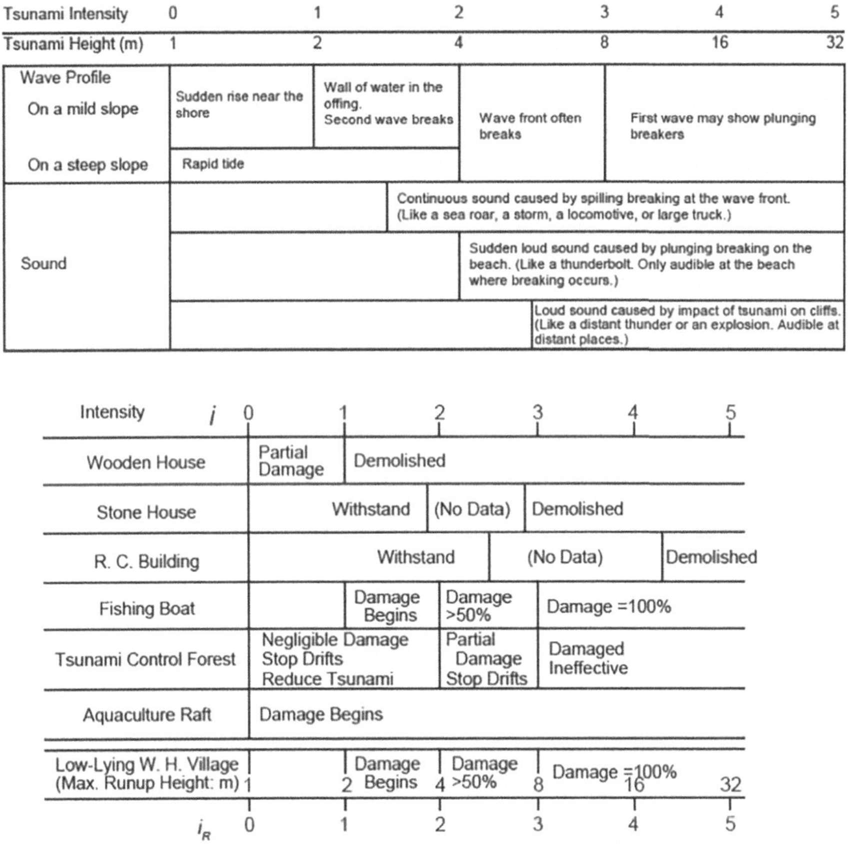

Fig. 7.3 Tsunami intensity and damage (Shuto 1993)

tsunami collides becomes extremely high. In recent years, new technologies have made large-scale tsunami experiments possible, and studies have begun to examine the effects of surge front wave pressure on buildings when the tip of a tsunami floods land. The vertical distribution of the surge front wave pressure clearly differs among breaking waves, stepped waves, and overflows, and when waves break in front of a building, a very large shock wave force is generated as a wave force.

The hydraulic characteristics of a tsunami include inundation, flow velocity (wave current), and wave force (hydrostatic force, hydrodynamic force, and shock wave force), which are predisposing factors that cause damage. There are many disaster cases in the past, which can be categorized by predisposing and triggering factors, as shown in Table 7.1. Damage is caused by three main types of predisposing factors related to tsunamis and is amplified by surrounding triggers.

Residential areas near the coast of Banda Aceh have experienced devastating damage. Photo 7.1 displays satellite images before and after the tsunami. Even though it is difficult to observe in satellite images, erosion control measures have been implemented on the sandbar coast facing the sea to protect it from high waves. However, the countermeasures were damaged by the action of the tsunami and caused the coastline to retreat. Furthermore, the land in the center of the sandbar has been completely submerged by erosion. In this way, tsunamis have the power to change topography.

Furthermore, it is very difficult to estimate the type of impact and the degree of damage, as impacts can vary greatly depending on topographical changes (e.g., sedimentation/erosion), the presence or absence of drifting debris, fires, etc., and their distribution. Further quantitative evaluation of the tsunami damage recorded during the 2011 Great East Japan Earthquake is progressing (Charvet et al. 2014; Leelawat et al. 2014). These findings serve as further references for evaluating tsunami damage as a complex disaster, and in particular, it is necessary to clarify the chain of events.

Table 7.1 Causes, effects, and damage of tsunami

Predisposing factors	Inducing factors	Impact	Damage
Flooding	Long-term inflow of seawater (including salt)	Long-term flooding, withering of marine plants, sparks on electric wires, batteries, etc., causing fires	Drowning, fire, agricultural damage
Flow and flow speed (wave current)	Scattering of drifting objects, soil, and combustible materials	Destruction of buildings and structures, erosion and sedimentation	Damage to houses and facilities, damage to infrastructure, environmental and ecological destruction
Wave force (hydrostatic force, hydrodynamic force, impulse wave force)	Movement of debris, etc.	Destructive force (increased destruction)	Damage to houses and facilities, damage to infrastructures

7.2.2 Characteristics and Causes of Human Casualties Caused by Tsunamis

This tsunami caused as many as 220,000 deaths and missing persons. Generally, a disaster that kills approximately 10,000 people is called a "colossal disaster," but this tsunami far surpassed that standard, making it the worst megadisaster in modern disaster history. There are three possible reasons for this major disaster:

① A large earthquake caused a very large tsunami to hit various places, including distant countries. This earthquake was a large ocean trench-type earthquake, and large tsunamis over 4 m hit various places. In areas where the tsunami height was greater than 4 m, major damage to houses occurred, and in areas where the tsunami height was greater than 8 or 10 m, the damage was catastrophic.

② The tsunami suddenly struck people who were not aware that a major earthquake had occurred and who had not been given an advance warning of the tsunami. Earthquake-induced damage, such as building collapse due to seismic vibrations, has occurred in the northern and western parts of Sumatra Island in Indonesia and surrounding areas, which are close to the epicenter, and in the Andaman Islands in India. Nevertheless, there was no damage from ground shaking in other countries that were far from the earthquake, such as Thailand. They only felt it slightly. Countries that do not experience earthquakes do not have earthquake or tsunami warning systems, so they are unaware of the earthquake and the tsunami, resulting in a large disaster.

③ Many people live on low-lying land near coasts, which have no coastal defense structures. The affected area was not severely affected by cyclones, the sea was relatively calm, and many people lived in comparatively weak buildings near the coast.

In particular, regions ② and ③ are different from Japan and other countries in the Pacific Ocean, where tsunamis occur relatively frequently and where many disasters, such as high waves and storm surges, are caused by typhoons. In addition, many tourists died along the Thailand coast because it was a resort area.

On the other hand, there are examples of people who were saved. It is said that an elephant being used on the coast began to scream for an hour prior to tsunami arrival, and just before that, it ran up into the hills with tourists on board, and many of the people who ran with it were saved. On Surin Island in Thailand, there is a tradition among the Moken people who live on Surin Island that they should flee to high ground or mountains after seeing the receded sea. It is said that there were only tourists remaining.

It is difficult to assess and predict human casualties caused by tsunamis. For historical tsunamis, the number of victims for each village was calculated, and the relationship between that number and the typical tsunami height was determined (Kawata 1997). This varies greatly depending on the evacuation system at the time, the presence or absence of a warning system, and the level of tsunami awareness. In other words, since there is a time lag between the occurrence of an earthquake and the arrival of a tsunami, the number of casualties can vary greatly depending on the evacuation behavior during that time, making it extremely difficult to estimate the damage function. Figure 7.4 shows the results of an attempt to evaluate human casualties as a fragility function in Banda Aceh. As expected, unlike the results for buildings in Fig. 7.3, there is a large variance.

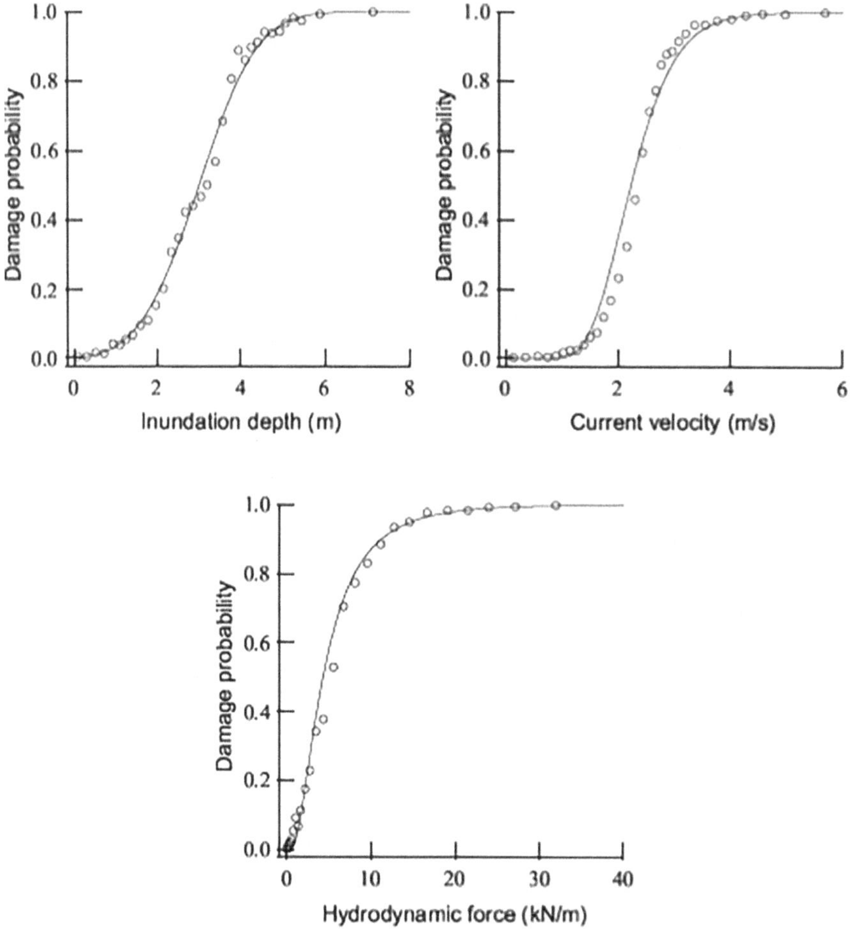

Fig. 7.4 Fragility functions (fragility curves) for building damage in terms of the inundation depth, current velocity, and hydrodynamic force (Koshimura et al. 2009)

7.2.3 Building Damage: The First Step in Estimating Complex Disasters

(1) Damage status of buildings: 3 classifications

In Sect. 2.1, building damage is considered from the perspective of tsunami forces, but the conditions for building destruction vary depending on the building, such as structural type and topographical conditions. Even if the structures are the same size, the quality of the materials used, the structural parameters, especially the methods of connecting columns and beams, and the structure of the walls, can vary by country or region, and the failure conditions are not uniform. For example, even if a building is made of reinforced concrete, the strength of the building can differ greatly depending on whether the walls are made of unreinforced brick or reinforced concrete. It would be complicated to classify these details and explain the conditions for destruction, which would be meaningless because the degree of damage varies greatly, and a detailed classification is not necessarily appropriate. Therefore, the extent of damage can be classified into three categories: major damage, moderate damage, and minor damage. The definitions of these disaster levels are as follows.

Major damage (□): A large portion of the pillars as well as the walls are destroyed. This also includes the state where all walls and pillars are destroyed. It is impossible to repair damaged buildings.

Moderate damage (△): Most of the pillars remain, but part of the wall has been destroyed. It is possible to reinforce and replace the pillars. It is possible to repair damaged structures.

Minor damage (○): Even if the windows have been destroyed, the wall remains. This includes cases where there is only water intrusion and no structural damage. The building can be reused again after minor repairs.

The relationships between the inundation depth and degree of damage for each building structure were categorized on the basis of past disaster cases. In Fig. 7.5, D. (□) represents major damage, P. D. (△) represents moderate damage, and W. (○) represents minor damage. The inundation depth refers to the side of the building where the flow acts (referred to as the front side of the building). However, the building structure, distance from the shoreline, and other regional characteristics are not considered.

There is only one known case of a reinforced concrete building suffering minor damage if the water depth is 5 m or less, moderate damage if the water depth is 5–8 m, and severe damage at 20 m (Fig. 7.5, top row). An example of moderate damage is damage to unreinforced brick buildings during the Indian Ocean tsunami. These buildings are thought to have a slightly weaker bearing capacity than ordinary buildings with reinforced concrete walls. With respect to major damage, there have been no cases of damage with flood depths of 820 m, so the limiting conditions are unknown.

There is no clear difference in the degree of damage to stone, brick, or block buildings due to differences in flood depth (middle of the figure). For example, if the water depth is 2–3 m, small, medium, or major damage can occur. According to the

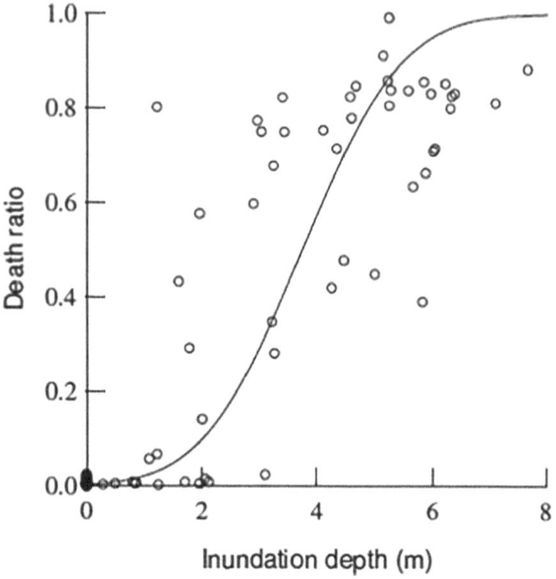

Fig. 7.5 Fragility function for tsunami casualties in terms of inundation depth (Koshimura et al. 2009)

results of damage cases prior to the Hokkaido Nansei-Oki Earthquake and Tsunami (1993), if the inundation depth was 3 m or less, there would be minor damage, and if the inundation depth was 7 m or more, there would be major damage. However, with the addition of disaster cases such as the East Java Earthquake and Tsunami (1994) and the Biak Island Earthquake and Tsunami in Indonesia (1996), this clear difference disappears. The additional damage cases include a building with a thin brick wall containing a small amount of thin wire that was severely damaged by flooding to a depth of 23 m. Stone, brick, and block buildings have different structures depending on the country and region, and their strength can vary widely.

During the 2004 Indian Ocean tsunami, a tsunami flowing like a large river flood was captured on video in Banda Aceh, North Sumatra, from the second floor of a house located approximately 2.3 km inland from the sea. The flow depth in and around the house was approximately 3.9 m. This video could be filmed because the house was not largely damaged by the earthquake or tsunami. This evidence shows that sturdy buildings can also be used as tsunami evacuation buildings. In addition, cases of intermediate damage due to small inundation depths of at most 0.25 m are thought to be caused by collisions with floating objects during tsunamis.

Subsequently, many buildings were damaged or destroyed during the Great East Japan Earthquake in 2011, and numerous field surveys were conducted to collect data on the extent of tsunami inundation and building damage in the affected areas. Data from surveys of more than 250,000 structures provided by Japan's Ministry of Land, Infrastructure, Transport and Tourism (MLIT) were used to organize the data into five damage categories and derive a damage function. The data include the damage levels, structural materials, number of floors, and building function of each location. This event provides us with one of the most complete datasets of tsunami events in history. Fragility functions have been derived from these data (Suppasri et al. 2013; Suppasri et al. 2015).

7.2.4 Damage to Infrastructure

Port and ship damage: Disaster damage in Banda Aceh, Indonesia.

Photo 7.1 shows a satellite image taken after the earthquake and tsunami occurred in Banda Aceh, Indonesia. The top of the image shows nearby port facilities, and the topography itself has changed significantly, leaving behind some breakwaters and quays. This tsunami also swept away large objects that would normally be unimaginable. The power generation ship (60 m long and 20 m wide) that was moored at this port was washed away by the tsunami approximately 3 km inland from the sea. The water marks left by the tsunami on the buildings near where the ship stopped are approximately 3 m above the ground, which is roughly the same as the ship's draft. It appears that the ship stopped when the bottom of the ship touched the ground. Large ships that drifted inland in other areas were flat-bottomed ships. It seems that pontoon boats are more likely to run upstream on land due to tsunamis (Fig. 7.6).

There was also a ferry terminal building at this port. The first floor was crushed, but the second floor and higher floors remained. The first floor was damaged by the earthquake only. This terminal building, which was still under construction, had a framework made of reinforced concrete columns, but the walls had not yet been installed. A 10 m high tsunami struck this area, but the tsunami passed through the pilot-style building without walls, and the building was spared from destruction by the tsunami. These findings indicate that pilot-style buildings can be effective as tsunami-resistant buildings if they can be built to withstand earthquake motion (Fig. 7.7).

Near Kruen Raya Port east of Banda Aceh, three empty oil tanks and nine full oil tanks were washed away by the tsunami, which reached 5 m above the ground. The oil tanks, 17 m in diameter and 11 m high were swept away approximately 300 m along the shoreline. The red arrow in the photo indicates the destination, but the three units did not move from their original position. Most likely, a push wave caused the three units to start moving inland, and the next wave caused them to move toward the coast. It is thought that they were pulled back, pushed, and pulled repeatedly, and finally stopped at this position. The drifting of these tanks may have caused significant damage to the surrounding houses (Photo 7.2).

Transport-related damage, especially railway damage: train disasters in Kahawa, Sri Lanka.

Traditionally, typical traffic-related damage has included damage to roads and automobiles. In the 2004 Indian Ocean tsunami, a train was washed away by a tsunami in the Kahawa district of Sri Lanka, killing more than 1,500 people, including those on board. The Kahawa district is located approximately 15 km north of Galle. Coconut trees are sparsely planted along the coast, and before the tsunami washed them away, one-story wooden and brick houses were scattered. The elevation of the area rises

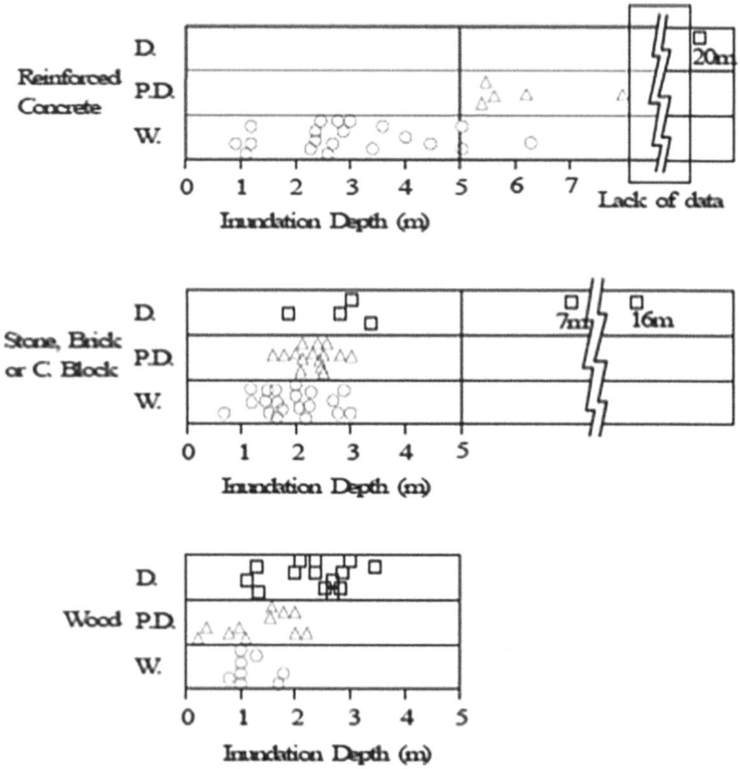

Fig. 7.6 Relationship between the degree of building damage and tsunami inundation depth (Matsutomi and Shuto 1994)

slightly from the shoreline and then gently descends to a continuous flat area. There is a railway line approximately 200 m inland from the coast, and the altitude near the railway line is approximately the same as the altitude at the coastline. It is thought that the tsunami that entered from the coast formed into tiered waves, broke up, and rushed at a higher speed than normal and was further accelerated by the downward slope and hit the train. It is thought that the situation resembled a jet current (Figs. 7.8 and 7.9).

An emergency stop was made by the train because of the first tsunami wave, but the train was subsequently washed away by a tsunami larger than 1 m. Sri Lanka had not been hit by a tsunami for more than 100 years. Therefore, both the crew and the passengers had no idea what was happening in response to the sudden flood. The height of the second tsunami can be estimated from the traces left on nearby reinforced concrete buildings. The building was located closer to the sea than the railroad tracks were, and it was approximately 200 m away from where the train overturned. There were traces of the tsunami on the building wall at a distance of 4.2 m from the ground. Since the wall was located behind the wall from the direction

Photo 7.1 Topographic changes before and after the tsunami (Banda Aceh, Indonesia) ©Digital Globe

of the tsunami, the maximum tsunami height may have been greater. The building now has markings indicating the locations of the tsunami traces. The height of the second tsunami was approximately the same as the height of the train, and the train that was hit by this tsunami was torn apart, including the power section, and swept away 70 m.

Fig. 7.7 Topography of Banda Aceh after the tsunami and the drifted power generation ship ©digital globe

Photo 7.2 Drifting oil tank near Kruen Raya Port (by the Aceh Province Office)

Fig. 7.8 Train disaster situation and train locations in Kahawa, Sri Lanka

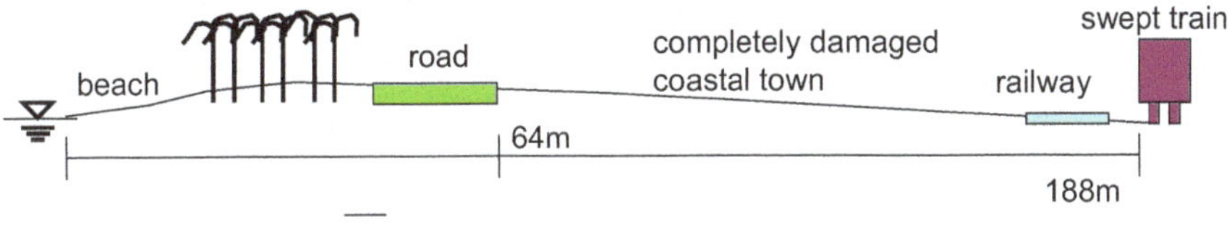

Fig. 7.9 Topographical cross section near the train disaster site in Kahawa

7.3 Factors Contributing to the Spread of Damage: Inundation and Drifting of Buildings, Ships, Etc.

One of the characteristics of a complex tsunami disaster is floating debris. In the past, it was noted that the movement and drifting of fishing boats and the remaining destruction of houses, cars, tanks, etc., can contribute to the spread of damage; however, this time, the scale of the impact was extremely large.

7.3.1 Drifting and Damage

During the 2004 Indian Ocean tsunami, floodwater was captured on video in the center of Banda Aceh, North Sumatra, with large amounts of floating debris, such as building debris, coastal trees, furniture, and vehicles. When tsunami inundation is accompanied by a large amount of drifting debris, the current velocity and the drifting speed of the drifting debris tend to decrease, and the depth of inundation increases due to friction between the debris and between the ground and debris. The video shows a strong current, which does not slow even to the center and is more than 3 km from the coast. The impact force of a tsunami flood flow accompanied by drifting debris is also unknown, and future investigations and research are needed (Photo 7.3).

Photo 7.3 After tsunami flooding with drifting debris (photographed by the author, 2006)

Typical examples of floating objects include driftwood and drifting ships. Since both of these items float on the ocean, there is a high chance that they will drift over very long distances. For example, many ports in Japan have coastal lumber yards, and when a tsunami occurs, the stored lumber is often washed away from the lumber yards, causing greater damage. However, no lumberyard has yet taken full-fledged measures to prevent tsunami spills, and no effective method for preventing such spills has been established. On the other hand, research on the behavior of driftwood is progressing, and it is now possible to quantitatively estimate the impact force from the inundation depth and driftwood specifications (thickness, length, weight, strength, etc.). Recently, the probability of multiple pieces of driftwood colliding simultaneously has also begun to be considered. When floating debris such as driftwood (or fishing boats) is trapped between buildings, the area blocked by the flow leads to an increase in the water level. This creates a large drag force on the building. This force can be much greater than the drag force acting on a building in the absence of debris, leading to its destruction. Other floating objects include aquaculture rafts, fishing gear, building debris, furniture, etc. In port areas, floating objects include containers, cars, oil tanks, etc.

7.3.2 Washed-Away Buildings: Undertows and Buoyancy

This section explains damage to buildings due to undertow and buoyancy, which has not been reported often in the past, even though buildings tend to be prone to drifting away. After a tsunami propagates in undulating waves on a seabed slope with a uniform slope, the force of the current acts on buildings and other objects. The hydraulic force acts when a tsunami runs up onto land, that is when it is a push wave, and it also acts during an undertow when seawater returns to the sea. Many cases of disasters also occur during undertows. Pushing waves act on buildings and structures on the coast, and in places where the topography is such that seawater flows in concentrated waves during an undertow, there are many cases of damage caused by the undertow. Furthermore, the structural strength of a building may be weakened against undertow waves. In other words, coastal seawalls are built to protect against waves that come from the sea, and the backs of embankments are not covered with concrete or other materials, so they may be weak. It has also been reported that a tsunami coming from the ocean can overtop a seawall and flow into the back side, where it will be accelerated on the slope behind it and create a very strong current.

Furthermore, structures on land, such as buildings, are normally designed without considering buoyancy. This tendency is likely noticeable in modern, highly airtight houses. In addition, buildings with basements may be subject to double buoyancy due to flooding, in addition to ground liquefaction.

7.4 Regional Damage

7.4.1 Urban Damage in Banda Aceh, Indonesia

Banda Aceh is a city in a flat area located at the mouth of the Aceh River, and, behind the coastline, there are aquaculture ponds and rice paddies spanning 11 km^2. Even in urban areas located approximately 2.5 km inland from the coast, the ground height relative to sea level at the time of the tsunami was less than 2 m. According to the testimony of a person who witnessed the tsunami and was on high ground on the west coast of Banda Aceh, the tsunami struck approximately 1520 min after the shaking caused by the earthquake. The height of the tsunami that hit the coast exceeded 10 m, killing approximately 61,000 people out of Banda Aceh's population of 264,000.

A tsunami height of 12 m was measured at a mosque near the coast, and it also left a mark at a height of 7 m at the mouth of the Aceh River to the east. This large tsunami caused catastrophic damage to an area up to 2–3 km from the coast and flooding up to 5 km, as shown in the example in Photo 7.4.

The tsunami that hit the Banda Aceh area was not just one wave. There are testimonies that there were at least three waves, and the second wave, which occurred approximately 15 min after the first wave, seemed to be the largest. According to residents who encountered the tsunami on the west side of the city, the first tsunami wave came from the northern coast, but the second wave, which was stronger, came from the west side. Although this west coast is facing the coast indirectly, there is a creek that leads to the north coast, and it is thought that this was caused by the tsunami moving up the creek. On the other hand, reports indicate that the first wave of the tsunami hit the northern coast almost from the north, and the second wave originated slightly further west. These results suggest that tsunamis that cause flooding exhibit

Photo 7.4 Tsunami inundation at Banda Ache and a boat on a house (photographed by the author, 2006)

complex behavior, depending on the topography, including other waterways, and the direction of the tsunami coming from offshore (Murata et al. 2010).

7.4.2 Damage at Tourist Destinations in Phuket, Thailand

Phuket Island is an example of a famous resort area in southern Thailand, and the Khao Lak coast on the mainland side is also undergoing tourism development, with many resort hotels located on the coast. At the time the tsunami arrived on December 26th, it was the middle of the tourist season from Christmas to New Year, and many tourists stayed there. Approximately 5,400 people were killed in southern Thailand, and among them, more than 2,000 foreign tourists were killed, with the majority coming from Northern Europe. It was apparent that the victims were people from countries that rarely experienced earthquakes or tsunamis, but the number of victims in Japan, which is an earthquake-prone country and has repeatedly been hit by tsunamis, was not small.

The tsunami hit the Phuket and Khao Lak coasts approximately 2 h after the earthquake occurred (Photo 7.5). The height of the tsunami trace was approximately 5 m at Patong Beach on Phuket Island (the ground elevation is approximately 3.5 m), approximately 10 m at Khao Lak Beach (the ground elevation is approximately 2.5 m), and approximately 10 m at Bang Nam Kem Fishing Port, which is located on the north side of the Khao Lak Coast (the ground elevation is approximately 6 m). The tsunami trace was large on the Khao Lak coast. The ground level was low, the terrain behind it was flat, and the lowlands extended inland, so the flooding from the tsunami spread approximately 1 km inland from the coastline. The people on Phuket

Photo 7.5 Tsunami damage near Khao Lak, Thailand (photographed by the author, 2006)

Island seemed to have sensed that an earthquake was occurring. However, because the shaking of the earthquake was not strong and because there was no information on the scale or location of the earthquake or any tsunami forecast in the two hours after the earthquake occurred and before the arrival of the tsunami, the situation was unclear, and Patong Beach seems to have had many people who went out and became victims.

The tsunami began at approximately 10 a.m. when the water began to recede offshore. Immediately afterward, what appeared to be the first wave hit, and the water reached a waist-deep level in urban areas facing the coast. The sea water soon receded offshore. At this point, people did not seem to be aware of the tsunami, but the fear that they felt when they were submerged up to their waists in the first wave caused them to evacuate. A few minutes later, a second, larger wave hit, and the height of this tsunami was as high as a person's head. Many of those who survived said that they would not have been able to evacuate if the first wave had been as large as the second wave, submerging them up to their heads.

Notably, a 10-year-old British girl who came from England saved her parents and tourists in the area on the Mai Khao coast of Phuket Island, and no one was killed. The reason is that it did not appear. Two weeks prior, the girl remembered from an elementary school geography class that "the phenomenon of the tide receding from the coastline and the formation of white foam is a precursor to a tsunami." When she shared her knowledge with her parents, they were able to inform the people around them and the hotel staff, which allowed them to evacuate before the tsunami hit. This area was one of the few beaches for which no deaths were reported.

On the other hand, it appears that the second tsunami wave was also large on the Khao Lak coast. Strangely, while there are many testimonies on Phuket Island in which many people began to evacuate after the first tsunami hit, there is no evidence that people on Khao Lak Beach began evacuating after the first wave. In fact, many people gathered on the shore after the first wave, and some even went to retrieve fish left behind in puddles on the ocean floor created by the receding waves. The second, larger tsunami washed away many people before they could evacuate. Therefore, it can be inferred that evacuation was delayed because the first tsunami on the Khao Lak coast was so small that people on the coast did not feel fear (Murata et al. 2010).

7.4.3 Damage in the Port City of Galle, Sri Lanka

In Sri Lanka, which is located on the west side of the island of Sumatra in Indonesia, the tsunami reached the east coast of the island approximately two hours after the earthquake, split into the north and south sides, and even reached the rear, causing impact and damage throughout the country. There was no impact from the earthquake, but the tsunami caused extensive damage, with approximately 35,000 casualties. In the past, there was almost no experience with tsunamis, and there were no tsunami warning systems or protective facilities.

We report damage caused by the tsunami in Galle, which is located in the southern part of the island where the World Heritage City is located. Photo 7.6 shows an image of the city center from the old town (fortress). According to the testimony of a local employee of the LPA (Sri Lanka Ports Authority) at the time of the tsunami, seawater receded not only inside the port but also outside the breakwater (approximately 10 m deep) before the tsunami struck. Her 200-ton (DWT) class dredger, which was moored there, has been washed ashore on the quay and is sitting there. On the quay in the foreground, pieces of wood that appear to have been washed ashore can be seen. The area behind the quay where the dredging boat was launched was severely damaged, and containers and other items were submerged in the water. A capsized fishing boat was floating. Additionally, a fishing boat was broken into two parts and washed ashore on the quay. Two supply boats lined up and launched approximately 50 m to the shore.

This shows that the tsunami caused extensive damage to structures and ancillary facilities. However, no damage was observed to the northern or southern masonry breakwaters. At Galle Port, the tsunami-damaged part of the seawall inside the port and removed the ground behind it, causing the access road into and out of the port to be cut off. The part of the shed where the cargo was temporarily stored was also damaged.

A sailor onboard a tugboat in Galle Port described the situation when the tsunami hit: "*The seawater in the port receded within 10 min after the sudden change in sea level began to occur. We hurriedly tried to berth the ship and go out of the port, but the ship hit the seabed inside the port and tilted diagonally. However, the sea level started to rise, and the ship was safe.*" The water depth in the port is 89 m, so since the bottom of the ship touched the seabed, accounting for the draft of the tugboat, the sea level must have receded by several meters. Ships are relatively safe if they

Photo 7.6 Photo near Galle, Sri Lanka (photographed by the author, 2006)

are out at sea, but if a ship is moored in a port, it is difficult to take it out to sea after a tsunami hits. The sailors spoke passionately about how they tried to protect themselves and the ship despite being suddenly attacked by a tsunami that they had never experienced before.

7.5 Toward Disaster Prevention and Mitigation: Understanding Complex Disasters

7.5.1 Creating a Chain Disaster Flow

Various types of damage were reported in the 2024 Indian Ocean tsunami, particularly in Banda Aceh, which includes urban areas and is a typical example of a complex disaster. Before the tsunami hit after a large tremor (earthquake), it was reported that the earthquake had an intensity of VII-VIII on the modified Mercalli intensity scale and 5–6 on the Japan seismic intensity scale (JMA). It was so strong that it was difficult to withstand. Many RC buildings were damaged by this shaking, and some had their first floors crushed. Furthermore, liquefaction, landslides, and ground changes (uplift and subsidence) occurred at the same time. Then, a tsunami struck, destroying ports and other facilities and displacing ships and cars. It is believed that these items drifted into inland urban areas and contributed to the spread of damage. A video taken in central Banda Aceh at the time shows that the tip of the tsunami first moved gently but then gradually accelerated into a torrent that engulfed cars, ships, and driftwood. Near Kruen Raya Port, which is located east of Banda Aceh, three of the nine oil tanks in total were washed away by the tsunami. These strong currents caused changes in the topography, especially in coastal areas, which caused significant erosion and affected vegetation such as mangroves. Afterward, long-term inundation with seawater caused major damage to agriculture and fisheries (particularly aquaculture).

These series of events are summarized in Fig. 7.10 as a chain disaster flow. In the figure, the black arrows indicate the main effects of the earthquake, and the red arrows indicate the main effects of the tsunami. First, the occurrence of an earthquake causes liquefaction, landslides, ground deformation, etc., in addition to a tsunami. At that time, the tsunami caused flooding and fires and topographical changes due to erosion and sedimentation. In the case of a tsunami, drifting debris is generated by flow and inundation, increasing the destructive external force. These various infiltrations cause damage to houses and buildings, infrastructure and traffic disruptions, agricultural damage to land, and fishery damage in the sea. In addition, this infiltration affects the environment, including vegetation such as mangroves, and ultimately causes human suffering. Furthermore, we must not forget damage to the economy and community or indirect damage, such as reputational damage to tourist destinations.

This complex flow increases in relation to the hazards of the tsunami (e.g., wave height, inundation depth/duration, flow speed/fluid force) and the fragility of the target. For each instance, the external force should be evaluated; damage occurs when a certain strength (standard value) is exceeded, and the resulting increase is thought to govern the extent of the damage. Here, hazard evaluation is possible to a certain extent analytically or by numerical simulation, accounting for the various conditions of the tsunami (e.g., wave height and period) and coastal topography. On the other hand, the strength (standard value), which is an important element of fragility, is estimated from past experience, but it is inevitably only an empirical evaluation. Therefore, it is essential to determine this strength in advance according to each facility and function. The reason that it is difficult to respond is that the level of development of facilities and functions differs in each region, and damage occurs and spreads from the lowest level.

During the 2011 Great East Japan Earthquake, this impact was large because disaster prevention facilities and equipment were already in place to a certain extent. Furthermore, a nuclear disaster occurred, which spread over a wide area and became long lasting. In addition, emergency earthquake warnings and tsunami warnings were in operation in Japan at the time, so this information was also included (Fig. 7.11). Suppasri et al. (2021) organized each item in the process (classification) of chain damage as follows: red frame: main cause (triggers); chain and chronological occurrence of shaking and tsunamis; yellow frame: impact and damage (the darker the color is, the greater the degree); blue frame: points where the impact is particularly amplified; warnings after the tsunami occurs, etc.; this concept of classification is also included here in Figs. 7.10 and 7.11. The degree of chain damage is classified from M0 to M4, with the Indian Ocean tsunami and the Great East Japan Earthquake corresponding to M4 and M5, respectively.

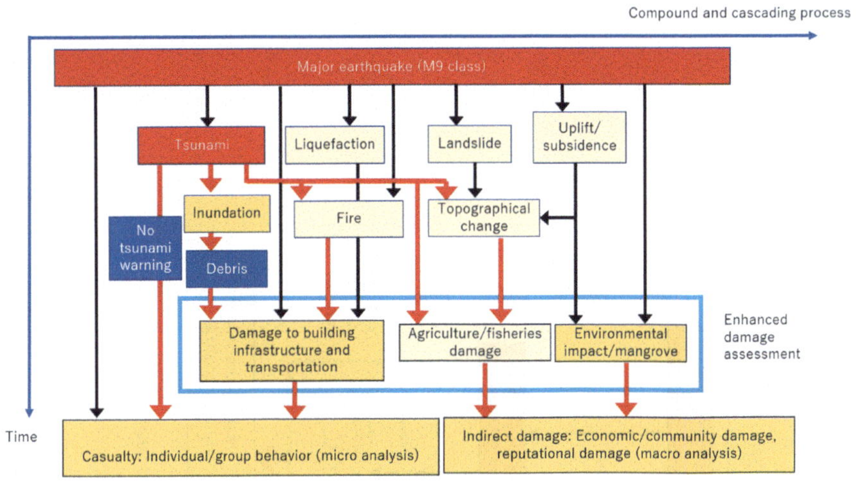

Fig. 7.10 Damage flow from the 2024 Indian Ocean tsunami as a wide-area, complex, and chain disaster

7.5.2 Promotion of Comprehensive Tsunami Disaster Prevention

Tsunami mitigation science is based on the foundation of existing tsunami engineering and earthquake engineering and is a new field that comprehensively combines elements such as estimating the history of tsunami occurrences, devising social infrastructure and building structures, and urban planning. It must be systematized as an academic field. This field must contribute to the construction of social systems that can reduce human and material damage and enable early recovery and reconstruction even in the case of infrequent large-scale disasters such as the Indian Ocean tsunami and the Great East Japan Earthquake. To this end, several research topics should be promoted as follows:

(1) Tsunami frequency and risk assessment

A method for investigating and evaluating the world's tsunami history from a geological perspective should be established. One of the challenges in assessing tsunamis, such as the Great East Japan Earthquake, was the limitations of historical data and information. Because evaluations have focused on investigating descriptions in ancient documents spanning hundreds of years, it has not been possible to address tsunamis that repeat over hundreds or thousands of years. In the future, it will be necessary to carry out tsunami deposit surveys via geological techniques such as boring, not only in Japan but also worldwide, to share this information and integrate this information with geomorphology and change studies.

(2) Protecting areas from tsunamis

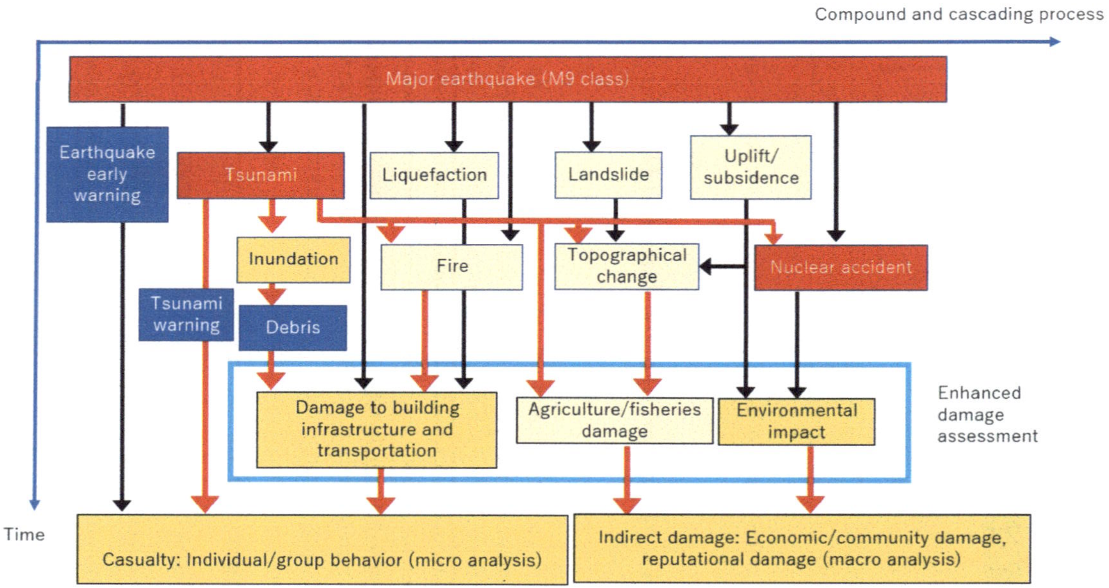

Fig. 7.11 Damage flow from the 2011 Great East Japan earthquake as a wide-area, complex, and chain disaster

To create a city that is tsunami resistant, it is necessary to improve social infrastructure and buildings to minimize land use regulations and damage based on numerical analysis that includes the behavior of tsunamis that run up on land. It is necessary to promote tsunami-resistant town development by promoting the selection of residential areas, street designs, hillside residential developments, and the construction of tsunami evacuation facilities.

In the construction of social infrastructure facilities and buildings that can withstand tsunamis, in addition to Indian Ocean tsunamis, there are also tsunami external forces and damage conditions obtained from the Great East Japan Earthquake, so we must scientifically elucidate these mechanisms and develop structures that can withstand tsunamis. It is essential to construct social infrastructure facilities such as buildings, bridges, and sea walls. We summarize our experiences and lessons learned and take measures to maintain functionality and quickly recover from tsunami-induced wave forces, collisions with floating objects, and flooding.

(3) Measures to protect lives

It is necessary to establish a system for the early understanding of a disaster situation over a wide area and for information collection and communication. For wide-area disasters that are expected to occur in the future, methods for collecting and communicating hazard and damage information and establishing a system for life-saving, emergency response, and emergency recovery are important. Furthermore, it is important to enhance disaster prevention education and inspect and maintain disaster prevention training. In Kamaishi city and Kesennuma city, where disaster prevention education was common before the earthquake, the death and missing rates of children and students were less than 1/10 of the death and missing rates for the total population, demonstrating the effectiveness of disaster prevention and evacuation drills. This was a case where the results were clearly evident. However, the number of deaths among children, students, etc., and teachers in coastal areas reached 625, and the damage to schools and social education facilities exceeded 10,000 (Ministry of Education, Culture, Sports, Science and Technology publication: August 22, 2011). In the event of a major earthquake, it is necessary to examine the effectiveness of disaster prevention education and training, including for cases of failure, and to consider how disaster prevention education and training should be conducted. In this environment, it is necessary to cultivate not only knowledge but also the ability to grasp the situation, recognize risks, and decide on appropriate responses.

7.5.3 Development and Progress of Tsunami Warnings

The first initiative for a tsunami warning system in the Indian Ocean started in January 2005 at the United Nations World Conference on Disaster Risk Reduction, which was held in Kobe city, Japan. UNESCO's Intergovernmental Oceanographic Commission (UNESCO/IOC) took the lead in establishing a tsunami warning system for the Indian Ocean. Since March 2005, the IOC has held coordination working meetings in Paris and Mauritius, whereas the IOC, in collaboration with other United Nations agencies, has set up a mission to investigate the current status and challenges of tsunami countermeasures in countries surrounding the Indian Ocean. In August 2005, the first Indian Ocean Tsunami Warning System (ICG/IOTWMS) conference, which was held by two neighboring countries, was held in Perth, Australia. However, this time, the preparations were completed in approximately half a year (Yamamoto 2023).

Until March 2013, the Japan Meteorological Agency and PTWC provided provisional "Indian Ocean Tsunami Monitoring Information" until the Indian Ocean warning system began stable operation. The goal was to establish a tsunami warning system that would determine the epicenter within 3 min of an earthquake and issue a tsunami warning within 5 min (BMKG 2012).

7.5.4 Mitigation Effects of Natural Forces (Vegetation)

The 2004 Indian Ocean tsunami survey reported cases where dense mangrove forests and other forest belts mitigated damage to houses and villages behind them. Because the wave force was reduced by the mangrove tree belt, tsunami damage also decreased in villages behind and within the mangrove tree belt.

Single trees or forests that are not very dense cannot be expected to weaken the power of tsunamis, but they can be expected to have effects related to human lives. The most direct effect is the escape effect, where people are saved by taking shelter in a tree. There are several examples of Indian Ocean tsunami events. A video aired as a news video in Indonesia and other countries showed people holding on to trees on the coast to avoid being swept away by the tsunami. In addition, students from Siakla University in Banda Aceh survived the tsunami by holding onto tall coconut trees along the coast. Furthermore, some people climbed trees and evacuated to the second floor of a hotel at Patong Beach on Phuket Island. In this way, even a single tree can be useful, such as by holding onto it to prevent being swept away or by traveling along the tree to evacuate to a higher location near a building. However, the roots of trees growing near the coast tend to be shallow, and young trees whose roots do not extend sufficiently horizontally are likely to be knocked down by a tsunami, making them unsuitable candidates for evacuation.

The tree (mangrove, pine tree, etc.) belt near the coast is called the green belt. Individual trees reduce the tsunami flow velocity, and as the flow velocity decreases, the fluid force (drag force) proportional to the square of the flow velocity also decreases. Since green belts are made up of trees growing in clusters, their tsunami suppression effects can be evaluated as the sum of the flow velocity reduction effects (and thus the drag reduction effects) of individual trees. Therefore, the total projected area of tree trunks in a flooded area, which can resist tsunami flow, is an important parameter that is a function of (tree trunk diameter × number of trees) × flood depth (Yanagisawa et al. 2009).

7.5.5 Simulation of Sediment Movement and Tsunami Flow

Research on damage caused by tsunamis is being conducted with a focus on human and residential damage related to the maximum wave height. However, recently, cases of unexpected disasters caused by tsunamis resulting from currents and sediment movement have also been studied. In fact, there have been reports of damage to ships and aquaculture rafts due to strong currents, as well as pre-excavation of structure foundations and collapse due to earth movement. In addition, recent tsunami surveys have confirmed the presence of a sedimentary layer of several tens of centimeters after a tsunami occurs, and the behavioral characteristics of tsunamis on land have been estimated from this information (Masaya et al. 2020).

Since the flow is a vector quantity, its direction changes significantly depending on the local topography. When prediction and reproduction are considered via numerical calculations, the discretization error due to differences in calculation grid spacing is larger than the water level (scalar quantity). Moreover, there are few examples of actual measurements and observations of tsunami flows, and studies on the accuracy and reproducibility of numerical calculations have been limited. However, since the fluid force and destructive force that cause damage are related to the flow velocity rather than the water level, this accuracy has a large effect on the results of damage estimation. Furthermore, people swept away by a tsunami can drift over a fairly wide area. It is very effective to estimate the drifting area in advance through numerical flow calculations. Considering the survival rate of victims in past disasters, it is necessary to conduct rescue activities within 72 h at most. Identifying the range of drifting people and ships will be useful for rescue and lifesaving operations.

7.5.6 Lessons from the 2024 Noto Peninsula Earthquake

The Noto Peninsula experienced a wide range of cascading hazards, such as uplift, liquefaction, landslides, fires, and tsunamis, and the damage from these hazards was notable. Strong shaking occurred, highlighting that it is important to retrofit buildings and promote earthquake resistance for both housing and public buildings, as many structures on the Noto Peninsula were destroyed. Although the direct impact on casualties was limited, unexpectedly large uplifts and widespread liquefaction caused further cascading impacts and difficulties in recovery and reconstruction. Several landslides blocked roads and rivers, causing many villages to become isolated. The spread of fire could not be eliminated because the tsunami warning deterred the firefighting process. There were almost no deaths caused by this tsunami because of the high tsunami awareness of both local residents and tourists throughout Japan after the 2011 Great East Japan Earthquake. Nevertheless, tsunamis caused by submarine landslides are another remaining problem for tsunami warning systems and evacuation responses.

7.6 Conclusion

The Indian Ocean tsunami caused not only human casualties but also a variety of other types of damage because it struck an area that did not have physical structures such as breakwaters and seawalls or soft measures such as advanced tsunami awareness, warning systems, and evacuation systems. This resulted in the widespread flooding of coastal lowlands; the destruction of wooden and brick buildings; damage to social infrastructure facilities such as coastal roads, bridges, railways, and ports; the erosion of coastal land; and the destruction of ships, oil tanks, and automobiles. In addition, farmland was damaged. A compound disaster occurs when multiple disaster events with different characteristics are closely related. This chapter summarized the actual situation in each region with respect to complex disasters in wide areas caused by earthquakes and large tsunamis, as discussed by Suppasri et al. (2021) and described as (1) hazards (predisposing factors), (2) impacts and damage, and (3) points that particularly increase the impact to express the relationships among events in a flow. The results were compared with the complex disaster flow chart of the subsequent 2011 East Japan Earthquake to deepen our understanding of the actual state of complex disasters, which were caused mainly by giant tsunamis.

References

BMKG (Meteorological, Climatological and Geophysical Agency) (2012) Tsunami early warning service guidebook for InaTEWS, 2nd ed. https://www.gitews.de/tsunami-kit/en/E3/tool/Tsunami%20Early%20Warning%20Service%20Guidebook%20for%20InaTEWS.pdf

Charvet I, Ioannou I, Rossetto T, Suppasri A, Imamura F (2014) Empirical fragility assessment of buildings affected by the 2011 Great East Japan tsunami using improved statistical models. Natural Haz 73:951–973. Published: 15 March 2014. https://doi.org/10.1007/s11069-014-1118-3

Choi BH, Siripong A, Sundar V, Wijetunge JJ, Diposaptono S (2005) Past runup survey of the December 26, 2004 earthquake tsunami of the Indian ocean, Sumatra Tsunami on 26 December 2004 Asian and Pacific Coast 2005, September 4–8, 2005, Jeju, Korea, pp 1–20

Choi BH, Imamura F (2005) Sumatra tsunami on 26th December 2004. Prof of the special Asia Tsunami Session at APAC 2005, Jeju-do, Korea, September 4–8, Hanrimwon Publishing Co., p 221

Imamura F (2024) Noto tsunami arrived in minutes: Did Japan Learn from 3/11? https://www.nippon.com/en/in-depth/d00986/

Hatori T (1984) On the damage to houses due to tsunamis. Bull Earthq Res Inst Univ Tokyo 59:433–439

Imamura F, Shuto N, Goto C (1988) Numerical simulation of the transoceanic propagation of tsunamis. In: Paper presented at the Sixth Congress of the Asian and Pacific Regional Division. Int Assoc Hydraul Res, Kyoto, Japan

Kawata Y (1997) Prediction of loss of human lives due to catastrophic earthquake disaster. Jpn Soc Nat Disaster Sci 16(1):3–13 (in Japanese)

Koshimura S, Oie T, Yanagisawa H, Imamura F (2009) Developing fragility curves for tsunami damage estimation using numerical model and post-tsunami data from Banda Aceh, Indonesia. Coast Eng J 51:243–273

Leelawat N, Suppasri A, Charvet I, Imamura F (2014) Building damage from the 2011 Great East Japan tsunami: Quantitative assessment of influential factors—A new perspective on building damage analysis. Nat Hazards 73(2):449–471. https://doi.org/10.1193/053013EQS138M

Lobkovsky LI, Mazova RKH (2007) Source mechanism of the tsunami of 2004 in the Indian Ocean: Analysis and numerical simulation, Izvestiya. Phys Solid Earth, 43(7):573–582. ISSN 1069-3513. https://doi.org/10.1134/S1069351307070063

Masaya R, Suppasri A, Yamashita K, Imamura F, Gouramanis C, Leelawat N (2020) Investigating beach erosion related with tsunami sediment transport at Phra Thong Island, Thailand caused by the 2004 Indian Ocean tsunami. Nat Hazard 20:2823–2841

Matsutomi H, Shuto N (1994) Inundation depth due to a tsunami and damage of houses. Proceedings of Coastal Engineering, JSCE 41:246–250 (in Japanese)

Murata S, Imamura F, Katoh K, Kawata Y, Takahashi S, Talayama T (2010) Tsunami: to survive from Tsunami, Advanced series on ocean engineering vol 32. World Scienctific, p 302

Papadopoulos GA, Imamura F (2001) A proposal for a new tsunami intensity scale. In: ITS 2011 Proceedings, session 5, No.5–1

Subarya C, Chlieh M, Prawirodirdjo L, Avouac JP, Bock Y, Sieh K, Meltzner AJ, Natawidjaja DH, McCaffrey R (2006) Plate-boundary deformation associated with the great Sumatra–Andaman earthquake. Nature 440. https://doi.org/10.1038/nature04522

Shuto N (1993) Tsunami intensity and disasters. Tsunamis in the World, Kluwer Academic Publishers, Dordrecht, pp 197–216

Suppasri A, Koshimura S, Imamura F (2011a) Developing tsunami fragility curves based on the satellite remote sensing and the numerical modeling of the 2004 Indian Ocean tsunami in Thailand. Nat Hazards Earth Syst Sci 11:173–189

Suppasri A, Goto K, Muhari A, Ranasinghe P, Riyaz M, Affan M, Mas E, Yasuda M, Imamura F (2015) A decade after the 2004 Indian Ocean tsunami- The progress in disaster preparedness and future challenges in Indonesia, Sri Lanka, Thailand and the Maldives. Pure Appl Geophys 172(12):3313–3341

Suppasri A, Kitamura M, Alexander D, Seto S, Imamura F (2024) The 2024 Noto Peninsula earthquake: Preliminary observations and lessons to be learned. Int J Disaster Risk Reduct 110:104611

Suppasri A, Charvet I, Macabuag J, Rossetto T, Leelawat N, Latcharote P, Imamura F (2015) Building damage assessment and implications for future tsunami fragility estimations. Handbook of Coastal Disaster Mitigation for Engineers and Planners, pp 147–178. https://doi.org/10.1016/B978-0-12-801060-0.00009-5

Suppasri A, Pakoksung K, Charvet I, Chua CT, Takahashi N, Ornthammarath T, Latcharote Pl, Leelawat N, Imamura F (2019) Load-resistance analysis: An alternative approach to tsunami damage assessment applied to the 2011 Great East Japan tsunami. Nat Hazards Earth Syst Sci, 19:1807–1822. Highlight article

Suppasri A, Muhari A, Syamsidik, Yunus R, Pakoksung K, Imamura F, Koshimura S, Paulik R (2018) Vulnerability characteristics of tsunamis in Indonesia: Analysis of the global centre for disaster statistics. J Disaster Res 13(6):1039–1048

Suppasri A, Maly L, Kitamura M, Syamsidik, Pescaroli G, Alexander D, Imamura F (2021) Cascading disasters triggered by tsunami hazards: A perspective for critical infrastructure resilience and disaster risk reduction. Int J Disaster Risk Reduct 66:102597

Mori N, Satake K, Cox D, Goda K, Catalan PA, Ho TC, Imamura F, Tomiczek T, Lynett P, Miyashita T, Muhari A (2022) Giant tsunami monitoring, early warning and hazard assessment. Nat Rev Earth & Environ 3:557–572. https://doi.org/10.1038/s43017-022-00327-3

Yamamoto M (2023) 100 Years of Tsunami Forecasting—tsunami forecasting is difficult. Earthq J 2023(76):72–78

Yanagisawa H, Koshimura S, Miyagi T, Imamura F (2009) Tsunami damage reduction performance of a mangrove forest in Banda Aceh, Indonesia inferred from field data and a numerical model. J Geophys Res: Ocean 115:C06032

Chapter 8
Science, Technology, and Innovation in Disaster Risk Reduction: *A Review of 20 Years*

Rajib Shaw

Abstract Significant advancement has been made over the past twenty years at global, regional, and national level on the use of science and technology in policy advocacy, decision making, and implementation of the global agenda as well as national plans on disaster risk reduction. There are formal structures of global and regional science advice, and over the years regional science and technology advisory groups have become prominent in making science advice and having science technology plan for global agenda. At national level, some countries have adopted a formal structure on science advice directly to the head of state and / or head of the national disaster management agency. The key challenge still remains on how to bring science into local level decision making in a more formalized way. This is more relevant in current days as well as in future in the perspective of complex risk landscape as well as emerging of different types of technologies. Science and technology in future need to be more localized, inclusive, as well as the youth and young generation should be involved in making new entrepreneurship and innovation with science and technology to solve critical local problems.

Keywords Global science advocacy · Regional science advisory group · National science platform · Local science advice · Inclusive science and technology

8.1 Introduction

Science, technology, and innovation have always there with mankind to cope with different types of disasters. However, its form has changed and evolved over time. While in the past, people were mostly depending on local and indigenous knowledge, which is possibly still valid in some parts of the world and with some communities, be it urban or rural, coastal or mountain. Over the years, with the advancement of science and technology, the form of science, technology, and innovation (STI) has

R. Shaw (✉)
Graduate School of Media and Governance, Keio University, Fujisawa, Japan
e-mail: shaw@sfc.keio.ac.jp

R. Shaw et al. (eds.), *Two Decades from the Indian Ocean Tsunami*, Disaster Risk Reduction, https://doi.org/10.1007/978-981-96-2669-4_8

changed to cope with disasters, be it geological or climate-related hazards. Over the past few years, the advancement of disruptive technology has also made deep impacts in the disaster field and will possibly change the course of STI in future.

Over the last 20 years, the global risk landscape has changed drastically, and the world has become more complex with many complex, compounded, and cascading events occurring frequently. One of the classic examples is to look at the Global Risk Outlook of the World Economic Forum (WEF) over the past several years. The report started in 2007 after the global economic turmoil, and WEF has categorized into different types of risks: economic, environmental, social, technological, geo-political, etc. A close look from 2007 shows that economic risks globally dominated in terms of likelihood and impacts over the first few years. However, over time, environmental risks (disasters, climate change, extreme weather, biodiversity losses, etc.) took top ranks both in terms of likelihood and impact. In 2020 (which is the analysis of the year 2019), the top five risks are all related to environmental risk (WEF 2020). This says that in a complex global landscape, addressing environmental risk is important to tackle the economic and other associated risks. The 2021 report (WEF 2021) shows infectious disease as one of the critical risks because of COVID-19, which impacted the whole world. However, two new arose in that year, which are: digital power concentration and digital divide. In the year 2020, due to COVID-19, work, education, health system, shopping changed from in-person to online mode. However, this can only be done with good digital penetration. Many parts of the world (like developing countries), many parts of the same country (like rural areas), many age group in the same country (like aged people) did not have proper digital connectivity or were not digitally native. Thus, the digital divide came as a new risk. The following years saw the emergence of other new risks like (cyber security, WEF 2022), the energy crisis, cost of living due to Russia-Ukraine war (WEF 2023), and mis- and dis-information (WEF 2024).

Therefore, the evolution of STI needs to be looked at in the evolving risk landscape. We are seeing more complex/compound disasters, where two or more disasters are happening together. Sometimes, we also see cascading disaster happening more often, where one hazard is triggering another disaster. Thus, the STI required for these complex emergencies needs to be looked at. This chapter provides a brief overview of some of the key issues and evolution over the last twenty years.

8.2 Science, Technology, and Innovation

Kameda (2009) described three types of technologies in disaster risk reduction. He termed these as: Implementation Oriented Technology (IOT), Process Technology (PT), and Transferable Indigenous Knowledge (TIK). Categorization of these three types was made through a series of consultation process at global, regional, national, and local level, which started in 2005 World Conference on Disaster Reduction in Hyogo, Kobe. IoT refers to the technology, which can be implemented on the ground in cooperation with different stakeholders. PT is more like social process in the communities, including community-based disaster risk reduction, education, awareness raising, etc. TIK refers to the local traditional and indigenous knowledge, the principles of which can be transferred in different locations and contexts. This was the first time that the social processes and local/traditional knowledge were recognized as technologies, which emphasized the importance of social science in the landscape of STI.

Over the past 20 years, conventional technologies have evolved significantly. Major breakthrough have been made in earthquake retrofitting techniques, especially in non-engineered construction (Shaw 2023). Several countries have adopted strengthening rural housing through simple and affordable retrofitting techniques. The high-rise buildings and engineered construction (including infrastructures) have developed through learning from different disasters. A significant improvement has been made in the early warning system. The Indian Ocean Tsunami has prompted the establishment of the Indian Ocean Early Warning System, led by UNESCO. This has made a significant impact on connecting Indian Ocean countries through a shared early warning system, similar to the Pacific Early Warning System, which was developed in 1960s, after the Chile Tsunami. There have also been improvements in the weather-related early warning system, not only in terms of its prediction (through advanced satellite monitoring), but also in bringing the early warning system to the last mile, bringing it to the people and communities through different media. In 2022, The United Nations (UN) Secretary General launched "Early Warning for All", where he appealed that all parts of the world should be connected through early warning system, and a specific action plan was launched for 2023 to 2027 for its implementation (UN 2022).

Sakurai and Shaw (2021), in their book, made a detailed analysis of emerging technologies in disaster risk reduction. Technologies have become an enabler for strong and resilient local communities and have influenced every day's practice and people's behavior. The book suggested to develop/categorize different technologies under existing, new, and emerging technologies, keeping in mind the dynamic evolution of technologies. The suggested framework focused on: pre-disaster, early warning and evacuation, and post-disaster scenario, using the IOT, PT, and TIK, mentioned earlier. The book also argued that while there has been significant development of emerging technologies over the past several years, the governance mechanism to support the emerging technologies is still lacking. Thus, there seems to be a gap in the speed of new technology development, and possible adjustments and changes need to be made in the governance regulations to implement them.

8.3 Role of Global Advisory Group and Initiative

Shaw (2020) has made a 30-year's review of the development of science, technology, and innovation in the field of disaster risk reduction. The year 2020 marked the 30 years from the first inception of the United Nations International Decade of Natural Disaster Reduction (IDNDR). In the post-Indian Ocean Tsunami, the year

2005 marked 10 years after the Great Hanshin Awaji Earthquake (popularly known as Kobe Earthquake).

In 2005, a global disaster framework called Hyogo Framework for Action (HFA) was adopted which laid out the path for the next 10 years until 2015. While the role of science and technology was mentioned in all five priorities of the framework, it is explicitly mentioned in Priority 2, "*Identify, assess and monitor disaster risks and enhance early warning*," and Priority 3, "*Use knowledge, innovation and education to build a culture of safety and resilience at all levels*." The Hyogo Framework explicitly mentioned the need to "*Strengthen the technical and scientific capacity to develop and apply methodologies, studies and models to assess vulnerabilities to and the impact of geological, weather, water and climate-related hazards, including the improvement of regional monitoring capacities and assessments*" (UNISDR 2005, p. 10).

During the UN IDNDR period (1990–1999), there was a global Science Technology Committee (STC) which used to provide technical advice to the UN and through UN to its member states on the importance of science technology in disaster risk reduction. This was later replaced by STAG (Science and Technology Advisory Group). The STAG played a pivotal role in disaster risk reduction (DRR) by providing expert guidance to the United Nations Office for Disaster Risk Reduction (UNDRR). STAG helped bridge the gap between scientific research and policy implementation to ensure that DRR strategies are evidence-based and effective. It consisted of scientists, technologists, and experts from various fields who contributed their expertise in understanding disaster risks, developing early warning systems, and promoting innovative solutions for resilience. Globally, STAG promoted the use of scientific data and technology in formulating disaster risk management plans. It ensured that science-based information was accessible to policymakers and the public, enabling informed decision-making in risk mitigation, preparedness, and recovery efforts. The group supported global DRR frameworks like the Sendai Framework for Disaster Risk Reduction by ensuring the integration of scientific knowledge in the assessment of risks and vulnerabilities. STAG also facilitated collaboration between governments, research institutions, and civil society to foster innovation in DRR, focusing on technological advancements such as artificial intelligence, remote sensing, and data analytics. Its efforts contribute significantly to reducing disaster risks, improving resilience, and safeguarding lives and livelihoods across the world (Shaw 2020).

Another global initiative in the field of science and technology in disaster risk reduction was IRDR (Integrated Research on Disaster Risk). IRDR is an international, interdisciplinary research program aimed at addressing the complex factors that contribute to disasters and reducing their global impact. Launched in 2010 by the International Science Council (ISC), the International Social Science Council (ISSC), and the United Nations Office for Disaster Risk Reduction (UNDRR), IRDR promotes a holistic approach to understanding disaster risks. IRDR's primary role in global risk reduction lies in fostering collaboration among scientific communities, policymakers, and practitioners to improve risk assessments, governance, and resilience-building strategies (Shaw 2020). By integrating natural, social, and engineering sciences, IRDR seeks to uncover the root causes of vulnerabilities, enhance the effectiveness of disaster response, and propose long-term risk mitigation measures. A key focus of IRDR is enhancing global capacity in disaster research through data sharing, knowledge transfer, and capacity-building initiatives. This approach is exemplified by its work on improving risk assessments, building resilient societies, and advancing risk communication strategies to ensure that scientific findings are applied to policy and practice. IRDR also contributes to global frameworks like the Sendai Framework for Disaster Risk Reduction by providing scientific insights and policy recommendations that promote evidence-based disaster risk management worldwide. Currently, the second phase of IRDR is ongoing from 2022 to 2031.

8.4 Emergence of Regional Advisory Groups

While we found that global advisory groups have played important roles in generating awareness of the need for science-based decision making, the importance of regional groups has been discussed over the past 10 years, especially after the Sendai World Conference on Disaster Risk Reduction in 2015. Asian region was the first to establish its regional science technology advisory group in May 2015, known as the Asian Science Technology Academia Advisory Group (ASTAAG), which later changed its name to Asia Pacific Science Technology Advisory Group (AP-STAG). Gradually, several other regional advisory groups were formed in Europe, the Arab-North African region, Americas, etc. These regional groups had much stronger footprints in the regions to advance the science, technology implementation in their respective regions (Shaw et al. 2016).

In Asia, to foster science technology widely and beyond scientist and academicians, a platform was created called Science Technology Conference for Disaster Risk Reduction, which is currently known as APSTCDRR (Asia Pacific Science Technology Conference for Disaster Risk Reduction). It started in 2016, hosted in Thailand, followed by 2018 in China, 2020 in Malaysia, 2022 in the Philippines, and 2024 in India. Figures 8.1 and 8.2 provides the detailed theme of the conference and number of participants. The key characteristic of the platform is that it does not include scientists and academics only, it is an open forum for any stakeholders (like civil society, private sector, media) including governments and UN agencies also who use science/technology for their own activities, and who can contribute to the advancement of science and technology. During each of these meetings, a Science Technology status report is created, which shows the advancement in different aspects of science and technology in the region (Fig. 8.3). In Fig. 8.3, there are four outcomes: data and knowledge, dissemination, monitoring and review, and capacity building. The numbers 1, 2, 3, and 4 denote Sendai Priority 1, 2, 3, and 4 respectively.

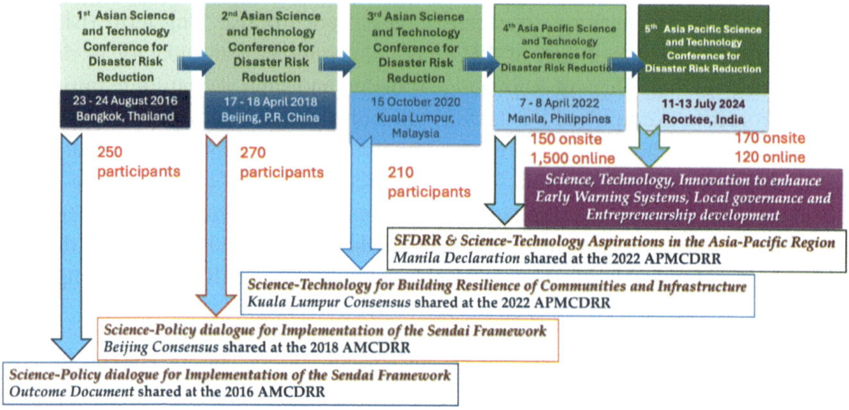

Fig. 8.1 Evolution of regional science technology platform in Asia–Pacific region

Fig. 8.2 Asia Pacific science technology status report

	2020					2022					2024				
OUTCOME	1	2	3	4	Average	1	2	3	4	Average	1	2	3	4	Average
OUTCOME 1: DATA AND KNOWLEDGE	2.84	2.97	2.85	3.00	2.90	3.08	3.06	2.95	3.14	3.06	3.20	3.10	3.00	3.21	3.13
OUTCOME 2: DISSEMINATION	2.96	3.12	3.00	2.82	2.98	3.10	3.15	3.04	3.04	3.08	3.25	3.23	3.05	3.13	3.17
OUTCOME 3: MONITORING AND REVIEW	2.68	3.15	2.97	2.48	2.81	3.03	3.09	2.93	2.83	2.97	3.14	3.19	2.98	2.98	3.07
OUTCOME 4: CAPACITY BUILDING	3.18	2.95	3.14	2.63	3.00	3.16	3.13	2.93	2.89	3.03	3.26	3.26	3.09	3.05	3.17
AVERAGE	2.94	3.05	2.93	2.77	2.92	3.09	3.11	2.96	2.98	3.03	3.21	3.20	3.03	3.10	3.13

Fig. 8.3 Advancement of different criteria in science technology in the region

8.5 National Science Platforms

To implement the HFA, at the country level, national platforms were established, which provided an excellent opportunity for interactions among different stakeholders, and science and technology communities played an important role in influencing decision making. National platforms served as multi-stakeholder mechanisms, coordinating efforts across government agencies, civil society, private sector, and academic institutions to enhance national disaster preparedness and resilience. The roles of the national platforms were critical at an early stage of the implementation of HFA priorities, as well as sensitize the national governments on the role of stakeholders in implementing risk reduction activities.

National platforms helped develop and align national disaster risk management policies with the priorities set out in the HFA. They facilitated coordination between various government agencies to ensure an integrated approach to DRR, reducing overlaps and filling gaps in policy implementation. These platforms were instrumental in conducting risk assessments at local and national levels, identifying vulnerabilities, and implementing early warning systems to enhance preparedness. They promoted the use of scientific research and data to inform decision-making. National platforms worked to strengthen local and national capacities for DRR by providing training, disseminating knowledge, and raising public awareness about disaster risks and preparedness strategies. They also played a role in monitoring the implementation of the HFA, collecting data on progress, and reporting it to international bodies like the UNDRR.

In the post-2015, there have been several initiatives to promote science-based decision making at national level. In Japan, the National Disaster Prevention Committee plays a central role in coordinating and overseeing disaster risk reduction (DRR) efforts at the national level. This committee operates under the Cabinet Office and works to ensure a comprehensive and strategic approach to disaster prevention, preparedness, and response. This committee members mostly come from academia and technical institutions. Malaysia is a classic example where a formal science-based decision making system emerged in post-2015, where a scientific expert panel on disaster risk reduciton was formed (Fig. 8.4) with participaiton from government technical agencies, universities, civil societies, and private sector, and provide direct technical advices to the NADMA (National Disaster Management Authority).

A similar yet different system is formed in the Philippines in the name of the National Resilience Council (NRC), which is composed of government, private sector and academia, and provides advice to the local governments.

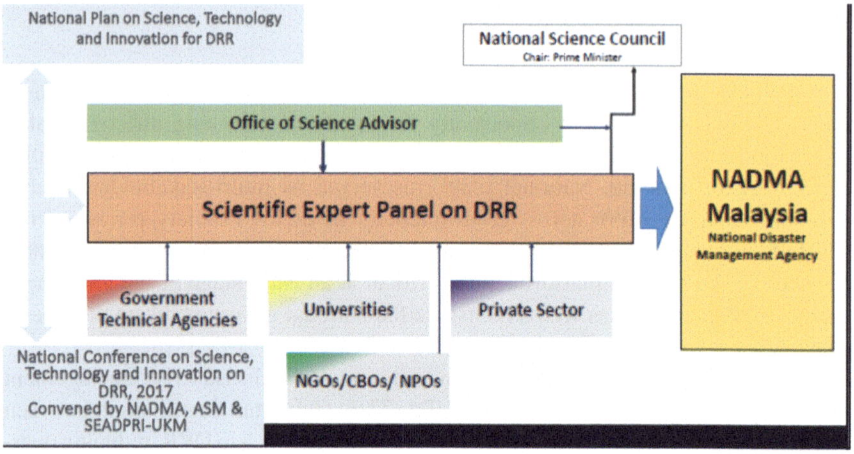

Fig. 8.4 Role of science and technology in disaster risk reduction decision making in Malaysia

8.6 Need for Local Science and Innovation Platforms

Although the disaster managed locally over the years, and importance of community-based and role of local governance has been discussed over the years, the role of science-based decision making in local governance has evolved significantly in post-2015. The changing hazard as well as risk patterns have urged tailoring strategies, policies, and actions to the specific needs, vulnerabilities, and capacities of local communities. Localizing DRR ensures that interventions are more effective, sustainable, and culturally appropriate, ultimately reducing the impact of disasters on lives and livelihoods. Disasters affect regions differently depending on geography, climate, and socio-economic conditions. Localization allows DRR strategies to be customized to the unique risks and challenges of each area. For example, coastal communities may prioritize tsunami preparedness, while mountainous regions focus on landslide mitigation.

Local communities often have traditional knowledge and coping mechanisms that have been developed over generations. Localization integrates this local expertise with scientific knowledge, making disaster plans more practical and relevant to the people affected. Linking science/technology with local knowledge has been found effective before, during, and after disasters. By involving local communities in the design and implementation of DRR strategies, localization fosters a sense of ownership and responsibility. When communities actively participate, they are more likely to implement and sustain DRR initiatives, enhancing resilience at the grassroots level. Local actors are often the first responders when disasters strike. Strengthening local capacities ensures quicker and more effective responses, reducing reliance on external aid. Localization enables communities to adapt recovery efforts based on their immediate needs and priorities. Localization ensures that marginalized groups, such as women, children, and people with disabilities, are included in DRR planning.

By addressing local inequalities and vulnerabilities, DRR becomes more inclusive and equitable.

Overall, localization enhances the relevance, sustainability, and effectiveness of disaster risk reduction efforts by ensuring that strategies align with the specific needs, risks, and capacities of local communities. However, there is still a lack of formal approaches to make local science advice linking local governments, local academic/technical institutions and local civil societies, and private sectors. In the critical and fast technological evolution bringing science benefits to the people and communities as well as local governments.

8.7 Post-script

The world is undergoing different changes in risk landscape, with the emergency of new risks, as explained in the first part of this chapter. At the same time, technological progress has been very fast, and today's emerging technologies are quickly becoming parts of essential technologies. A new governance mechanism (which I argue as adaptive governance) is required to apply these technologies in crisis situation. Japan has been promoting the concept of Society 5.0, which is a technology-driven and human-centric society. We tried to make some analysis on the role of Society 5.0 and digital transformation and its implications for disaster risk reduction (Kanbara et al. 2022). Over a couple of years, Artificial Intelligence (AI) has become more accessible and popular with the new Generative AI interfaces. In this context, as a postscript, I argue for two points:

To use new technologies properly in disaster risk reduction, especially at the local level, capacity enhancement is extremely important. In many cases, the local governments, especially the smaller towns and municipalities may not have people who are digitally native. Therefore, the challenge is how the scientists and technical groups, who are mostly based in the larger cities can help smaller cities and municipalities to gain from the advancement of the technologies. There are different remote and technologies that can make emerging technologies more accessible and inclusive. The future growth of science and technology needs to be more inclusive and affordable.

The other point is entrepreneurship development in the field of disaster risk reduction. While we see a strong development of start-ups, which are mostly concentrated in the high-tech domains, social entrepreneurship/social innovation start-ups, especially in the field of disaster risk reduction are very rare. We need to develop platforms/support system for youth and young scientists to promote more social innovation and address local critical challenges in disaster risk reduction.

References

Kameda H (2009) Implementation technology for disaster reduction. In: Shaw R, Krishnamurthy R (eds) Disaster management: global challenges and local solutions. University Press, Chennai, India, pp 206–219

Kanbara S, Shaw R, Kato N, Miyazaki H, Morita A (2022) Society 5.0, Digital Transformation and disasters: Past, present and future. Springer, p 217

Sakurai M, Shaw R (2021) Emerging technologies for disaster resilience: Practical cases and theories, Springer, p 260

Shaw R (2020) Thirty years of science and technology, and academia in disaster risk reduction and emerging responsibilities. Int J Disaster Risk Sci. https://doi.org/10.1007/s13753-020-00264-z

Shaw R (2023) Repositioning earthquake risk reduction: Implications to global risk landscape. J Disaster Res 18(6):649–655

Shaw R, Izumi T, Shi P (2016) Perspectives of science and technology in disaster risk reduction of Asia. Int J Disaster Risk Sci 7:329–342

UN (2022) Early warning for all. Available at and accessed on 14th Sep 2024. https://www.un.org/en/climatechange/early-warnings-for-all

UNISDR (2005) United Nations International Strategy for Disaster Reduction: "Hyogo Framework for Action" 2005–2015. UNISDR, Geneva

WEF (2020) The global risks report 2020. Published by World Economic Forum. Available from https://www.weforum.org/reports/the-global-risks-report-2020/. Accessed on 23rd of June 2024

WEF (2021): The Global Risks Report 2021, Published by World Economic Forum, Available from https://www.weforum.org/reports/the-global-risks-report-2021/. Accessed on 23rd of June 2024

WEF (2022) The global risks report 2022. Published by World Economic Forum. Available from https://www.weforum.org/reports/the-global-risks-report-2022/. Accessed on 23rd of June 2024

WEF (2023) The global risks report 2023. Published by World Economic Forum. Available from https://www.weforum.org/reports/the-global-risks-report-2023/. Accessed on 23rd of June 2024

WEF (2024) The global risks report 2024. Published by World Economic Forum. https://www.weforum.org/reports/the-global-risks-report-2024/. Accessed on 23rd of June 2024

Chapter 9
Progress in Ecosystem-Based Approach for Disaster Risk Reduction Over the Last 20 Years

Ambika Dabral, Krishnakali Ghosh, Ranit Chatterjee, and Rajib Shaw

Abstract The Hyogo Framework for Acton: 2005–2015, implicitly upheld the Ecosystem-based Approach (EbA) by calling for "sustainable use and management of ecosystems" and "integrated environmental and natural resource management approaches" that imbibe Disaster Risk Reduction (DRR) under one of its priorities for action: reducing the underlying risk factors. A decade later, the Sendai Framework for Disaster Risk Reduction (SFDRR) 2015–2030 explicitly specifies "ecosystem-based approaches". Guided by this and various other global instruments including IPCC Assessment Reports, Agenda 2030, and Convention on Biological Diversity, the EbA has gained much prominence in policy, research and practice. Globally, there have been increasing initiatives by various actors for leveraging EbA for DRR. The current chapter, through an extensive literature review, brings out the evolution in inclusion and uptake of the EbA for DRR in techno-legal instruments and practices along with the key lessons learned and limitations that still prevail.

Keywords Ecosystem-based Approach (EbA) · Disaster Risk Reduction (DRR) · Sendai framework · Resilience · Nature-based Solutions (NbS)

9.1 Introduction

For long Disaster Risk Management (DRM) has predominantly used engineered and grey solutions with a focus on prevention and reduction of hazard. However, the paradigm shift from managing disasters to managing disaster risks necessitated a holistic understanding of risk in all its dimensions rather than merely focusing on hazards. This required that the researchers and practitioners engaged with Disaster Risk Reduction (DRR) imbibe a multi-dimensional (socio-cultural,

A. Dabral (✉) · K. Ghosh · R. Chatterjee · R. Shaw
RIKA Institute, New Delhi, India
e-mail: ambika@rikaindia.com

R. Shaw
Graduate School of Media and Governance, Keio University, Tokyo, Japan

R. Shaw et al. (eds.), *Two Decades from the Indian Ocean Tsunami*, Disaster Risk Reduction, https://doi.org/10.1007/978-981-96-2669-4_9

economic, ecological, physical, institutional, etc.) and all-encompassing approach for effective comprehension and management of components and drivers of risks and the complex interactions therein. Ecosystem-based Approach (EbA) is appropriate in this regard as approaches that embrace ecological conservation while also nurturing the sociocultural and economic well-being of the local communities. These very dimensions are critical in shaping the disaster risk (or resilience) in a given context. EbA works on the principle that healthy ecosystems are better placed to resist, reduce, cope and adapt to adverse effects of disaster and climate risks (Ommer et al. 2022).

EbA has over the decades continued to gain popularity and acceptance with the growing recognition of the tangible and non-tangible co-benefits they offer to people and the planet. Global commitments, agreements and frameworks along with the academic discourse including the Millennium Ecosystem Assessment, IPCC Assessment Reports, Agenda 2030 and Convention on Biological Diversity (CBD), the Sendai Framework, the Paris Climate Agreement have provided a huge impetus in this regard.

9.2 Methodology

The chapter is informed by a desk-based review of secondary literature. The literature included research papers, outcome documents of major conferences & conventions, techno-legal and financial instruments (commitments, policies, guidelines, plans) of various global, regional, inter-governmental and national organisations/bodies, and reports on relevant initiatives by different actors. Keyword search has been used on Google Scholar, Science Direct and Prevention Web to identify the related literature. After that, the snowballing approach on the cited literature was used to further identify more relevant literature. Additionally, the web portals of associated organisations/ stakeholders were also studied to map and study the relevant reports on different initiatives undertaken by them.

9.3 Literature Review

9.3.1 Origin of EbAs

The United Nations Environment Programme (UNEP) was established in 1972 amidst the growing concerns of ecosystem degradation and loss/threat to varied species of flora and fauna. A pertinent challenge at that time was the lack of appropriate economic policies and instruments that would promote conserving and managing natural resources and biodiversity sustainably. The need for the same was further brought forth by the World Commission on Environment and Development of 1987 (the Bruntland Commission). Through its report, "Our Common Future", the Commission underscored the need for economic development and policies that uphold the environment at their core. Further, the introduction of the CBD at the UN Conference on Environment and Development, Rio de Janeiro in 1992 and further ratification by other nations led to an amplified discourse on the importance of biodiversity conservation. This also enhanced the understanding and awareness of the inter-relatedness of the loss of biodiversity and adverse impacts/disruptions in the associated ecological functions and services and hence the effects on human health and socio-cultural and economic wellbeing (Convention on Biological Diversity 2020). The Convention also paved the way for the "ecosystem approach" as a suitable framework for action in this regard. Such an approach considers all products and services that biodiversity offers in a given ecosystem (ibid). Towards this end, the Global Environment Facility (GEF) was created to support international cooperation and fund initiatives targeted at addressing the loss of biodiversity, thinning of the ozone layer, deterioration of international waters and adverse impacts of climate change.

9.3.2 Uptake of EbA for Disaster Risk Reduction (DRR)

Utilisation of the EbA to mitigate disaster risk is not a recent development. Since time immemorial (even when terminology like EbA did not exist), indigenous and traditional communities across the globe have been leveraging different ecosystem services such as those provided by mangroves (to coastal and riverine communities) and forests to protect themselves and their respective habitats and symbiotically reap their socio-economic and cultural benefits. However, the global impetus triggered by the era of environmental movements (1970–1980s) significantly contributed to the subsequent formalisation and integration of the EbA in DRR during the 1990 and 2000s. In the current period (post-2000s), EbA is well established and considered a viable solution to bank on for not only mitigating disaster risks but also enhancing societal and ecological resilience against disaster and climate risks. Thus, currently, the focus is rather on mainstreaming the EbA in DRR. This section talks about the progress in the uptake of the EbA for DRR in research, techno-legal instruments, financial instruments and practices.

9.3.2.1 Research

Heightened research interest in the topic can be gauged considering the consistent increase in publications on the topic from 2005 to 2024 (Fig. 9.1). Using the keyword search for "ecosystem approach, disaster risk reduction" on Science Direct, 14,810 literature were found. The figure shows a significant annual increment (in %) in the number of publications in 2007 (compared to 2006), 2012 (compared to 2011) and 2021 (compared to 2020).

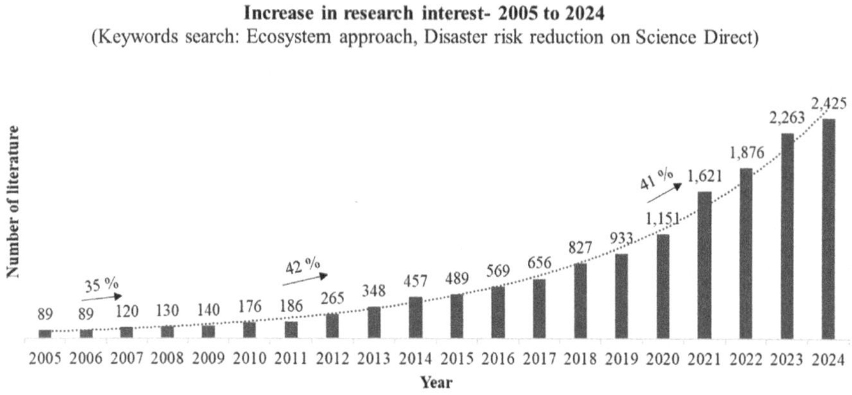

Fig. 9.1 Increase in research interest (2005–2024) in EbA for DRR

Coincidentally, these and the immediately preceding years have witnessed some major disaster events across the globe. For example, 2020–21 witnessed the unprecedented COVID-19 pandemic and its repeated waves which brought the world and varied services to almost a standstill. Besides, the pandemic period also saw the occurrence of other disaster events leading to compound disaster scenarios. The World Disaster Report 2020 informs that over 100 disaster events were experienced globally within 6 months of the declaration of the pandemic in March 2020 (IFRC 2020). Similarly, 2011 saw the occurrence of the Great East Japan Earthquake and Tsunami and 2010 saw the deadly Haiti earthquake. The economic losses by natural hazard-induced disasters in 2011 were over USD 350 billion; the maximum that the decade recorded (IFRC 2012). The year 2005 is considered a year of disasters, following the devastating Asian Tsunami (December 2004), the year witnessed Hurrican Stan and Katrina which also led to massive mudslides and floods along with humanitarian emergencies caused by conflicts in different parts of Africa (Oxfam 2005). These large-scale disasters and associated damage and loss may have triggered an increased research interest in EbA for DRR as an alternative to conventional DRR measures whose limitations were exposed by some of these disasters. Further, it is interesting to note that out of these 14,810 pieces of literature, 12,810 have been published in the last decade (2015–2024) while only 2000 in the preceding decade of 2005–2014. This can likely be due to the momentum gained and the clarion calls for uptake of and investment in sustainability and resilience by the three global commitments in 2015 on DRR, climate action and sustainable development.

A keyword-based review of some recent relevant research papers indicates the key themes popular in the current academic discourse on EbA (Table 9.1). Some of the striking themes studied by these papers include ecosystem services, nature-based solutions, cost-effectiveness, co-benefits, non-market benefits, disbenefits, equity, sustainability, resilience, governance, data technology, public acceptance and perception and ecosystem service dependency, among others.

Table 9.1 Key themes covered by literature published during 2015–2024

Pre-dominant themes	Literature
Cost-effectiveness/benefit–cost analysis/ economic viability	Janzen et al. (2024), Chaba et al. (2022)
Cluster analysis	Tyllianakis et al. (2022)
Indicator framework	Tyllianakis et al. (2022)
Environment impact assessment	Ommer et al. (2022)
Pre-assessment	Ommer et al. (2022)
Sustainability	Chaba et al. (2022), Ommer et al (2022)
Equity	Chaba et al. (2022)
Resilience/disaster resilience/urban resilience/ island resilience/resilient societies	Chaba et al. (2022), Tyllianakis et al. (2022), Talubo et al. (2022), Ruckelshaus et al. (2020), Crosweller and Tschakert (2021), Faivre et al. (2018)
Public acceptance/public perception	Anderson et al. (2022)
Data technology	Ruckelshaus et al. (2020)
Governance/leadership	Mundaca and Heintze (2024), Crosweller and Tschakert (2021)
Stakeholder engagement	Trogrlic et al. (2024)
Co-benefits/non-market/co-benefits/ multi-beneficial solutions	Chaba et al. (2022), Ommer et al. (2022)
Disbenefits/ecosystem disservice	Ommer et al. (2022)
Socio-ecological systems	Talubo et al. (2022)
Ecosystem services/ecosystem functioning/ natural capital	Janzen et al. (2024), Tyllianakis et al. (2022), Ommer et al. (2022), Ruckelshaus et al. (2020), Mundaca and Heintze (2024), Faivre et al. (2018)
Nature-based solutions	Tyllianakis et al. (2022), Ommer et al. (2022), Anderson et al. (2022), Mashiyi et al. (2023), Faivre et al. (2018)
Green infrastructure	Faivre et al. (2018)
Ecosystem service dependency	Mundaca and Heintze (2024)
Ecosystem-based adaptation	Tyllianakis et al. (2022)
Disaster risk reduction/disaster risk management	Ommer et al. (2022), Anderson et al. (2022), Mashiyi et al. (2023), Ruckelshaus et al. (2020), Trogrlic et al. (2024), Faivre et al. (2018)
Climate change/climate change adaptation	Crosweller and Tschakert (2021), Faivre et al. (2018)

With respect to the hazard-specific focus of the research studies on EbA, a slight tilt towards hydrometeorological hazards can be observed. Studies including (Janzen et al. 2024), Mashiyi et al. (2023), (Anderson et al. 2022; Ruckelshaus et al. 2020) focused on hydrometeorological hazards such as floods, coastal floods, erosion, among others. Besides, many researchers such as (Janzen et al. 2024; Chaba et al. 2022; Talubo et al. 2022) and others have gone beyond a mere focus on hazards and have dealt with contributing conditions of vulnerabilities too. A few studies such as (Chaba et al. 2022; Trogrlic et al. 2024) considered multi-hazard and multiple risks while others like (Tyllianakis et al. 2022) also studied climate change risks.

Studies suggest that EbA for DRR supports not only mitigating hazards but also improving the conditions of exposure and vulnerabilities (Janzen et al. 2024). Thus, EbA tends to address the systemic interconnectedness prevailing within and across different components and drivers of risks. On the contrary, conventional solutions centred around technology/engineering often target the hazard component of risk only while being guided by a hazard-by-hazard approach (Kelman 2017 and UNDRR 2019 as cited in Chaba et al. 2022). The solutions guided by the EbA, thus, have at their core, the comprehensive well-being of socio-cultural, economic (local livelihood) and ecological systems. Thus, over the decades and with the ever-increasing complex manifestations of the climate and human-induced disaster risks, ecosystem-based solutions have become all the more relevant for not just ecological conservation and restoration but also for mitigating and managing risks in built environments. EbA has been increasingly seen and used as a solution to sustainable urbanisation (Tyllianakis et al. 2022) by catering to many of the current urban and peri-urban challenges including extreme temperatures, urban heat islands, urban floods, wastewater management, air and water pollution, etc.

However, some literature also indicates potential "disbenefits" or undesired impacts of EbA. These may include the potential creation of habitats favouring vector growth (such as mosquitoes) and, a hike in real estate prices potentially leading to social segregation, among others (Ommer et al. 2022). Similarly, during the current discourse, EbA is also often linked with the limitations of having a long gestation period (particularly for green solutions such as plantations) before the benefits can be realised, having associated uncertainties of risk reduction results and being very context/site-specific (Anderson et al. 2022).

9.3.2.2 Techno-Legal Instruments

The Hyogo Framework for Action has 2 explicit mentions of "ecosystem". Firstly, in the context of the promotion of using and managing ecosystems sustainably and secondly, for uptake of integrated approaches for the management of "fragile ecosystems" (UNISDR 2005). The Sendai Framework explicitly uses "ecosystem (s)" and "ecosystem-based approach to DRR" at a total count of 7. While the explicit mentions may not have seen a significant increase in terms of absolute numbers, there is a positive expanse in focus, scope and contexts in which the terminologies have been used in the SFDRR in contrast to the Hyogo Framework. For example, the scope of SFDRR encompasses beyond sustainable use, conservation and management of fragile/declining ecosystems to also leveraging ecosystem functions and services to reduce disaster risks and build socio-economic and ecological resilience. Further, their coverage in the SFDRR is closely linked to two of its priorities for action, namely, understanding disaster risks (even at the ecosystem level) and strengthening risk governance (using EbA) (UNDRR 2015). Implicitly, SFDRR through the use of related terminologies like "environment" and "environmental" and promotion of closely associated approaches such as "gender-equitable", "universally accessible", "all-of-society", "all-of-state", "multi-hazard" and "multi sectoral" further pushes the underlying principles of EbA. Further, EbA, particularly Nature-based Solutions (NbS) and ecosystem-based adaptation, and related measures have also been included in the Nationally Determined Contributions (NDCs) by a majority of the signatories of the Paris Agreement as an appropriate pathway towards climate change mitigation and adaptation (Tyllianakis et al. 2022). Similarly, EbA has also been upheld as suitable for guiding adaptive governance towards fulfilling sustainability goals (ibid).

At the national level, a nation's risk profile and its development and climate change adaptation priorities are shaped in part by its national plans, the distribution of its resources and the institutional mechanisms that oversee resource distribution. Policies include the strategic principles for solving problems or identifying lines of action addressing issues such as maintenance and conservation of ecosystem services and linking those to vulnerability reduction and strengthening of adaptive capacities. The policies should be anchored in indigenous, local and scientific knowledge to ensure better adaptive capabilities and scope for addressing climate uncertainties. Despite the importance of ecosystem health and services to diverse sectors, they are not usually adequately addressed in development plans and strategies. As a result, it is critical to build policies and assign stakeholder roles in order to collaborate closely and establish cooperation structures that allow parts of biodiversity and climate change to be integrated into national policies. EbA can be incorporated into national policies and plans through periodic reviews and updates, which can then be included in the National Adaptation Plans.

Legal frameworks are crucial when it comes to encouraging and supporting ecosystem-based adaptation to climate change. Some of the important legal mechanisms include strategic planning for sustainable land and resource use; governance frameworks for adaptive management; regulation of current land and resource usage; environmental impact assessment and approval of proposed development activities. Limited resources; inadequate awareness of the complex impacts of climate change; government capacity to administer legal frameworks; capacity for monitoring and enforcement; and governance issues, such as political instability and corruption, are some of the obstacles to the effective implementation of laws across the world. E.g. The Philippines classifies mangroves as public-owned assets and the resultant sequestered carbon is owned by the government. Ownership of mangroves can be transferred to a private body or local community through legal tools like lease agreements and/or community-based management of forests.

9.3.2.3 Financial Instruments

Over 60% of countries' NDC pledges prioritise ecosystem-based adaptation, proposing conservation, restoration, agroforestry and community-led alternatives. Most NDCs that mention ecosystem-based solutions require additional climate financing for successful implementation (GIZ 2018). EbA is frequently financed using a combination of mechanisms. To address the lack of a specialised/dedicated EbA financing mechanism, the following instruments are chosen for their ability to provide climate finance in a manner appropriate to EbA.

- Risk transfer through insurance against the potential damages by extreme climate events is one of the effective instruments for strengthening community resilience. Climate adaptation can be incentivised through risk insurance systems by lowering the risk of policy payouts and increasing attention. Furthermore, insurance mitigates natural investment barriers by offering protection against underperformance risks. International financing organisations often support these plans, reducing risk for national policymakers. These have the potential to assist developing countries in implementing EbA. E.g. The Caribbean Catastrophic Risk Insurance Facility (CCRIF) and the fisheries and aquaculture industries have created a sovereign insurance product. Further, as per its policy on fisheries and the associated protocol for CCA and DRM in the sectors of fisheries and aquaculture, the sovereign product supports governments in fulfilling the mandate of the regional mechanism on fisheries while enhancing fishing communities' resilience by supporting them to rebuild in aftermath of catastrophes.
- Grant finance which involves linking payouts to either upfront or results-based indicators for implementation. E.g. The NbS Invest project, a US$10 million project funded by the Global Environment Facility funds climate mitigation and adaptation in the world's Least Developed Countries (LDCs).
- Equity finance involves a financial source that ties its requirements and stipulations to the EbA outcome metric.
- Debt finance involves a fund aligning its debt terms and conditions with EbA measures. E.g. Coastal Louisiana uses Environmental Impact Bonds (EIBs) with incentives that are tied to the success and performance of the projects. Thus, EIBs represent a form of debt financing that operates on a pay-for-success model, i.e. the repayment on the bond purchased by an investor depends on the successful attainment of a pre-determined environmental outcome.
- Concessional loans offer more favourable terms than market loans, including lower interest rates, grace periods, or a mix of the two.
- Market development tools such as access/price premium to green markets have the potential to boost value and improve access to markets for eco-friendly products and services.

9.4 Analysis and Discussion

9.4.1 Existing Challenges and Opportunities

Firstly, despite the holistic gains offered by the EbAs for DRR, the mechanisms for assessing its effectiveness are still found to be majorly focussing on aspects of mitigation of hazards (Janzen et al. 2024). This undermines the contribution of EbAs in limiting the conditions of exposure and vulnerabilities; thereby enhancing resilience. A commonly used tool in this regard is that of Benefit–Cost Analysis (BCA). More often than not, the BCA tool disregards the consideration of a multitude of benefits the EbAs offer through different ecosystem services and upholds societal equity (Chabba et al. 2022).

Secondly, a critical observation by researchers is that more often than not, environment and conservation fields have witnessed varied approaches gaining popularity and increased uptake followed by them soon getting replaced by a newer approach (Lesisher 2015). While the evolution of the EbAs and their increased acceptance over the decades is not yet signalling in that direction, more is needed to be done at the varied governance levels (beyond mere prescribing mentions) to ensure that EbAs and solutions are effectively mainstreamed and practised across jurisdictions and sectors (Tyllianakis et al. 2022).

Thirdly, being context-specific, the appropriateness and effectiveness of EbAs depend on local socio-ecological conditions which necessitates in-depth and temporal assessment of the processes and systems thriving at the landscape level. However, the literature suggests often the decision-making behind the selection of EbAs for a given context/ site/hazard is not adequately backed by landscape-level risk assessment and knowledge systems. This also creates barriers to knowledge and acceptance and hence effective uptake of such measures by local decision-makers (UNDRR 2021). Willing decision-makers who recognise and appreciate the multiple co-benefits of EbAs, also often face the lack of actionable frameworks, tools and systematic databases that can help them identify the most appropriate ecosystem-based solutions for their given context out of a suite of solutions they are aware of.

9.4.2 Way Ahead and Role of Different Stakeholders

To further expand the uptake of the EbAs for DRR, the identified research and praxis gaps must be bridged and an enabling environment must be created via supporting policies and financial mechanisms. These instruments must also look into the critical aspects highlighted in the earlier sections. Chabba et al. (2022) and Faivre et al. (2018) underscore the need for more scientific evidence, particularly from developing countries, for showcasing resilience building and economic viability of the EbAs along with their risk reduction benefits across different ecosystems. This is important as different ecosystem-based mitigation measures may provide varying degrees of expected outcomes for different local risk conditions, ecosystem characteristics and climatic variability. Thus, researchers and practitioners, through such studies and evidence can better inform the conventional evaluation tools that undermine local variability and non-economic (socio-cultural and ecological) co-benefits offered by the EbAs for DRR. Evaluation tools so upgraded should also be able to address the existing gap highlighted by UNDRR (2021) and Ommer et al. (2022) in terms of lack of assessment of potential limitations, undesired negative outcomes (disbenefits/ co-harm) along with critical factors that may undermine achieving the full potential of the EbAs.

Further, vulnerable communities and local authorities are among the key stakeholders who are often involved/engaged in collaborative conceptualisation, designing, execution and monitoring of the EbAs. Thus, achieving the expected outcomes of EbAs also relies on the acceptability and sense of ownership of these critical stakeholders. To this effect, the importance of putting in place simplified yet innovative and robust tools and strategies to enhance their understanding and awareness, demystifying the prevailing misconstruction while also showcasing the multi-dimensional co-benefits through demonstrative and experiential models cannot be overstated (Anderson et al. 2022). Likewise, detailed studies are also required to better gauge the outlook of the local stakeholders on EbAs and the placement of their preferences (and why) for a given pool of ecosystem or engineered/ grey solutions for mitigating and managing risks (ibid).

The emerging technologies supporting the timely, accurate and repeated data collection and management of large databases on socio-ecological ecosystems and their varying local contexts can be further leveraged to enable the local decision-makers and other stakeholders with the right pool of data and information for uptake and mainstreaming of EbAs (Ruckelshaus et al. 2020). With the increasing traction of concepts like ESG, net zero, circular economy and CSR in businesses, private sector and corporate foundations are also likely to lean towards EbA as a suitable pathway towards these compliances and commitments. However, the future of EbA and its mainstreaming in DRR and CCA substantially rests on the degree and quality of collaborative efforts by different actors to address the existing barriers and bottlenecks in its uptake and scaling up (Fig. 9.2).

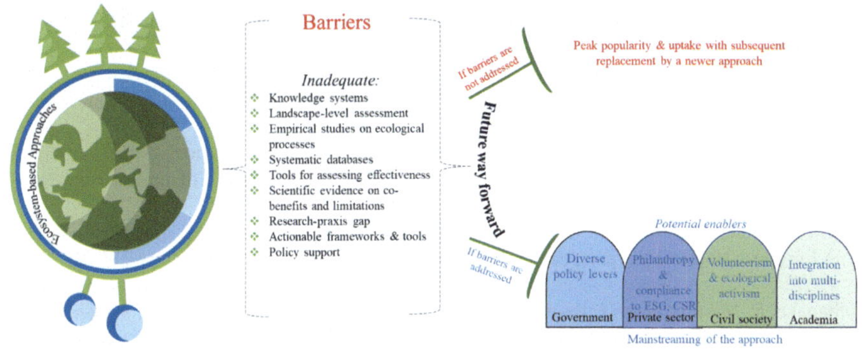

Fig. 9.2 Future way forward of EbA

References

Anderson C, Renaud F, Hanscomb S et al (2022) Green, hybrid, or grey disaster risk reduction measures: What shapes public preferences for nature-based solutions? J Environ Manage. https://doi.org/10.1016/j.jenvman.2022.114727

Chabba M, Bhat M, Sarmiento J (2022) Risk-based benefit-cost analysis of ecosystem-based disaster risk reduction with considerations of co-benefits, equity, and sustainability. Ecol Econ. https://doi.org/10.1016/j.ecolecon.2022.107462

Crosweller M, Tschakert P (2021) Disaster management and the need for a reinstated social contract of shared responsibility. Int J Disaster Risk Reduct. https://doi.org/10.1016/j.ijdrr.2021.102440

Convention on Biological Diversity (2020) Sustaining life on earth: How the convention on biological diversity promotes nature and human well-being. https://www.cbd.int/doc/publications/cbd-sustain-en.pdf

Faivre N, Sgobbi A, Happaerts S et al (2018) Translating the Sendai framework into action: The EU approach to ecosystem-based disaster risk reduction. Int J Disaster Risk Reduct. https://doi.org/10.1016/j.ijdrr.2017.12.015

GIZ (2018) Finance options and instruments for Ecosystem-based Adaptation. Overview and compilation of ten examples. Authors: Kiran Hunzai, Thiago Chagas, Lieke ‘t Gilde, Tobias Hunzai, Nicole Krämer. Deutsche Gesellschaft für Internationale Zusammenarbeit (GIZ) GmbH, Bonn

International Federation of Red Cross and Red Crescent Societies (2020) World disasters report 2020: Come heat or high water. Geneva. https://www.ifrc.org/sites/default/files/2021-05/20201116_WorldDisasters_Full.pdf

International Federation of Red Cross and Red Crescent Societies (2012) World disasters report 2012: Focus on forced migration and displacement. https://www.ifrc.org/sites/default/files/2012_WDR_Full_Report.pdf

Janzen S, Balzer J, Merk F et al (2024) Moving towards a comprehensive evaluation of ecosystem-based disaster risk reduction: The example of agroforestry for flood risk reduction. Nat-Based Solut. https://doi.org/10.1016/j.nbsj.2023.100104

Leisher (2015) How useful to biodiversity conservation are ecosystem services-based approaches? In: Martin-Ortega J (ed) Water ecosystem services, a global perspective. Cambridge University Press, Cambridge, UK, pp 65–70

Mundaca L, Heintze J (2024) Banking on ecosystem services. Ecol Econ. https://doi.org/10.1016/j.ecolecon.2024.108284

Ommer J, Bucchignani E, Leo L (2022) Quantifying co-benefits and disbenefits of nature-based solutions targeting disaster risk reduction. Int J Disaster Risk Reduct. https://doi.org/10.1016/j.ijdrr.2022.102966

Oxfam International (2005) 2005: Year of disaster https://reliefweb.int/report/indonesia/2005-year-disasters

Ruckelshaus M, Reguero B, Arkema K et al (2020) Harnessing new data technologies for nature-based solutions in assessing and managing risk in coastal zones. Int J Disaster Risk Reduct. https://doi.org/10.1016/j.ijdrr.2020.101795

Talubo J, More S, Saroj D (2022) Whose resilience matters? A socio-ecological systems approach to defining and assessing disaster resilience for small islands. Environ Challenges. https://doi.org/10.1016/j.envc.2022.100511

Trogrli'c R, Reiter K, Ciurean R (2024) Challenges in assessing and managing multi-hazard risks: A European stakeholders perspective. Environ Sci Policy. https://doi.org/10.1016/j.envsci.2024.103774

Tyllianakis E, Martin-Ortega J, Banwart S (2022) An approach to assess the world's potential for disaster risk reduction through nature-based solutions. Environ Sci Policy 136:599–608. https://doi.org/10.1016/j.envsci.2022.07.021

United Nations International Strategy for Disaster Reduction (2005) Hyogo framework for action 2005–2015: Building the resilience of nations and communities to disasters. World Conference on Disaster Reduction

United Nations Office for Disaster Risk Reduction (2015) Sendai framework for disaster risk reduction 2015–2030. Geneva

United Nations Office for Disaster Risk Reduction (2021) Nature-based solutions for disaster risk reduction: Words into action. Geneva

Chapter 10
The Shelter Sector Two Decades After the Indian Ocean Tsunami

Iftekhar Ahmed

Abstract The 2004 Indian Ocean Tsunami was one of the most massive disasters in recent history and shelter was the most impacted sector. With a generous outpouring of international aid, a plethora of humanitarian agencies undertook extensive shelter recovery programs in the affected countries. The approaches followed defined the trajectory of this sector because this was the first such large-scale undertaking. Strategies for coordination and a staged approach to shelter recovery were some of the elements that influenced the sector, evident after some of the major disasters since then, such as in Bangladesh, Haiti, Philippines and Nepal. However, problems were also evident in these contexts in applying the tsunami lessons, particularly in addressing the shelter needs of poor and vulnerable people. During a couple of decades after the 2004 tsunami, two key trends emerged: increasingly a central government-led shelter recovery process and the self-recovery paradigm espoused by humanitarian agencies. With a changing global context beset with conflicts and financial crises, there is no longer a deep purse of funding for the shelter sector. Two decades after the tsunami, the sector is at a crossroads presenting a quandary for its future trajectory.

Keywords Cluster system · Indian Ocean Tsunami · Owner-driven approach · Recovery · Shelter

10.1 Introduction

This chapter presents a review of the critical issues relating to shelter that have evolved over the past couple of decades since the Indian Ocean Tsunami of 2004. The chapter is concerned with disasters caused by natural hazards with a focus on the post-disaster context when most shelter interventions are implemented, as reflected in the post-tsunami experience. Shelter interventions in response to human conflict are not covered here, nonetheless, that requires a separate treatise beyond

I. Ahmed (✉)
University of Newcastle, Newcastle, Australia
e-mail: ifte.ahmed@newcastle.edu.au

R. Shaw et al. (eds.), *Two Decades from the Indian Ocean Tsunami*, Disaster Risk Reduction, https://doi.org/10.1007/978-981-96-2669-4_10

the scope of this chapter because of the preponderance of recent global conflicts and the consequent need for shelter.

At the outset, the term 'shelter' needs to be clarified with regard to the context of this chapter. The term has been largely used in humanitarian agency circles, often in post-crisis situations after disasters or conflicts, to imply support for accommodation for the affected populations. Its roots are reflected in early publications by Davis (1978) and UNDRO (1982); this past paradigm was practised in scattered places around the world without an established global coordination framework before the 2004 Indian Ocean Tsunami, which changed the game, although the 'shelter' term persists. Furthermore, the term has been and is applied primarily in the Global South context, whereas in the Global North 'temporary housing' is more commonly used. Quite simply, shelter is connected to the provision of post-crisis support by humanitarian agencies in the Global South, mostly beyond or sometimes in partnership with national governments, and as discussed later in this chapter, increasingly government-led shelter interventions as well. Also, all the countries impacted by the 2004 tsunami were in the Global South, therefore, this chapter mainly focuses on this context in line with the application of the 'shelter' term.

Following the 2004 tsunami, people from around the world provided generous support—donations exceeding USD 7 billion for recovery projects only in Indonesia (Masyrafah and McKeon 2008). Indonesia was the most impacted country with more than 150,000 homes destroyed or damaged (UNORC-IAS 2009), and a large proportion of the international aid was allocated for shelter recovery there (Masyrafah and McKeon 2008). Among all the sectors, shelter was the most severely impacted (Marti 2005; Masyrafah and McKeon 2008). More than 140,000 new houses were built (Deprez and Labattut 2005) by more than 120 International Non-Governmental Organisations (INGOs) and local NGOs, with support from different United Nations (UN) agencies; the government also implemented shelter projects (Masyrafah and McKeon 2008). Similarly, Sri Lanka, the second most affected country, lost more than 80,000 houses (Suppasri et al. 2015), and extensive shelter recovery programs were necessary. There were various shelter programs of different types implemented by different agencies in the other affected countries including India, the Maldives and Thailand. The scale of the devastation and massive recovery interventions led to chaos and a random flurry of activities with multiple and diverse agencies on the ground on numerous projects acting without a uniform coordination mechanism. A number of lessons emerged from this experience that informed the shelter sector in subsequent decades, which is discussed in this chapter.

10.2 Materials and Methods

The author specialises in post-disaster shelter recovery and over the last few decades has been engaging extensively in this field through research, teaching and professional activities. After the 2004 tsunami, several research projects and training missions allowed gaining first-hand experience and knowledge relating to shelter after the disaster, which was followed through in the subsequent couple of decades. This chapter is derived from this experience and also some of the related publications were reviewed. Furthermore, this experience has allowed identify the key and extensive literature in relation to the topic, which is reviewed to produce a comprehensive narrative.

To arrive at a structure for this chapter, thematic analysis (Braun and Clarke 2019; Caulfield 2019) of the above materials was conducted. It was found that a case study approach (Yin 2017) would be suitable for the chapter structure and to draw out the main arguments. Thus, a selection of case studies of major disasters since the 2004 tsunami were selected, again drawing on the author's experience in relation to these disasters.

10.3 Key Shelter Strategies After the 2004 Tsunami

The 2004 tsunami experience, particularly in Indonesia and Sri Lanka, led to efforts for better coordination and a structured approach to shelter provision, as discussed subsequently in this chapter. These approaches were later applied within the tsunami-affected countries as well as in other countries. Two key aspects that emerged from the tsunami experience in the shelter sector—cluster system and staged shelter provision—are discussed below, highlighting their advantages and challenges.

10.3.1 Cluster System

The lessons gained from the chaotic situation after the 2004 tsunami led to the development of the Humanitarian Cluster System, led by the UN, a key coordination mechanism that came into place in 2005 soon after the 2004 tsunami (Global Shelter Cluster n.d.). Shelter is one of the sectors among eleven other sectors included in the cluster system (UNOCHA n.d.), coordinated by the International Federation of Red Cross and Red Crescent Societies (IFRC) and the United Nations High Commission for Refugees (UNHCR) (Global Shelter Cluster 2018). This has led to the formation of the Global Shelter Cluster based in Geneva, which promotes establishing national shelter clusters after a disaster in respective countries, which are deactivated after shelter response, that is emergency shelter assistance (ESA), is completed and early recovery is initiated (see for example, Global Shelter Cluster 2017).

In essence, the cluster system aims to bring coordination between the different agencies that are involved in post-disaster response and recovery in each sector, such as shelter, so that their interventions are harmonised. Through a series of interactions and reporting mechanisms, a level of harmony is sought between the interventions by a diverse set of agencies, a common set of principles and standards are agreed upon, and agencies agree in principle to avoid competition and attempt to outdo each other, which was an issue after the 2004 tsunami. For example, the Sphere Standards for disaster response (Sphere Association 2018) are agreed to be adhered to. The cluster system mostly includes humanitarian agencies such as INGOs and to some extent some local NGOs. The government is usually not included in the cluster system and acts independently on its own shelter interventions. However, the government can sometimes impose stipulations that need to be followed by humanitarian agencies. For example, according to the Indonesian government, a post-disaster shelter has to be 36 m^2 (Ahmed and O'Brien 2011).

Since the 2004 tsunami, a sequence of massive disasters continued to occur around the world, where the cluster system has been followed, although elements of the chaos evident after the 2004 tsunami persisted. It is mainly applied in Global South countries; there are no examples of its application in Global North countries because shelter recovery in such countries is mainly led by the government, unlike the Global South whether a plethora of INGOs and other international agencies engage to fill the public sector resource gap in these countries, although governments have begun to adopt a central role in some Global South countries, as will be discussed later.

10.3.2 Staged Approach to Shelter Provision

Another paradigm that emerged after the tsunami is the staged approach to shelter provision, beginning with temporary shelter, then transitional shelter (T-shelter), and then, ideally, leading to permanent housing. Here the distinction between 'shelter' and 'housing' becomes apparent. 'Housing' implies more permanence and durability beyond the short-term emergency shelter stage. While 'shelter' is the blanket term, it essentially applies to temporary and transitional shelters, where humanitarian agencies are most active. Permanent housing is much more expensive, and it is often difficult for humanitarian agencies to fund its construction. However, because of the large funding available after the 2004 tsunami, humanitarian agencies were able to engage extensively in the provision of permanent housing.

10.3.2.1 Temporary Shelter

Subsequent to the tsunami, when the typical emergency shelter approach involved provision of family tents, agencies have begun exploring other options such as support for repair and retrofit through provision of toolkits and training, hosting (affected people hosted in someone else's dwelling) and rental subsidy (Opdyke et al. 2017); tents are expensive in the Global South context, there is often no local supply and they need to be brought in from elsewhere, adding extra cost and time. Facility conversion is another temporary shelter approach that has evolved, where disaster-affected people are provided short-term accommodation in public buildings such as schools and religious buildings. Space constraints, exposure to adverse climatic conditions and limited privacy make tents an unsuitable option, a lesson gained that triggered efforts to apply other innovative ESA approaches such as rental subsidy and hosting. For example, people affected by the 2004 tsunami by devastation to their island settlement in the Maldives were hosted by a community on another island with support from the Red Crescent Society until they could be provided permanent housing (Ahmed et al. 2020); here tents were not an option because of the destruction of their island and an innovative solution had to be found.

10.3.2.2 T-shelter

Typically, a T-shelter is a small shed-like temporary dwelling, or sometimes a row of rooms in a barrack style allocated to separate households, expected to provide accommodation for a few years before permanent housing becomes available. T-shelters were generally uncommon in Indonesia after the tsunami, but they were widely built in Sri Lanka (see Fig. 10.1) and have been applied later in different places. A key criticism of T-shelters is that because building permanent housing is a cost-intensive and protracted process, people end up staying in T-shelters for much too long in a state of uncertainty (Global Shelter Cluster 2020; Sanderson et al. 2014). Not being sturdy and often built of lightweight materials, they are exposed to subsequent disasters, which have been recorded for example in Haiti and the Philippines (Deprez and Labattut 2017; Global Shelter Cluster 2017). Additionally, instead of moving from temporary shelter to early recovery, T-shelters can block that process and reduce the imperative to move rapidly to building permanent housing and allowing the recovery process to be fulfilled.

Even for T-shelters, significant funds are required, so agencies sometimes apply other transitional approaches instead of constructing a full dwelling. An example

Fig. 10.1 T-shelters in Sri Lanka, which were found to be occupied five years after the 2004 tsunami

Fig. 10.2 A 'starter' T-shelter built after the 2006 earthquake in Yogjakarta, Indonesia

was found in Yogjakarta, Indonesia, after the 2006 earthquake (Wilderspin et al. 2008), when humanitarian agencies assisted people in building a sort of transitional shelter with a low wall made of bricks salvaged from the debris of the earthquake, with bamboo walls above and roofing also of salvaged clay tiles (see Fig. 10.2). This was then upgraded incrementally by the homeowner as and when funds were available to eventually have a permanent house. In the interim, the bamboo walls served well because they were treated against decay and insect attack, and also included bracing to resist wind and seismic stress. Such a 'starter house' approach designed for upgrading to permanent housing can offer a more effective solution instead of a single-use T-shelter.

10.3.2.3 Permanent Housing

Providing permanent housing is capital-intensive and often protracted because of the complexities of procurement and implementation processes. To address that, an approach which also emerged after the tsunami is that of the 'core shelter', where the agency assists a household to build a sturdy and durable room to which extensions can be added incrementally over time to build a complete house. In the interim, the 'core' provides protection against hazards. This is somewhat similar to the Yogjakarta 'starter house' approach mentioned earlier in Sect. 3.2.2. Extensions built by owners often do not follow hazard-resistant guidelines, nonetheless, the core provides a level of safety even if the other parts of the dwelling are impacted. However, mobilising funds for subsequent extensions can be challenging for the owners and eventually building a permanent house can be a long-term process and the shelter needs of the household continue to suffer.

Support for permanent housing can only be provided where there is land tenure security, and the beneficiary owns the land or has a long-term lease. Thus, renters and informal settlement dwellers are excluded from such support—often the people who need it the most. This problem became evident in Indonesia after the 2004 tsunami (ACHR 2006; Greenblott 2007; Nazara and Resosudarmo 2007), and through the advocacy of human rights groups, there was an effort to address it. However, it was found that out of the more than 140,000 permanent houses built in Aceh, only about 3,000 were built for people who did not own land (ReliefWeb 2009). This is sometimes a guise for evicting informal settlements—for example, in India after the tsunami, the Slum Clearance Board, provided coastal informal fishing settlements with inland 4-storey walk-up housing, with suboptimal results. Distance from the livelihood source, and the very small size of poorly constructed apartments (22 m^2), led to selling them off to seek shelter options more suitable to the people's needs (Shaw and Ahmed 2010).

It should be noted that, while the Sphere Handbook (Sphere Association 2018) provides guidelines for temporary and transitional shelter, there are no internationally accepted guidelines for permanent housing. Although Priority 4 of the Sendai Framework of Disaster Risk Reduction places priority on post-disaster recovery (UNISDR, 2015), there is no guidance there specifically on shelter. There are many country-specific building codes and other shelter-related guidelines, and the Shelter Cluster produces such guidelines after a disaster in a country (see for example, Flinn and Beresford 2009; Shelter Cluster Philippines 2016), which bring value to shelter recovery initiatives, but these tend to be largely for temporary or transitional shelter, seldom for permanent housing. This is thus an area that requires institutional attention at the global level.

10.4 Case Studies

After the 2004 tsunami, a number of major disasters impacted different countries, and some of the approaches that emerged after the 2004 tsunami found their way into shelter interventions after these disasters, as discussed in the case studies below. One key element that became prominent in the 2004 tsunami was the engagement of humanitarian agencies, particularly INGOs, and this became common in the subsequent disasters, with the Shelter Cluster of the newly formed cluster system playing a lead role in the shelter sector. The case studies below are mostly from Asia, where most of the major disasters occurred, and also, Asia being the most disaster-affected continent the focus here has been placed on Asian case studies.

10.4.1 Cyclone Sidr (2007) and Cyclone Aila (2009), Bangladesh

One of the major disasters that occurred soon after the 2004 tsunami was Cyclone Sidr in 2007 which struck the southwestern coastal areas of Bangladesh, followed by Cyclone Aila two years later in 2009. These almost back-to-back cyclones caused widespread devastation to the typically lightweight shelter in the region. Sidr was categorised as a 'super cyclone' with powerful wind speed and destroyed or damaged more than 1.5 million dwellings (Ministry of Food and Disaster Management 2008). Aila had a lesser intensity yet devastated more than 550,000 dwellings (Bangladesh Red Crescent Society 2011), and a high storm surge breached the coastal embankment and inundated vast tracts of land, causing extended displacement of people from their homes for more than two years (IFRC 2010). The devastation caused by the combination of these two cyclones was much more significant than any of the subsequent cyclones that hit the country.

Two key elements that emerged after the 2004 tsunami were evident in the recovery after the cyclones. Firstly, the engagement of the Shelter Cluster, named locally as the Shelter Working Group (SWG), composed of humanitarian INGOs and local NGOs, led by the United Nations Development Programme (UNDP), with the intention of coordination to avoid the chaos that ensued in some of the tsunami-affected countries. To this purpose, the SWG drafted a set of minimum standards for 'core shelter' that SWG member agencies were expected to follow (Flinn and Beresford 2009), the second carry-over from the tsunami. However, it is not clear to what extent these guidelines were followed. It was reported as variable—by the time the guidelines were produced, some agencies had already started their projects (Kabir 2009). There is also some confusion between core and transitional shelters (Kabir 2009), it seems that shelters were randomly assigned these classifications. In any case, those that were considered as core shelters (Flinn and Beresford 2009), were only minimally more durable and sturdier than pre-existing shelter typologies and were not built as a hazard-resistant central core unit as discussed above in Sect. 3.2.3, the key characteristic of a true core shelter.

Perhaps one of the only actual interpretations of the core shelter concept was by UNDP, where single-room shelters with a safe refuge loft were built of durable materials such as brick and good quality metal sheet roofing (Ahmed 2016) (see Fig. 10.3). These obviously came at a higher cost and limited the number of people who could be supported. Nonetheless, these could serve as mini-cyclone shelters for the wider community. There were however design problems of inadequate protection from rain and local salinity. Another initiative supported by UNDP after Aila of durable shelters raised on a reinforced concrete (RC) frame also faced similar problems of rain penetration and salinity, and eventually had to be upgraded with brick and metal sheet roofing (Ahmed 2021).

Thus, it appears that the vocabulary that emerged after the 2004 tsunami in humanitarian circles was transferred to other contexts without adequate understanding and was rather loosely interpreted. While the shelter sector did develop a coordination

Fig. 10.3 A core shelter built by UNDP, Bangladesh

mechanism, in terms of physical shelter outcomes, lessons were still being learned in the early years after the tsunami.

10.4.2 Earthquake (2010), Haiti

After the 2010 earthquake in Haiti, more than 10,000 humanitarian agencies, mostly INGOs, descended there leading it to be called a 'republic of NGOs' (Baptiste 2015, p. 1). Despite the cluster system in place, with so many agencies gathered in one place, there was bound to be chaos, replicating a similar situation as after the 2004 tsunami.

Following on from the 2004 tsunami experience, T-shelters were built extensively in Haiti after the earthquake; more than USD 500 million was spent to build more than 100,000 T-shelters (Phelps 2015). Problems as discussed earlier in Sect. 3.2.2 were evident here, particularly prolonged occupation of the shelters without prospects of early recovery and permanent housing. Conceived to be short-term accommodation, adequate services were lacking, which led to unhygienic conditions and the spread of diseases such as cholera. Similarly, extensive temporary camps were developed and after some years of extended occupation, these camps became informal settlements and developed long-term roots (authors field notes, 2013). Shops and businesses grew in and around, and the settlements became permanent. Removing them without adequate shelter options could potentially lead to conflict, hence they were allowed to stay; before the earthquake, about 90% of the capital city, Port au Prince, was informally built, so this phenomenon was generally accepted. Again, health hazards proliferated. The outcome was a perpetuation, and even increase, of pre-existing vulnerabilities. The T-shelters and temporary camps were built for short-term use without following hazard-resistant construction and safe siting, and hence experienced massive devastation in cyclones later (Hurricane Tomas 2010; Hurricane Matthew 2016; Deprez and Labattut 2017; McDonald, 2010).

A development that arose after the 2004 tsunami, particularly in Sri Lanka, was that of the 'owner-driven' shelter recovery approach, where households rebuild their homes with technical and financial support from agencies. Compared to the 'donor-driven' approach consisting of contractor-built shelter recovery, the owner-driven approach leads to high beneficiary satisfaction (Dias et al. 2016; Ratnayake and Rameezdeen 2008), however, there are some disadvantages too such as the need for monitoring and quality control, time management and availability of adequate construction skills (Aubrey 2008; Ingirige et al. 2008). This approach was also followed in Haiti after the earthquake, where agencies drew on the lessons gained after the tsunami. For example, a recovery project supported by the INGO Cordaid in partnership with various other agencies, particularly Build Change (specialising in hazard-resistant design and construction) utilised a community-based, owner-driven model (Charlesworth and Ahmed 2015) (see Fig. 10.4). Here, in addition to rebuilding new homes, existing buildings were retrofitted to earthquake-resistant standards. Thus, utilising the positive aspects of the owner-driven process, technical support to ensure quality and hazard-resistance allowed moving forward beyond the 2004 tsunami experience. Importantly, the 'one-size-fits-all' approach typical of most donor-driven shelter recovery projects was avoided and the varying needs of different households on different lot sizes were addressed, an important lesson starting to be followed by agencies by learning from previous sub-optimal experiences of mass-produced shelters after the 2004 tsunami.

Fig. 10.4 A house built in the owner-driven mode, Haiti

10.4.3 *Typhoon Haiyan (Yolanda) (2013), Philippines*

A somewhat newer development that became evident in the Philippines after Typhoon Haiyan, or Yolanda as locally called, is the increasing involvement of the government in shelter recovery. Here, in line with the experience after the 2004 tsunami, the government stated that shelter recovery would be owner-driven, but actually, the procurement process allowed only large construction companies or contractors to build the housing (Arroyo and Astrand 2019a). The target for permanent housing of the humanitarian sector led by the Shelter Cluster was reportedly just 1% of the overall shelter response, whereas repair and retrofit support was 70% (Shelter Cluster Philippines 2014). On the other hand, the government undertook extensive permanent housing projects, although still falling short of the vast need where more than a million homes were impacted by the disaster (Global Shelter Cluster 2017).

Perhaps one of the most well-known strategic decisions made after the 2004 tsunami was that of the 'buffer zone' in Sri Lanka, that initially restricted new construction within 100 m of the coastline to reduce the risk of future tsunami impacts, later reduced to 35 m due to pressures from protests. A similar policy decision was also made after Yolanda, with a 40-m wide 'no build zone' (NBZ), later changed to a 'no-dwell zone' (NDZ), along the coast, with government projects relocated inland (Schofield et al. 2017). However, available land and also adequately serviced land were difficult to find (Arroyo and Astrand 2019a; Global Shelter Cluster 2017; National Housing Authority 2015; Shelter Cluster Philippines 2016). Exacerbated by regulatory issues and corruption, it led to delay and uncertainty in provision of permanent housing (Arroyo and Astrand 2019a; Global Shelter Cluster 2017; National Housing Authority 2015), thereby causing people to live in a temporary and transitional shelter for an extended duration of time (Opdyke et al. 2017).

Thus, many people disregarded the NBZ restriction and informally established coastal settlements (Arroyo and Astrand 2019a; Flinn et al. 2017; Schofield et al. 2017; Opdyke et al. 2017; Shelter Cluster Philippines 2016). Most of them were poor and could only afford low quality building materials, and poorly built informal settlements developed along the coast (Arroyo and Astrand 2019b; Doyle 2016; Opdyke et al. 2017) (see Fig. 10.5). People in these communities had coastal livelihoods such as fishing and therefore continued living on the coast. Thus, many permanent housing projects in relocation sites remained abandoned (Ranada 2023). To build on their own in the NDZ, people had to obtain high interest loans and became trapped into long-term debt (Shelter Cluster Philippines 2014). Through this process, the pre-existing vulnerabilities before Yolanda were reproduced, and perhaps even augmented, and people in these informal settlements are exposed to future typhoons, which can only be expected to increase in frequency, magnitude and intensity because of climate change.

Fig. 10.5 Informal settlements in the NDZ, Philippines (*Source* Courtesy of Ivette Arroyo, Lund University, Sweden)

10.4.4 Gorkha Earthquake, Nepal, 2015

As in post-Yolanda Philippines, after the Gorkha Earthquake in Nepal, the government assumed a central role in shelter recovery, despite the plethora of humanitarian agencies and international aid that arrived in the country. The aid was channelled and disbursed through the government, though in this case, it was for an owner-driven shelter recovery process, a perpetuation in the essence of the 2004 tsunami experience. Significantly, there were three tranches of funding disbursal, which was dependent on following the national building codes for earthquake-resistant construction at each of the three stages, a measure to reduce future disaster risk (Schofield et al. 2017).

However, implementation of the owner-driven process ran into many hurdles. The technical support and information required to apply the codes effectively were scanty, especially in the remote areas affected by the earthquake. Together with bureaucratic barriers, this caused delays in the recovery. Escalation of costs of building materials and labour shortages contributed further to the problem. The amount of the grant money was far below the funds required to build a house; in some places, it was estimated to be more than tenfold (Schofield et al. 2017; von Meding et al. 2017). Thus, households found it difficult, if not impossible to access enough funds to rebuild their home and were often driven to borrow money at high interest rates, landing into life-long debt, as in the case of the Philippines discussed above in Sect. 4.3. A key issue is that sometimes people used the grant money to meet other household priorities, and hence were not able to receive the remaining funds, thereby being compelled to live in damaged or incomplete dwellings with exposure to future

Fig. 10.6 Households such as this in Nepal could not complete their house after the earthquake because they did not fulfil the government requirements

disaster risks (von Meding et al. 2017) (see Fig. 10.6). Once again, pre-existing vulnerabilities were replicated, or magnified in this process.

Other shortcomings in the shelter recovery process were also noted. The owner-driven shelter recovery could only take place on land owned by the beneficiary, whereas a significant proportion of the earthquake-affected people were landless. Particularly lower caste communities, or 'Dalits', were deprived of receiving any recovery assistance, as well as single mothers and widows. Thus, the people who needed shelter recovery support the most, did not receive it (Ahmed et al. 2020). The Nepal case study clearly demonstrates the perils of a centralised governmental shelter recovery intervention in a Global South country, similar in some ways to the Philippines case above (Table 10.1).

10.5 Emerging Paradigms

In the changing global context, new frameworks, such as the Sendai Framework for Disaster Risk Reduction and the Sustainable Development Goals, have emerged; the world was rocked by COVID-19; and now, two decades on the world is a different place with a changing nature of disasters due to climate change (such as heatwaves), conflicts and consequent human displacements, placing new demands on the shelter sector.

The timeline since the 2004 tsunami is depicted in Fig. 10.7, where the main drivers and trends of the transformations in the shelter sector are presented, drawing

Table 10.1 Key characteristics of the case studies

Country	Disaster	Year	Number of houses destroyed and damaged	Approaches applied based on 2004 tsunami experience
Bangladesh	Cyclone Sidr and Cyclone Aila	2007 2009	>1.5 million >550,000	Shelter cluster; core shelter; transitional shelter
Haiti	Earthquake	2010	>300,000	Shelter cluster; transitional shelter; owner-driven approach
Philippines	Typhoon Haiyan (Yolanda)	2013	1,012,790	Shelter cluster; core shelter; transitional shelter; coastal 'buffer' zone
Nepal	Gorkha earthquake	2015	>750,000	Shelter cluster; transitional shelter; owner-driven approach

from the preceding discussions and case studies, and the apparent prospects for the future. Driven by the post-tsunami frenetic shelter recovery activities emanating from the plethora of funding available at that time, and the consequent chaos and lack of coordination, a variety of innovations emerged, particularly during the period of 2005–2015. This was indeed a significant period for the shelter sector, driven by global humanitarian initiatives embodied in the shelter cluster, in response to some massive disasters such as those mentioned in the case studies. This expansion in the sector subsequently seems to have contracted, leading to governments assuming an increasingly central role, and with the dwindling purse of international aid, humanitarian agencies have been led to innovate, such as the shelter self-recovery mode discussed below. Subsequent global challenges and an even further shrinkage of funding have led to uncertainty for the future of the shelter sector. Within two decades since the 2004 tsunami, two key paradigms almost at opposite ends of the spectrum of shelter intervention modalities have emerged relating to post-disaster shelter recovery, which is discussed below.

10.5.1 Government-Led Shelter Interventions

After the 2004 tsunami support for and implementation of shelter recovery consisting of rebuilding thousands of shelters was largely undertaken by humanitarian agencies and some private sector agencies. The Indonesian government also implemented some shelter projects, but these were not occupied and abandoned because of allegations of poor design and construction (Ahmed and O'Brien 2011). However, as evident from the case studies above from the Philippines and Nepal, governments are stepping up and assuming a central role in the post-disaster shelter sector, and shelter interventions by humanitarian agencies do not extend much beyond ESA.

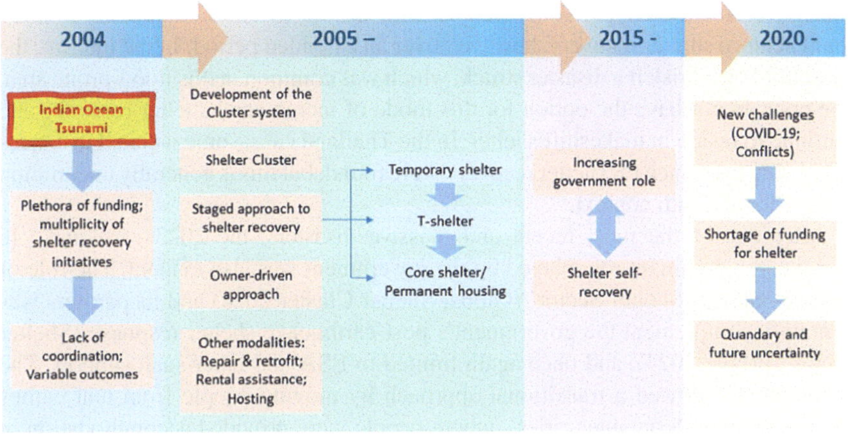

Fig. 10.7 Timeline since the 2004 tsunami with critical developments and emerging trends in the shelter sector

Even in Indonesia where after the 2004 tsunami the role of humanitarian agencies in the shelter sector was extensive and widespread, it has changed in recent times. For example, in a field study after the 2018 earthquake in Lombok (Ahmed and Purnomo 2024), it was found that shelter recovery was mainly undertaken by the Indonesian government with a low level of involvement by INGOs that mostly provided ESA and construction training (IFRC 2020; Lines et al. 2022). This reflects a shift from the previous paradigm led by humanitarian agencies. Nonetheless, as in the case of the Philippines and Nepal, similar discrepancies in the government-led shelter recovery were evident—delays, complexity of regulatory and bureaucratic processes, scarcity of labour and materials, establishing landownership, etc. (Tsulis Iq'bal et al. 2022; Lines et al. 2022; Pribadi et al. 2020; Sunarti et al. 2021; Widyayanti et al. 2020). Thus, again, people ended up in for an extended duration in temporary shelter, experiencing uncertainty, suffering and anxiety. Furthermore, the most vulnerable people were not prioritised, resulting in the poor facing further obstacles in securing adequate shelter. The grant money was not sufficient to rebuild a complete house, thus leading people to seek loans from informal moneylenders. Five years later, even though the government program was officially completed, many homes remained incomplete, and despite efforts to incorporate earthquake-resistant construction, due to lack of monitoring and technical support, such construction was widely ignored, making people vulnerable to future earthquakes (Ahmed and Purnomo 2024), thus reproducing pre-existing vulnerabilities.

A similar pattern of government-led shelter recovery was evident in Thailand after the massive flood of 2011 (Ahmed et al. 2023). Once again, the amount of funding support provided by the government was not adequate to rebuild a complete house; in one case, it was found that 75% of the cost was borne by the household. This could only be done gradually in increments as and when funds were available and in the interim, the household had to live in an incomplete house. Thus, even with governmental support only people with access to other sources of funding could achieve shelter recovery, however, over an extended period. In the interim, the household faced risk if a disaster struck, which was common in this flood-prone area. The poor do not have the option for this mode of incremental shelter recovery and continue to reside in makeshift shelter. In the Thailand case, some isolated instances of utilising insurance for shelter recovery were found, but this is generally uncommon in the Global South context.

After one of the most recent and massive disasters, the 2023 earthquake in Türkiye, the dominant role played by the government was also evident. The role of the now-named 'Shelter Sector' (Global Shelter Cluster 2024) and its partners was to mainly complement the government's post-earthquake shelter response (Shelter Sector Türkiye 2023), and once again limited to ESA and other such support. The government followed a transitional approach by moving people from tent camps and establishing 'container cities', where people were provided accommodation in shelters built of shipping containers. These obviously can only function over the short term with limited thermal comfort, services, space and privacy, and difficulty of access to livelihoods and social amenities. Although work on building permanent housing was proceeding (Kizilkoyun 2023), it would take a long time to provide for the large number of people displaced by the disaster; in May 2024, almost 500,000 people were living in container shelters, tents or makeshift shelters (Global Shelter Cluster 2024). This is another example where people continue to remain in transitional shelters for a protracted duration of time. Given the extensive disregard for seismic building codes that led to the massive devastation by the earthquake (Ahmed 2023), it would be essential to ensure that would not be the case in the new permanent housing, but at this point, it is not clear if the government has that capacity; there is hardly any reference to that in the literature. The government's dominant role was evident from reports of the lack of coordination with the Shelter Sector partners, and also, the low engagement of local NGOs and predominance of INGOs (Shelter Sector Türkiye 2023), perhaps another reason for the government's central role in shelter provision.

10.5.2 Shelter Self-Recovery

Gaps in institutional support, be it from humanitarian agencies or the government, or both, are often large and many people initiate their self-recovery. An estimate indicates that provision of shelter support remains under 30% of the affected households in the first year after a disaster (Parrack et al. 2014). For households to recover sufficiently on their own, it is possible for only those who have access to funds and other resources. For many others, they can only rebuild makeshift or incomplete shelters, and face the risks of future disasters.

An emerging paradigm in this context is for humanitarian agencies to support the self-recovery process, usually termed as 'shelter self-recovery'. This can take various forms including provision of toolkits and shelter materials, and training on safe construction (Flinn et al. 2017; Parrack et al. 2014; Schofield et al. 2017; World Habitat 2017). This is a practice largely followed by humanitarian agencies, and governments rarely follow it; it is almost the other end of the spectrum of government-led, town-down shelter interventions. Although in some cases, as seen in Lombok and Thailand mentioned above, governments do provide funding for households to repair and rebuild their homes, this is different from the humanitarian shelter self-recovery approach that combines a variety of inputs including materials and capacity-building support. The self-recovery paradigm is a shift from the provision of completed T-shelters or permanent housing, common after the 2004 tsunami, and is akin to the owner-driven modality. The global context has changed, and agencies no longer have the generous funds available after the 2004 tsunami. New demands, particularly conflicts, have drawn resources away from recovery after disasters caused by natural hazards, making shelter self-recovery a feasible option.

The positive aspects of shelter self-recovery have been widely highlighted (see for example, Parrack et al. 2014), especially the opportunities for community participation, and there are notable achievements, such as an award-winning shelter self-recovery project in post-Yolanda Philippines (World Habitat 2017). However, there are also challenges. Similar to the owner-driven projects in Nepal and Sri Lanka discussed earlier, beneficiaries need to own land or have tenure security to be assisted in this mode, usually possible in rural areas, leaving out urban informal settlements and landless people. Additionally, it is often difficult for people to find extra money to supplement the humanitarian support and build a safe and durable shelter (World Habitat 2017). Thus, people borrow money and fall into long-term debt, as evident from such projects in the Philippines after Yolanda (Shelter Cluster Philippines 2014; Schofield et al. 2017).

Despite efforts to incorporate safe construction, it was found that these dwellings built through the self-recovery approach were comparable to pre-disaster poor-quality shelter, hence vulnerability was perpetuated. The concept of 'building-back-better' is widely proclaimed by humanitarian agencies, and is embodied in the Sendai Framework for Disaster Risk Reduction (UNISDR 2015), however, in the self-recovery model, households often do not have the funds to achieve that, and despite efforts to build local construction capacity, skill shortages prevail, as found in post-Yolanda Philippines; households seemed to be 'building back worse' (Shelter Cluster Philippines 2016, p. 21). Most of the support is targeted at rural and lightweight structures, using natural building materials (Ahmed and Parrack 2022), and not applicable in the urban context or building with manufactured materials such as masonry and concrete (Ahmed and McDonnell 2020; Shelter Cluster Philippines 2016), a growing pattern and aspiration in the built environment of the Global South.

10.6 Conclusion

The 2004 tsunami heralded a significant transformation of the shelter sector because of the opportunity for humanitarian agencies to build large numbers of shelters in the affected countries, made possible by the generous availability of international aid and philanthropic funds. It allowed forging of new strategies, modalities and approaches to shelter provision, often gained from the post-tsunami shelter intervention experience. It led to the formation of the humanitarian cluster system and a staged approach to shelter provision, among other lessons distilled from the experience. These were subsequently transferred to other post-disaster contexts with a view to achieving coordination among diverse agencies and a systematic approach to shelter recovery interventions. But varying outcomes became evident and despite well-meaning intentions, these approaches were met with contextual challenges. Constraints relating to time, funding, political will and even corruption continue to undermine these initiatives. Thus, the question arises whether the 2004 tsunami experience with its massive shelter interventions has actually resulted in a positive, long-term legacy.

Two decades after the tsunami, the sector is at a critical crossroads presenting a quandary for its future trajectory. Case studies of the different major disasters after the 2004 tsunami indicate the transforming landscape of the shelter sector. Where the sector was predominantly led by humanitarian agencies, mostly INGOs, after the 2004 tsunami, and continued to some extent even five years after the 2010 earthquake in Haiti, gradually a decade or so later a surge in governmental engagement and dominance of the shelter sector emerged. Since the time of the 2004 tsunami, there has been rapid economic growth in many Asian countries, including Indonesia, making it possible for governments there to undertake shelter interventions on their own without external support. If this would be the trend of the future, then questions arise on issues of accountability and transparency. While the Shelter Cluster is bound by global mandates and protocols, national governments can act outside such mechanisms and are often free to do as they desire within their own national boundary. Thus, the key coordination mechanism that emerged from the 2004 tsunami experience, the humanitarian cluster system, would then fade into the background, with humanitarian agencies limited to only provision of ESA.

Such a development seems even more plausible in the global context of diminished funding for shelter, with the demand placed on the sector by significant conflicts and consequent displacement, requiring shelter support. This takes away the imperative to support people to build safe, permanent housing and humanitarian agencies become confined to only supporting temporary forms of shelter. It is also a sustainable development issue beyond post-disaster shelter recovery—a significant proportion of the people in the world live in inadequate and vulnerable shelter, and the impact of disasters on them is severe. Investing in safe shelter as a sustainable development activity is imperative to achieve disaster resilience, and such anticipatory action can reduce the pressures for undertaking post-disaster shelter recovery interventions.

References

ACHR (Asian Coalition of Housing Rights) (2006) Tsunami update. ACHR, Bangkok
Ahmed I (2016) Developing disaster resilient rural housing in coastal Bangladesh. In: proceedings of the sustainable futures conference on architecture and construction in the Global South, Kenya, pp 186–196
Ahmed I (2021) Post-disaster housing reconstruction in the Global South: An example from Bangladesh. Know Disasters 3:42–44
Ahmed I (2023) Key building design and construction lessons from the 2023 Türkiye Syria earthquakes. Architecture 3(1):104–105
Ahmed I, McDonnell T (2020) Prospects and constraints of post-cyclone housing reconstruction in Vanuatu drawing from the experience of tropical cyclone Harold. Prog Disaster Sci 8:1–9
Ahmed I, O'Brien D (2011) Donor-driven housing, owner-driven needs. In: Kaufman N (ed) Pressures and distortions: City dwellers as builders and critics: Four views. Raphael Vinoly Architects PC, New York, pp 237–337
Ahmed I, Parrack C (2022) Shelter self-recovery: The experience of Vanuatu. Architecture 2(2):434–445
Ahmed I, Purnomo AG (2024) Ready for the next earthquake? Housing reconstruction in Lombok, Indonesia. J Media Commun Sci 7(1):1–6
Ahmed I, Maund K, Gajendran T (2020) Disaster resilience in South Asia: Tackling the odds in the sub-continental fringes. Routledge, Abingdon (UK)
Ahmed I, Pal I, Chonlasin V (2023) Complexities of post-disaster recovery: An example from Thailand. Asian Curr
Arroyo I, Astrand J (2019a) Housing recovery outcomes after typhoon Haiyan in the Philippines: a critical perspective. J Crit Rism 18(2):142–168
Arroyo I, Astrand J (2019b) User involvement in housing recovery after typhoon Haiyan from a capability perspective. Int J Disaster Risk Reduct 38:1–14
Association S (2018) The sphere handbook: Humanitarian charter and minimum standards in humanitarian response. Sphere Association, Geneva
Aubrey D (2008) Response to rapid change: Post-tsunami shelter in Sri Lanka. Open House International 33(2):32–39
Baptiste N (2015) Are foreign NGOs rebuilding Haiti or cashing in? Retrieved on 25 July 2024 from https://archive-yaleglobal.yale.edu/content/are-foreign-ngos-rebuilding-haiti-or-cashing
BDRCS (Bangladesh Red Crescent Society) (2011) Early recovery response for Cyclone Aila affected most vulnerable people of Khulna District, Bangladesh. BDRCS, Dhaka
Browne K (2015) Standing in the need: Culture, comfort, and coming home after Katrina. University of Texas Press, Austin
Braun V, Clarke V (2019) Thematic analysis: a reflexive approach. University of Auckland. Retrieved on 26 February 2025 from https://www.psych.auckland.ac.nz/en/about/thematic-analysis.html
Caulfield J (2019) How to do thematic analysis. Retrieved on 8 July 2024 from https://www.scribbr.com/methodology/thematic-analysis/#:~:text=Thematic%20analysis%20is%20a%20method,meaning%20that%20come%20up%20repeatedly.&text=Reviewing%20themes,Defining%20and%20naming%20themes
Charlesworth E, Ahmed I (2015) Sustainable housing reconstruction: Designing resilient housing after natural disasters. Routledge, Abingdon (UK)
Davis I (1978) Shelter after disaster. Oxford Polytechnic Press, Oxford
Deprez S, Labattut E (2017) Impact evaluation: Improve the living conditions for the most vulnerable populations affected by the earthquake. Handicap International, Lyon, France
Dias NT, Keraminiyage K, DeSilva KK (2016) Long-term satisfaction of post disaster resettled communities: The case of post tsunami—Sri Lanka. Disaster Prev Manag 25(5):581–594
Doyle D (2016) Philippines has built only 1% of homes promised after Typhoon Haiyan. New Humltarian, 1–6

Flinn B, Beresford P (2009) Post-Sidr family shelter reconstruction Bangladesh. UK Department for International Development, London
Flinn B, Schofield H, Morel LM (2017) The case for self-recovery. Forced Migr Rev 55:12–14
Global Shelter Cluster (2017) Shelter projects 2015–2016: Case studies of humanitarian shelter and settlement responses. International Organization for Migration, Geneva
Global Shelter Cluster (2018) Shelter & settlements: The foundations of humanitarian response. Global Shelter Cluster, Geneva
Global Shelter Cluster (2020) Shelter projects—Haiti: 16 case studies. International Organization for Migration, Geneva
Global Shelter Cluster (2024) 2024–05 factsheet—Türkiye earthquake 2023. Global Shelter Cluster, Geneva
Global Shelter Cluster (n.d.) The cluster approach. Retrieved on 25 July 2024 from https://sheltercluster.org/about-us/pages/cluster-approach#:~:text=Following%20a%20review%20of%20international,%2C%20accountability%2C%20responsibility%20and%20partnership
Greenblott K (2007) Shelter programming: Learning from the Asia tsunami response. World Vision, London
IFRC (International Federation of Red Cross and Red Crescent Societies) (2010) Cyclone Aila, Bangladesh. Final Report. IFRC, Geneva
IFRC (International Federation of Red Cross and Red Crescent Societies) (2020) Lombok earthquake 26 months update (report). IFRC, Geneva
Ingirige B, Haigh H, Malalgoda M, Palliyaguru R (2008) Exploring good practice knowledge transfer related to post-tsunami housing (re-)construction in Sri Lanka. J Constr Dev Ctries 13(2):21–42
Kabir R (2009) Post Cyclone-Sidr family shelter constrution in Bangladesh: Documentation of plans and processes. Shelter Working Group, Dhaka
Kizilkoyun F (2023) Over 46,000 houses built in quake zone. Hurriyet Daily News. Retrieved on 15 August 2024 from https://www.hurriyetdailynews.com/over-46-000-houses-built-in-quake-zone-188018
Lines R, Walker F, Yore R (2022) Progression through emergency and temporary shelter, transitional housing and permanent housing: A longitudinal case study from the 2018 Lombok earthquake, Indonesia. Int J Disaster Risk Reduct 75:1–14
Marti RZ (2005) The 2004 Hurricanes in the Caribbean and the tsunami in the Indian Ocean. United Nations Publications, New York
Masyrafah H, McKeon JMJA (2008) Post-tsunami aid effectiveness in Aceh: Proliferation and coordination in reconstruction. Brookings Institution, Washington D.C
Ministry of Food and Disaster Management (2008) Cyclone Sidr in Bangladesh: damage, loss and needs assessment for disaster recovery and reconstruction. Government of Bangladesh, Dhaka
McDonald T (2010) Deadly storm menaces Haiti's tent cities. Retrieved on 25 July 2024 from https://www.abc.net.au/news/2010-11-05/deadly-storm-menaces-haitis-tent-cities/2325804
National Housing Authority (2015) Typhoon "Yolanda" permanent housing program. National Housing Authority, Manila
Nazara S, Resosudarmo B (2007) Aceh-Nias reconstruction and rehabilitation: Progress and challenges at the end of 2006. Asian Development Bank Institute, Tokyo
Opdyke A, Javernick-Will A, Koschmann M (2017) Typhoon Haiyan: Shelter case studies. Mortenson Centre in Engineering for Developing Countries, Boulder (CO, USA)
Parrack C, Flinn B, Passey M (2014) Getting the message across for safer self-recovery in post-disaster shelter. Open House Int 39(3):47–58
Phelps PM (2015) What the Haiti earthquake taught us about disaster response. World Economic Forum. Retrieved on 26 February 2025 from https://www.weforum.org/agenda/2015/01/what-the-haiti-earthquake-taught-us-about-disaster-response/
Pribadi KS, Pradoto RG, Hanafi EA, Rasmawan AB (2020) Lombok earthquake, one year later: Housing sector recovery. In: E3S web of conferences,vol 156, pp 1–10

Ranada P (2023) A decade later: 15% of Yolanda houses unfinished, thousands unoccupied. Retrieved from on 3 July 2024 https://www.rappler.com/newsbreak/investigative/super-typhoon-yolanda-housing-projects-unfinished-thousands-unoccupied-decade-later-2023/

Ratnayake RMGD, Rameezdeen R (2008) Post disaster housing reconstruction: Comparative study of donor driven versus owner driven approach. In: Keraminiyage K, Jayasena S, Amaratunga D, Haigh R (eds) Post disaster recovery challenges in Sri Lanka, pp 51–62. University of Salford, Salford, UK

ReliefWeb (2009) Indonesia: Waiting for permanent homes. Retrieved on 26 July 2024 from https://reliefweb.int/report/indonesia/indonesia-waiting-permanent-homes

Sanderson D, Sharma A, Kennedy J, Burnell J (2014) Lost in transition: Principles, practice and lessons from Haiti for urban post-disaster shelter recovery programs. Asian J Environ Disaster Manag 6(2):131–151

Schofield H, Lovell E, Flinn B, Twigg J (2017) Barriers to urban shelter self-recovery in Philippines and Nepal. J Br Acad 7(s2):83–107

Shaw J, Ahmed I (2010) Design and delivery of post-disaster housing resettlement programs: Case studies from Sri Lanka and India. Australian Research Council, Canberra

Shelter Cluster Philippines (2014) Analysis of shelter response. Shelter Cluster Philippines, Philippines

Shelter Cluster Philippines (2016) Typhoon Haiyan (Yolanda) shelter response outcome assessment: final report. Shelter Cluster Philippines, Philippines

Shelter Sector Türkiye (2023) Lessons learned workshop report 2023: Shelter Sector, Türkiye, Kahramanmaras earthquakes response. International Federation of Red Cross and Red Crescent Society, Geneva

Simon Deprez S, Labattut E (2005) Post-tsunami housing reconstruction in Aceh. Groupe URD, Paris

Sunarti E, Gunawan E, Widiyantoro S, Marliyani GI, Ida R (2021) Critical point on housing construction, resilience and family subjective welfare after disaster: Notes from the Lombok, Indonesia, earthquake sequence of July-August 2018. Geomat Nat Haz Risk 12(1):922–938

Suppasri A, Goto K, Muhari A, Ranasinghe P, Riyaz M, Affan M, Mas E, Yasuda M, Imamura F (2015) A decade after the 2004 Indian Ocean Tsunami: The progress in disaster preparedness and future challenges in Indonesia, Sri Lanka, Thailand and the Maldives. Pure Appl Geophys 172:3313–3341

Tsulis Iq'ba KA, Saputri US, Roxandi A (2022) Delay analysis using in-depth interview method: Case study post-disaster reconstruction in Lombok. ASTONJADRO: CEAESJ, 11(2):418–425

UNDRO (Office of the United Nations Disaster Relief Coordinator) (1982) Shelter after disaster: Guidelines for assistance. UNDRO, Brussels

UNISDR (United Nations International Strategy for Disaster Reduction) (2015) Sendai framework for disaster risk reduction 2015–2030. UNISDR, Geneva

UNOCHA (United Nations Office for the Coordination of Humanitarian Affairs) (n.d.) We coordinate. Retrieved on 25 July 2024 from https://www.unocha.org/we-coordinate#:~:text=The%20humanitarian%20cluster%20system%20is,organizations%20to%20respond%20to%20crises

UNORC-IAS (Office of the UN Recovery Coordinator for Aceh and Nias) (2009) Tsunami Recovery Indicator Package (TRIP): The third report for Aceh and Nias. UNORC-IAS, Banda Aceh

Von Meding J, Shrestha HD, Kabir H, Ahmed I (2017) Nepal earthquake reconstruction won't succeed until the vulnerability of survivors is addressed. The Conversation

Widayanti HH, Yuniarman A, Yynianti SR (2020) The level of satisfaction in construction of post-earthquake houses in Tanjung sub-district, North Lombok Regency. Earth Environ Sci 447:1–8

Wilderspin I, Barnham J, Gill G, Ahmed I, Lockwood H (2008) Evaluation of disaster risk reduction mainstreaming in DG ECHO's humanitarian actions. European Commission Humanitarian Aid Office, Brussels

World Habitat (2017) Post-Haiyan self-recovery programme. Retrieved on 8 July 2024 from https://world-habitat.org/world-habitat-awards/winners-and-finalists/post-haiyan-self-recovery-housing-programme/#outline

Yin RK (2017) Case study research and applications: Design and methods. Sage Publications, Thousand Oaks (USA)

Chapter 11
Enhancing Community Resilience: Lessons from Disaster Responses and Policies for Effective Community-Based Disaster Risk Reduction

Mikio Ishiwatari and Daniel P. Aldrich

Abstract The catastrophic 2004 Indian Ocean Tsunami, claimed over 200,000 lives and caused widespread devastation across multiple nations. The extent of the damage could have been mitigated, if communities had knowledge of tsunamis, understood their risks, prepared for such natural hazards, and left for higher grounds before the tsunami arrived. The event underscored the critical importance of community-based disaster risk reduction strategies and encouraged efforts worldwide to strengthen community resilience against coastal disasters. This article evaluates progress in community-based disaster risk reduction following the Indian Ocean Tsunami by analyzing the efforts of community response to major coastal disasters in various countries. We focus on the integration of indigenous knowledge with disaster management efforts, the role of education and public awareness programs, the challenges of sustaining initiatives, and the importance of multi-stakeholder involvement. By synthesizing lessons learned from diverse coastal disasters, this study contributes to the development of more effective and sustainable community-based disaster risk reduction strategies. It identifies the remaining issues and proposes essential policies that prioritize community resilience.

Keywords Disaster education · Indigenous knowledge · Multi-stakeholder involvement · Risk communication · Vulnerable group

M. Ishiwatari (✉)
Meiji University, Tokyo, Japan
e-mail: Ishiwatari1219@gmail.com

D. P. Aldrich
Northeastern University, Boston, USA

R. Shaw et al. (eds.), *Two Decades from the Indian Ocean Tsunami*, Disaster Risk Reduction, https://doi.org/10.1007/978-981-96-2669-4_11

11.1 Introduction

Local communities are on the front lines of disaster management, making the strengthening of their resilience a key component of disaster risk reduction (DRR) (Aldrich and Meyer 2015; Cutter et al. 2010). The catastrophic 2004 Indian Ocean tsunami (IOT) underscored the critical importance of community-based DRR (CBDRR) strategies. Prior to the IOT, most people in the world except for a few countries such as Japan and Chile, had limited knowledge about tsunamis. As a result, more than 200,000 people, including foreign tourists, died in the IOT (Aldrich 2012). The extent of the damage could have been mitigated, if communities had knowledge of tsunamis, understood their risks, prepared for tsunamis, and moved to higher grounds before the arrival of the tsunami.

This devastating event served as a wake-up call for the international community, prompting a significant shift in DRR approaches. In the aftermath of the IOT, countries around the world, not just those directly affected, intensified their efforts to strengthen CBDRR initiatives. This shift recognized the crucial role that informed and prepared communities play in mitigating the impacts of disasters.

This study aims to track the development of CBDRR following the IOT by reviewing community responses to major coastal disasters and their efforts to enhance resilient capacity. It examines how lessons from disasters were used for improving community-based DRR, with a particular focus on tsunamis and coastal hazards such as storm surges and cyclones.

The methodology for this study involves a comprehensive review of recent academic literature, government reports, and policy documents related to CBDRR. The study encompasses a range of case studies of major tsunami and coastal disasters from different geographical contexts, including Indonesia, the United States, Myanmar, Chile, Japan, and the Philippines, providing a comprehensive global perspective on CBDRR evolution. This study provides valuable insights to policymakers, professionals, and community leaders in their ongoing efforts to build resilient communities in the face of increasing climate-related risks and natural hazards.

11.2 Community Efforts and Response to Major Coastal Disasters

This section reviews how communities respond to coastal disasters by using lessons from previous disasters domestically or internationally. Also, it examines how disasters triggered enhancing CBDRR.

11.2.1 2004 IOT

The 2004 IOT served as a pivotal moment in CBDRR, highlighting both the devastating consequences of inadequate preparedness and the life-saving potential of indigenous knowledge. The practice on Simeule Island, Indonesia, highlights that CBDRR integrated with local knowledge can save people's lives. Local communities escaped to higher grounds from the 2004 tsunami, because they had transferred indigenous knowledge of the previous 1907 tsunami disaster to the next generations (Rahman et al. 2018).

Recovery programs enhance community resilience by involving communities in recovery planning (Harahap 2020). In addition, authorities conducted school education and public awareness programs. For example, the Indonesia Institute of Science with support from UNESCO and the Tsunami Disaster Mitigation Center promoted school-based disaster preparedness programs The program sought to make school facilities safer and more prepared for disasters through establishing networks with communities and government organizations (Oktari et al. 2018). Mercy Malaysia, an international NGO, also supported establishing the Tsunami Resource Center to implement disaster education programs at the Universitas Syiah Kuala in 2007 (Zubir and Amirrol 2011). Scientists of the French NGO Planet Risk collaborated with Indonesian organizations to conduct public awareness programs producing educational tools. Among other lessons was the fact that people in Java successfully escaped from an oncoming tsunami to higher ground in 2006 (Morin et al. 2008).

However, the sustainability of the education program remains a challenge. From 2007, some 18 aid agencies and NGOs supported disaster education in Aceh. But as resources dwindled and attention turned to other disasters, the framework was difficult to sustain (Ishiwatari 2018; Oktari et al. 2015). Sustainable community resilience needs to enhance knowledge of risks and empower communities through participatory processes over time, not just for a year or two (Stephan et al. 2017).

The effectiveness of public awareness programs was further demonstrated during the 2010 tsunami in Mentawai Island, Indonesia, People living and working in the high-risk area escaped from the tsunami thanks to drills and knowledge gained through awareness programs (Esteban et al. 2013).

11.2.2 Hurricane Katrina in the United States

Hurricane Katrina in 2005 exposed significant gaps in disaster preparedness and response in the United States. When Hurricane Katrina hit New Orleans in August 2005, one-quarter of the population did not evacuate, and some 1,500 people died (Aldrich 2012). The disaster disproportionately affected vulnerable groups, such as African Americans, the poor, the elderly, and the handicapped, these groups faced the hardships of slow rescue, poorly planned evacuations, and low quality temporary housing (Colten 2008; Aldrich and Crook 2008).

On the other hand, the disaster also demonstrated the effectiveness of certain preparedness measures. The massive evacuation of nearly 1 million residents was aided by public risk communication and education, media and government advice, hurricane forecasting, and state-of-the-art traffic management (Boyd et al. 2009). Following this disaster, communities across the U.S. began to recognize to become more proactive in managing disasters, and community resilience emerged as a popular concept for improving disaster preparedness, response, and recovery (CRSI 2011).

11.2.3 Myanmar Cyclone Nargis

During the Cyclone Nargis in 2008, which caused almost 140,000 deaths, communities had different levels of resilience against the cyclone. Communities that had exposure to storm-warning information before the 2008 disaster were better prepared for it (Min et al. 2020).

Following the 2008 cyclone, Myanmar formulated an action plan for DRR, which focuses on CBDRR. However, the country faced difficulties in promoting policies and strategies for community support and education (Smith and Chan 2018).

11.2.4 Chili Tsunami

In Chile the 2010 earthquake killed over 500 people (Carlin et al. 2014), while the tsunami itself took the lives of 124 residents, some 2% of population in at-risk coastal area. This 2% mortality rate is smaller than 4% in Great East Japan and Tsunami in March 2011 and the 12% seen in the 2004 IOT (Ranghieri and Ishiwatari 2014).

The relatively low casualty rate can be attributed to several factors. People traditionally shared knowledge on tsunami, worked with fishermen, and successfully ignored false tsunami warning. Good practices by community efforts were reported in El Morro community, where all community members escaped to a nearby hill. Fishermen's experience with the ocean and the collective memory of past disasters saved their lives. Further, residents worked collectively to help each other in their community for five days without support from outside (Moreno et al. 2019).

The national emergency agency recognized the necessity of strengthening community resilience and conducted 7 drills between 2010 and 2015. In addition, authorities worked to plan evacuation routes and enhanced capacities and governance at the community level (Herrmann-Lunecke and Villagra 2020). On the other hand, Rahman and Kausel (2013) found that community participation in DRR planning and the activities of training and drills were limited at Dichato a coastal community in Chile with a population of 4000. Other studies in Chile have similarly concluded that community participation in the recovery process was also limited (Valenzuela et al.

2019). The reconstruction efforts mainly covered reconstruction of physical infrastructure with limited assistance for rebuilding the social environment and social infrastructure, such as parks, libraries, and support for NGOs.

11.2.5 Great East Japan Earthquake and Tsunami

The 2011 Great East Japan Earthquake and Tsunami, which claimed over 20,000 lives in the Tohoku region, exposed both the limitations and advancements in Japan's tsunami risk management strategies (Suppasri et al. 2012). This catastrophic event, occurring on a millennial scale, demonstrated the constraints of existing technological measures in the face of extreme natural phenomena.

The disaster revealed several critical shortcomings in preparedness and response mechanisms. Initial tsunami warnings, which were revised latter, estimated lower heights of tsunami than actual one. Some local hazard maps showed smaller risk areas than those inundated by the tsunami (Aldrich 2019). The overreliance on structural defenses, particularly tsunami dikes, may have fostered a false sense of security among coastal communities, potentially undermining evacuation efforts (Ranghieri and Ishiwatari 2014). Historical lessons in community planning proved their value, as settlements relocated to higher ground following previous tsunamis survived, while those rebuilt in high-risk coastal areas suffered losses (Murao 2015).

The disaster highlighted a disconnect between tsunami awareness and effective evacuation behaviors. Vulnerable populations, particularly the elderly and individuals with disabilities, experienced disproportionately high mortality rates, indicating inadequacies in evacuation planning for these groups.

However, the disaster also showcased the effectiveness of certain education programs, such as those in Kamaishi, which saved many lives (Matsura and Shaw 2015). This highlighted the importance of continuous education and drills in maintaining community preparedness.

11.2.6 Typhoon Haiyan in the Philippines

Typhoon Haiyan in 2013 revealed significant disparities in community preparedness for storm surges across the central region of the Philippines. In San Francisco municipality, Cebu, the local government conducted annual drills, ensuring people know how to move efficiently to evacuation sites. The Mamanwa indigenous peoples used their local-indigenous knowledge to the shock caused by Typhoon Haiyan (Cuaton and Su 2020). Their knowledge helped them in constructing emergency evacuation shelters and storing relief foods, as well as forecasting the typhoon. These activities ensured food security and physical safety in their communities and contributed to quick recovery.

In contrast, many other communities had limited understanding of storm surges. They responded poorly based on experience with past minor typhoons, creating more vulnerability to storm surges (Dalisay and Guzman 2016; Esteban et al. 2015).

11.2.7 International Influence

The 2004 IOT catalyzed various international initiatives in CBDRR. The Intergovernmental Oceanographic Commission of UNESCO developed the Tsunami Ready Recognition Program to enhance community's preparedness for tsunami disasters. This voluntary program requires participating communities to meet 12 indicators. Once achieved, IOC recognizes the communities as "tsunami ready." Communities in Costa Rica received this status by creating hazard maps and conducting evacuation drills through community participation (Chacon-Barrantes et al. 2021).

IOT also drove tsunami preparedness programs in California in the United States. The California Tsunami Program produced hazard maps and conducted evacuation drills, public education, and community workshops from 2006 until 2008. The state created a Tsunami Warning System to provide information on inundation areas, wave heights, and arrival times, and to encourage people to evacuate. (Dengler et al. 2011). At the national level, federal agencies worked with states to start the TsunamiReady program in 2001 to promote community-based efforts against tsunamis. The program activities cover risk assessment, planning, education, and warning communications.

11.3 Discussion

The review of community responses to major coastal disasters since the 2004 IOT reveals several critical aspects of CBDRR (Table 11.1). It was found that knowledge is crucial in mitigating damage and that the major disasters including IOT triggered community resilience programs not only in affected countries but also in other countries with coastal disaster risks.

11.3.1 Key Points from Analysis

(i) **Sustainability and Scalability**: As Shaw (2014) pointed out the sustainability and scaling up of CBDRR initiatives remain challenges. Many programs, particularly those implemented by international NGOs or aid agencies, face difficulties in maintaining their impact once external supports end. This was evident in the case of Aceh, Indonesia, where disaster education programs struggled to continue after the departure of supporting organizations (Ishiwatari 2018;

Table 11.1 Summary of cases

	Indonesia	US	Myanmar	Chile	Japan	Philippines
Positive effect						
Preparedness		✓			✓	✓
Effect of education					✓	
Local and indigenous knowledge	✓			✓	✓	
Started new programs	✓	✓	✓	✓		
Negative effect						
Limited knowledge	✓		✓		✓	✓
Impact on vulnerable group		✓			✓	

Oktari et al. 2015). It is crucial to develop strategies that embed CBDRR initiatives within local governance structures and community practices, ensuring their sustainability beyond initial implementation periods.

(ii) **Multi-stakeholder Involvement and Community Participation**: The success of CBDRR efforts heavily relies on the involvement of various stakeholders. Government support is indispensable for providing financial resources, technical expertise, and legislative frameworks (Andrew 2018; Ishiwatari 2012). The case studies demonstrate that when local governments actively participate in DRR initiatives, such as in San Francisco municipality, Cebu, during Typhoon Haiyan, the outcomes are more positive. The experiences from Chile, Indonesia, and Japan emphasize the need for greater community involvement in the recovery and reconstruction processes. While infrastructure rebuilding often takes precedence, the social aspects of recovery, including community empowerment and participation, are equally crucial for long-term resilience building.

(iii) **Integration of indigenous and local knowledge**: The cases of the Mamanwa indigenous peoples in the Philippines, the Simeule Island community in Indonesia, and the Tohoku region in Japan highlight the value of integrating indigenous and local knowledge into DRR strategies. These examples demonstrate how traditional practices can enhance disaster preparedness and response. However, as seen in the Japanese case, even a wealth of knowledge about tsunamis may not connect with actual evacuation actions. This raises questions about how to effectively integrate historical disaster experiences, indigenous knowledge, and scientific approaches into community practices and decision-making processes. The process of learning and applying lessons is not always straightforward. As Egner et al. (2014) note, knowledge does not automatically translate into action due to complex social arrangements. There is a need for further research on how to effectively combine indigenous knowledge with modern scientific approaches to create more robust and culturally appropriate CBDRR frameworks.

(iv) **Education and Public Awareness**: The review underscores the importance of continuous education and public awareness programs. Successful examples, such as the school-based disaster preparedness program in Japan and the public awareness initiatives in Java and Mentawai Island, demonstrate the potential of these approaches. However, the challenge lies in maintaining the relevance and engagement of such programs over time, particularly in areas that do not frequently experience disasters.

(v) **Accurate risk communication**: The case of Japan highlights the importance of accurate risk communication and hazard mapping. Underestimated tsunami warnings and inaccurate hazard maps led to inadequate responses in some areas. CBDRR strategies should prioritize the development and dissemination of accurate risk information and ensure that communities understand and act upon this information.

(vi) **Addressing Vulnerabilities**: The cases of Hurricane Katrina and the Tohoku Tsunami highlight the disproportionate impacts on vulnerable groups. CBDRR strategies must explicitly address these vulnerabilities, ensuring that marginalized communities are not left behind in preparedness and response efforts.

11.3.2 Policy Recommendation

As the global disaster landscape becomes increasingly complex due to climate change, urbanization, and socio-economic disparities, CBDRR must evolve to meet these challenges. This study has identified several crucial efforts that need to be institutionalized to mobilize resources and ensure long-term sustainability. DRR strategies at all levels must include key components, such as education, public awareness, consideration of vulnerable groups, risk communication.

As key supporters and sustainers of community-level initiatives, local governments' capacities should be enhanced through training, resources, and policy support. This empowerment enables them to better coordinate with communities and implement long-term DRR strategies.

Government agencies should incorporate comprehensive disaster education programs into school curricula and develop engaging, accessible materials for all age groups. Local governments should foster partnerships between communities and educational institutions. These collaborations can lead to practical, context-specific DRR initiatives and provide opportunities for intergenerational learning and knowledge transfer. The efforts could contribute to building a culture of preparedness. Clear, culturally appropriate, and timely communication strategies should be developed to disseminate risk information and early warnings to all community members, including those with limited access to traditional media. Community participation in developing and implementing these communication strategies is key to ensuring that information translates into actionable measures on the ground.

To enhance these efforts, stakeholders should leverage mobile apps, social media platforms, and other digital tools. These technologies can be particularly effective in reaching younger generations, facilitating community engagement in DRR activities, and disseminating real-time information during emergencies. However, it's important to ensure that these digital approaches complement, rather than replace, traditional communication methods to avoid excluding less tech-savvy community members.

CBDRR initiatives must ensure the active involvement of marginalized groups, including women, the elderly, people with disabilities, and indigenous populations, in both the planning and implementation phases. This inclusive approach ensures that diverse perspectives and needs are addressed, leading to more effective and equitable outcomes.

11.4 Conclusion

This comprehensive review of CBDRR initiatives since the 2004 IOT reveals significant progress in enhancing community resilience against coastal disasters. However, it also highlights persistent challenges that need to be addressed to create more effective and sustainable strategies.

Key findings from this study underscore the importance of integrating indigenous and local knowledge with CBDRR strategies. The need for continuous education and public awareness programs to maintain community preparedness, especially in areas that do not frequently experience disasters, has been demonstrated. A major challenge involves the sustainability of initiatives after the departure of external support. The study also emphasizes the crucial role of multi-stakeholder involvement, especially local government participation. Additionally, the importance of accurate risk communication to ensure appropriate community responses has been highlighted, along with the necessity of addressing vulnerabilities of marginalized groups.

While CBDRR has developed significantly since the IOT, continued efforts are needed to create resilient communities. As climate-related risks continue to increase, the importance of effective, sustainable, and inclusive strategies cannot be overstated. It is through these concerted efforts that communities can hope to face future disasters with greater resilience and reduced vulnerability.

Acknowledgements This study is supported by the Japan–US Global Partnership of Japan Foundation grant program, "Government and Market in Disasters: Capacity Building, Burden Sharing and Insurance."

References

Aldrich DP (2012) Building resilience: Social capital in post-disaster recovery. University of Chicago Press, Chicago

Aldrich DP, Crook K (2008) Strong civil society as a double-edged sword: Siting trailers in Post-Katrina New Orleans. Polit Res Q 61(3):379–389

Aldrich DP, Meyer MA (2015) Social capital and community resilience. Am Behav Sci 59(2):254–269
Andrew M (2018) Revisiting community-based disaster risk management. In: Reducing disaster risks. Routledge, pp 42–52
Boyd E, Wolshon B, Van Heerden I (2009) Risk comunication and public response during evacuations: the New Orleans experience of Hurricane Katrina. Public Perform Manag Rev 32(3):437–462
Carlin RE, Love GJ, Zechmeister EJ (2014) Natural disaster and democratic legitimacy: The public opinion consequences of Chile's 2010 earthquake and tsunami. Polit Res Q 67(1):3–15
Chacón-Barrantes S, Murillo-Gutiérrez A, Rivera-Cerdas F, Rossel BA (2021) Advances in tsunami preparedness at the beginning of the Ocean Decade: the Costa Rica case. Ocean Coast Res 69(suppl 1):e21039
Colten CE, Kates RW, Laska SB (2008) Community resilience: Lessons from new orleans and hurricane katrina. CARRI Report 3(1):2–4
Cuaton GP, Su Y (2020) Local-indigenous knowledge on disaster risk reduction: Insights from the Mamanwa indigenous peoples in Basey, Samar after Typhoon Haiyan in the Philippines. Int J Disaster Risk Reduct 48:101596
Cutter SL, Burton CG, Emrich CT (2010) Disaster resilience indicators for benchmarking baseline conditions. J Homel Secur Emerg Manag, 7(1)
Dalisay SN, De Guzman MT (2016) Risk and culture: the case of typhoon Haiyan in the Philippines. Disaster Prev Manag: Int J 25(5):701–714
Dengler L, Goltz J, Fenton J, Miller K, Wilson R (2011) Building tsunami-resilient communities in the United States: An example from California. TsuInfo Alert 13(2):1–14
Egner H, Schorch M, Voss M (2014) Introduction: Can societies learn from calamities? In: Learning and calamities. Routledge, pp 1–23
Esteban M, Tsimopoulou V, Mikami T, Yun NY, Suppasri A, Shibayama T (2013) Recent tsunamis events and preparedness: Development of tsunami awareness in Indonesia, Chile and Japan. Int J Disaster Risk Reduct 5:84–97
Esteban M, Valenzuela VP, Yun NY, Mikami T, Shibayama T, Matsumaru R, Takagi H, Thao ND, Leon MD, Oyama T, Nakamura R (2015) Typhoon Haiyan 2013 evacuation preparations and awareness. Int J Sustain Futur Hum Secur 3(1):37–45
Community and Regional Resilience Institute (CRSI) (2011) Community resilience system initiative steering committee final report: A roadmap to increased community resilience: CRSI, Washington DC
Harahap GY (2020).Instilling participatory planning in disaster resilience measures: Recovery of tsunami-affected communities in Banda Aceh, Indonesia. Bp Int Res Exact Sci (BirEx) J 2(3):394–404
Herrmann-Lunecke MG, Villagra P (2020) Community resilience and urban planning in tsunami-prone settlements in Chile. Disasters 44(1):103–124
Ishiwatari M (2012) Government roles in community-based disaster risk reduction. In: Shaw R (ed) Community-based disaster risk reduction. Bingley, Emerald Group Publishing Limited, pp 19–33
Ishiwatari M (2018) How can recovery be linked with long-term development?: the case of Indonesia. In: Crisis management beyond the humanitarian-development nexus. Routledge, pp 162–184
Matsuura S, Shaw R (2015) Exploring the possibilities of school-based recovery and community building in Toni District, Kamaishi. Nat Hazards 75:613–633
Min ZY, Bhaktikul K, Aroonsrimorakot S, Sucharitakul S, Tabucanon AS, Eko Siswoyo B (2020) Factors related to coastal communities' water-related natural disaster awareness, preparedness, resilience and recovery in three Cyclone nargis affected areas in the Ayeyarwaddy Delta Region, Myanmar
Moreno J, Lara A, Torres M (2019) Community resilience in response to the 2010 tsunami in Chile: The survival of a small-scale fishing community. Int J Disaster Risk Reduct 33:376–384

Morin J, De Coster B, Paris R, Flohic F, Le Floch D, Lavigne F (2008) Tsunami-resilient communities' development in Indonesia through educative actions: Lessons from the 26 December 2004 tsunami. Disaster Prev Manag: Int J 17(3):430–446

Murao O (2015) Recovery after Sanriku Tsunamis in 1896 and 1933, and transition of housing Location before the 2011 Great East Japan earthquake and tsunami. Challenges, potentials and future, Tohoku Recovery, pp 93–105

Oktari RS, Shiwaku K, Munadi K, Shaw R (2015) A conceptual model of a school–community collaborative network in enhancing coastal community resilience in Banda Aceh, Indonesia. Int J Disaster Risk Reduct, 12:300–310

Oktari RS, Shiwaku K, Munadi K, Shaw R (2018) Enhancing community resilience towards disaster: The contributing factors of school-community collaborative network in the tsunami affected area in Aceh. Int J Disaster Risk Reduct 29:3–12. Environ & Nat Resour J 18(3)

Rahman MS, Kausel T (2013) Coastal community resilience to tsunami: A study on planning capacity and social capacity, Dichato, Chile. IOSR J Humanit Soc Sci 12(6):55–63

Rahman A, Sakurai A, Munadi K (2018) The analysis of the development of the Smong story on the 1907 and 2004 Indian Ocean tsunamis in strengthening the Simeulue island community's resilience. Int J Disaster Risk Reduct 29:13–23. https://doi.org/10.1016/j.ijdrr.2017.07.015

Ranghieri F, Ishiwatari M (2014) Learning from megadisasters: lessons from the Great East Japan Earthquake. World Bank Publications

Shaw R (2014) Disaster risk reduction and community approaches. Community practices for disaster risk reduction in Japan, pp 3–20

Smith AD, Chan EY (2018) Disaster risk reduction in Myanmar: a need for focus on community preparedness and improved evaluation of initiatives. Disaster Med Public Health Prep 12(4):422–426

Stephan C, Norf C, Fekete A (2017) How, "Sustainable" are post-disaster measures? Lessons to be learned a decade after the 2004 Tsunami in the Indian Ocean. Int J Disaster Risk Sci 8:33–45. https://doi.org/10.1007/s13753-017-0113-1

Suppasri A, Muhari A, Ranasinghe P, Mas E, Shuto N, Imamura F, Koshimura S (2012) Damage and reconstruction after the 2004 Indian Ocean tsunami and the 2011 Great East Japan tsunami. J Nat Disaster Sci, 34(1):19–39

Valenzuela VPB, Samarasekara RSM, Kularathna AS, Perez GCC, Norikazu F, Crichton RN, Esteban M (2019) Comparative analysis of tsunami recovery strategies in small communities in Japan and Chile. Geosciences 9(1):26

Zubir SS, Amirrol H (2011) Disaster risk reduction through community participation. WIT Trans Ecol Environ 148:195–206

Chapter 12
Evolution of Disaster Education in the Schools: The Experience of Asia and the Pacific Region

Ma. Theresa P. Pamaong, Ayse Yildiz, and Rajib Shaw

Abstract Disaster education is now more important than ever in promoting disaster resilience, especially in schools, where young people may be provided with the necessary information and skills to reduce risks and effectively respond to disasters. The study investigates the evolution of disaster education in Asia and the Pacific, analyzing various approaches and pedagogical techniques used to improve disaster risk reduction (DRR) in school curricula. This will help evaluate the current practices, challenges, gaps, and new approaches that aim to establish more efficient and engaging learning environments through case studies from Asia and the Pacific region. The results indicate that although there has been significant advancement, there is still a need for continuous innovation and collaboration among educators, policymakers, and communities to build a resilient future. This paper enhances the discussion on disaster education by offering valuable ideas and perspectives on improving disaster education in schools. The aim is to ensure that children and young people are well-informed and empowered to act in disaster.

Keywords Disaster education · Disaster risk reduction · Resilience · Children and young people

12.1 Introduction

The Hyogo Framework for Action (HFA) priorities for action 3 stipulate, "use knowledge, innovation, and education to build a culture of safety and resilience at all levels. (UNDRR 2007) This is one of the examples that emphasizes the value of disaster education as a crucial component to enhance a country's resilience. It is clear from previous studies that schools play an essential role in this initiative (Yildiz and

Ma. T. P. Pamaong (✉) · R. Shaw
Graduate School of Media and Governance, Keio University, Fujisawa, Japan
e-mail: matesapam@keio.jp

A. Yildiz
University of Leicester, Leicester, UK

R. Shaw et al. (eds.), *Two Decades from the Indian Ocean Tsunami*, Disaster Risk Reduction, https://doi.org/10.1007/978-981-96-2669-4_12

Shaw 2024). The investment in education is relatively low compared to the expense of rebuilding infrastructure following a disaster. Education is crucial to a disaster-resilient society (Wahlström 2009, p. 157). However, achieving this requires a significant and sustained commitment from various organizations, government agencies, and practitioners. As Shaw et al. (2011) discussed, disaster education should go beyond teaching about disasters and campaigns for risk awareness; it should guide people to discover potential solutions. This transformative aspect empowers the individual to build a resilient society. Establishing this empowerment is a crucial skill that a child or young people should achieve in disaster education. Understanding these concepts is very important, especially in the Asia Pacific region, which remains the most disaster-prone region in the world, where 2 million people have lost their lives to disasters since 1970. ESCAP (2023) reports that in 2022, over 140 disasters struck the Asia–Pacific region, which resulted in over 7,500 deaths, affected over 64 million people, and caused economic damage estimated at US$ 57 billion. Accordingly, Chan et al. (2020) surveyed children and youth living in the Asia Pacific countries on how they learn about climate change and disaster risks. It showed that formal education is a significant channel for learning about climate change and disaster risk for the younger age cohort.

This chapter aims to evaluate disaster education from the implementation of the HFA in 2005 up to 2015. As a case study, this chapter highlights the experiences of schools in selected countries in Asia and the Pacific region: China, Japan, India, Indonesia, and the Republic of Korea. The selection of the countries was based on the Asia–Pacific Disaster Report 2019, 2021, and 2023, which was based on understanding the regional "riskscape" or regional risk landscape. The 2019 Asia–Pacific Disaster Report has ranked China, Japan, India, Indonesia, and the Republic of Korea as the countries in the Asia Pacific with the greatest risks regarding multi-hazard absolute average annual loss (AAL) (ESCAP 2019). Similarly, Fig. 12.1 highlights the top five countries in the Asia Pacific region and the economic impact of multi-hazards, particularly the average annual loss and adaptation costs from 1995–2014.

The study describes the selected countries' status in disaster education and emerging pedagogies for disaster education in schools. Next, the challenges and

Country	Average Annual Loss (million US$)	Adaptation Costs (million US$)
China	352,611.406	75,342.403
India	151,593.634	30,693.954
Japan	131,040.808	14,570.289
Indonesia	41,555.782	8,080.308
Republic of Korea	28,467.26	6,077.154

Fig. 12.1 The top 5 countries in the Asia Pacific region and the economic impact of multi-hazards from 1995–2014. *Source* Accessed through the Risk and Resilience Portal of the ESCAP and compiled by the first author

gaps in schools will be discussed to establish innovative disaster education concepts. Lastly, an analysis of how these phenomena revolutionize disaster education from the onset of HFA. This research applied a desk review of secondary resources from journal articles, reports, and documents. The following keywords were used: disaster, education, disaster risk reduction, and selected country names.

12.2 Country Experiences

12.2.1 China

China's location is in the convergence of the Eurasian, Pacific, and Indian Ocean plates. This caused the country to experience several major large-scale disasters, such as earthquakes and floods. Notably, many of China's provinces are at high flood risk. The World Bank Group and The Asian Development Bank (2021) have noted that the flooding has contributed to the country's average annual losses. As a result, the country's disaster prevention and mitigation plans focused on reducing economic loss and mortality (World Bank 2020).

In addition, Zhong et al. (2021) highlighted that efforts in disaster education have been incorporated into China's educational system in the aftermath of the Wenchuan earthquake in 2008. Zhu and Zhang (2017) argue that after the earthquake, more people became aware of the disaster issues and put more effort into learning about disaster prevention and reduction. The Chinese government has introduced some measures to incorporate disaster risk reduction in schools by focusing on practical drills to improve students' awareness and knowledge about hazards based on lessons and experiences from past disasters. It is important to note that the implementation of disaster education efforts is higher in elementary and secondary schools than in colleges (Zhu and Zhang 2017). Similarly, Zhong et al. (2021) assessed the impacts of disaster education on children's flooding perceptions in Sichuan City. They found that the internet, television, schools, social networks, parents' involvement, and participation in disaster prevention activities are important to improving children's disaster awareness.

Moreover, the Ministry of Civil Affairs (2012) directs local departments to develop localized disaster prevention and school-based curricula by leveraging modern technologies to enhance student's understanding of disaster education. Another important effort from the government was the integration of disaster prevention and reduction knowledge into teacher training. Likewise, Zhu and Zhang (2017) revealed in their study that teachers suggested developing special teaching materials for teachers to enhance their disaster education knowledge.

On the other hand, the formal education that schools provide has focused its content on knowledge and lacks attractiveness (Zhu and Zhang 2017). Educational content is crucial to conveying the message of disaster education. They also mentioned that educational content creation in disaster education is limited due to the limited educational resources and inexperienced teachers in elementary and secondary schools, particularly in rural regions.

Meanwhile, the country plays an active role in technological advancement at the high school level. For example, Song et al. (2022) showed that a web hazard map using a GIS application increased students' skills in reading maps, and students preferred to use the application for the printed map. However, students experienced some issues, i.e., network and technical. Students also have found it more difficult to identify dangerous areas, such as their residence and local surroundings. The GIS application is available in Chinese and Japanese and was also implemented in Japan, a disaster-prone country that prioritizes disaster education.

12.2.2 Japan

Japan's geographical location in the Pacific Ring of Fire makes the country vulnerable to various disasters. Large-scale typhoons and earthquakes have caused profound damage and casualties in the country. These experiences provided huge lessons to help the country enhance and strengthen its disaster management strategies and policies (Cabinet Office 2023). These lessons are also incorporated into disaster education at Japanese schools, particularly in the "*period of integrated study*" (Kagawa et al. 2012). The Ministry of Education, Culture, Sports, Science and Technology (MEXT) implemented policies to support teachers by providing reference materials for safety guidance and evacuation drills to enhance their knowledge and skills to prepare for disasters. (Cabinet Office 2009). They also developed more age-appropriate, systematized programs that apply to the current curriculum guidelines. Teaching disaster-prevention education was highlighted after the 1995 Great Hanshin-Awaji Earthquake (Chinoi 2007). Disaster education in Japan aims to enhance individuals' capacity building based on "three helps," namely self-help, mutual help, and public help (Sakurai and Sato 2016). These concepts are proactively being promoted in the various trainings conducted by different government agencies in the country to promote disaster education.

Also, research studies have been conducted to investigate the effective use of science and technology in disaster education. Shaw (2020) describes that the role of science and technology has evolved from understanding risk to enhancing resilience. The Ministry of Land, Infrastructure, and Transport (MLIT) is active in operating websites dedicated to disaster education so that students and residents can access educational materials and lectures. The Fire and Disaster Management Agency (FDMA) has been introducing disaster prevention courses designed to teach people about it.

The Cabinet Office published the "Basic Framework for Promoting a Nationwide Movement for Disaster Reduction—Actions with Added Value to Security and Safety" in 2006 to promote nationwide commitment to reduce disaster risks (Cabinet Office 2013). This prompted various strategies for community participation to generate information sharing related to disasters and suggested that all educational institutions should develop disaster education that promotes learning lessons from past disaster experiences in their regions. Kagawa et al. (2012) discovered that past disaster experiences have prompted students to develop an important competency: a proactive attitude in responding to the changing environment by taking action to save lives from natural hazards. Chinoi (2007) indicated that many previous disaster education programs in the country have emphasized acquiring basic knowledge. A clear purpose and specifics on drills, training, and forums have been established recently.

It is also important to note that, as Sakurai and Sato (2016) describe, disaster education in Japan is taught only at a limited number of schools by passionate teachers. They also argue that some Japanese field experts already focus on handling domestic disasters, which can highly influence the pedagogy practices in the country. On the other hand, pre-service teacher training programs for disaster education can bridge this gap. Kawasaki et al. (2022) argue that introductory training about post-disaster situations for pre-service teachers can enhance their disaster-related knowledge and contribute to schools and communities after disasters.

12.2.3 India

India's socio-economic vulnerability and geo-climatic conditions have made the country vulnerable to disasters, such as droughts, floods, and tropical cyclones. NDMA (2021) has reported that out of 36 states and Union Territories (UTs) in the country, 27 of them are prone to disasters. India's disaster management approach has shifted from reactive, relief-centered search and rescue operations to a proactive, holistic approach incorporating preparedness, prevention, mitigation, and risk resilience (The Asia Foundation 2022). One of the most important initiatives in India was conducted by the Central Board of Secondary Education (CBSE), which included disaster management in the social sciences subject in the secondary education curriculum since 2004 (Ministry of Home Affairs 2011). This was introduced in classes VI to VIII from 2009–2010 (Gupta, 2022). In higher education, the Indian government has set up a technical committee for developing approach methodology and has issued all universities the task of integrating and institutionalizing disaster management education (Ministry of Home Affairs 2013). It materialized in creating the Integrated Disaster Management Program (IDMP), which started in 2009 at the Symbiosis International University (SIU). However, the National University Grants Commission (UGC) has yet to implement the program at the national level for all colleges and universities (Gupta 2022).

Meanwhile, the Ministry of Home Affairs (2011) developed a strategy focusing on the segments of society who never joined formal schooling where specific learning and teaching aids should be provided. Moreover, Das et al. (2013) conducted a capacity-building program on the role of teachers in disaster education. They discovered that teachers in the City of Delhi encounter various issues when teaching the subject. Some teachers felt that social science and science teachers could provide more correct answers about disaster management. The teachers have emphasized that physical infrastructure, such as ramps and open spaces, can facilitate earthquake preparedness. They also expressed that the lack of emergency resources and systematic management makes it difficult for them to implement disaster preparedness and mitigation plans.

12.2.4 Indonesia

Indonesia is in the Pacific Ring of Fire and is one of the world's biggest archipelagos. The country is vulnerable to earthquakes, volcanoes, landslides, and tsunamis. The 2004 Aceh earthquake and Indian Ocean tsunami remain among the deadliest. (The World Bank 2019). The Indonesian government has reformed its laws, policies, and institutions to manage disaster risk better since the 2004 Indian Ocean Tsunami. The government of Indonesia now has contingency plans for every disaster-prone city, which identify the vulnerabilities, outline the relief response, and build overall preparedness (CFE-DM 2018). One initiative to build preparedness is through disaster education. Desilia et al. (2023) said that significant progress in disaster education in Indonesia began after the 2004 Indian Ocean Tsunami. It marked various educational reforms that led to the launch of Act Number 23 of the Year 2007, focusing on the disaster education Circular Letter, No. 70a/MPN/SE/2010, about DRR mainstreaming in education. Selby and Kagawa (2012) described that in Indonesian school education, the subject area of physical education and health (PEH) includes the largest number of subsections and units related to disasters and safety. Indonesia has multiple hazards where students are ideally supposed to have a fundamental knowledge of how and why it happens. However, the current disaster education program has no scientific reference or knowledge of the school curriculum.

Meanwhile, the National Agency for Disaster Management (Badan Nasional Penanggulangan Bencana) (2009) has reported that another challenge is that the integration of disaster risk reduction in the existing curriculum is considered burdensome for the students at each level. Astuti et al. (2021) have pointed out that the main challenges for current DRR curriculum integration were teacher training and teaching resources, which can be solved by creating DRR training or programs and establishing collaboration among government agencies. Desilia et al. (2023) emphasize that teachers play an important role in disaster education. Their readiness and willingness to promote it can be done by incentivizing and empowering them. The National Agency for Disaster Management (Badan Nasional Penanggulangan Bencana) (2009) articulates that sufficient human resources are expected to encourage the regions and schools to be more creative and innovative in mainstreaming disaster risk reduction. Amri et al. (2017) argue that forming inter-agency collaboration and receiving support from a high level is crucial. Consequently, this creates better dynamics for them to work together and commit to improving the teaching and learning environment. Building commitment is important to developing effective and efficient regional disaster risk reduction programs (National Agency for Disaster Management (Badan Nasional Penanggulangan Bencana) (2011).

Moreover, at the grassroots level, Amri et al. (2017) describe in their study during a teaching training program in Jakarta that the participants vent out that government agencies, DRR, and education agencies still view children as passive participants. This creates problems regarding the alignment of goals and compromise between the top-down and bottom-up approaches.

12.2.5 Republic of Korea (South Korea)

The Republic of Korea (South Korea) is in a region vulnerable to disasters, especially typhoons and heavy rains (National Atlas of Korea n.d). The Korean government has established the Master Plan for National Safety Management (MPNSF) every five years to establish systematic and integrated disaster management (Ministry of the Interior and Safety 2022).

The Ministry of the Interior and Safety (2009) reported that the country introduced programs to strengthen disaster risk reduction, such as developing fire and disaster management science as school subjects and constructing safety experience facilities. Meanwhile, Korean government agencies and ministries have carried out disaster and safety education programs, which the Korean government plans to integrate into a mandatory and systematic disaster and safety education and training program. Moreover, the Republic of Korea, government (2013) highlighted legislative mechanisms need to be in place for professional training in disaster and safety education. In line with this, the National Disaster Management Institute focuses on teaching disaster managers and officials despite the limited educational institutes and experts on the subject matter.

The Sewol Ferry disaster on April 16, 2014, further highlighted the conversation of disaster and safety in South Korea. (Park et al. 2023). It raised an urgent social and political response on how safety should be discussed. The Ministry of Education at that year announced "Comprehensive Measures for Safety in Education," where standards for school safety were created, and the educational contents were suitable for elementary, middle, and high school students. Park et al. (2023) emphasized the importance of safety education in the curriculum, which can be delivered to the students by providing them with creative experiences to enrich their knowledge and skills. The said disaster has imposed reforms on the curriculum, such as science. However, implementing disaster safety has created a different perspective among science teachers. Park et al. (2023) pointed out that teachers saw disaster safety as a topic to be discussed with the students during their spare time once all the topics about physics, chemistry, biology, and earth science are covered. This brings forth the issue of disaster safety as an optional topic that is not prioritized in classroom topics.

On the other hand, safety education is promoted among local and international students living in the country. The study by Ryoo and Cheung (2021) underscored that international students are considered vulnerable groups due to the language barrier and cultural differences they encounter while living in the country. Students have described learning about safety education through video and smartphone applications. Notably, the participants raised topics that can be taught in safety education classes, such as meteorological disasters, criminal fraud, and sexual violence safety.

12.3 Implications of Hyogo Framework for Action (HFA)

Figure 12.2 shows the implication of the Hyogo Framework for Action (HFA) in China, India, Indonesia, Japan, and the Republic of Korea. Notably, China, Indonesia, Japan, and the Republic of Korea integrated Disaster Risk Reduction (DRR) into their curriculum from the primary to the professional level. However, India does not integrate DRR in the primary curriculum. This reflects the country's priority of determining who can learn disaster education. All selected countries have integrated DRR concepts into teachers' professional development, particularly teacher training programs. Teacher training programs in China, India, and Japan have focused on advancing the skills of teachers in disaster education. Notably, Indonesia's teacher training enhances teachers' capability and serves as an avenue to spur partnerships with non-government partners. Also, the legislative mechanisms in the Republic of Korea to strengthen teacher training can contribute to integrating disaster education as a key priority in policymaking.

Meanwhile, disaster education programs are focused on building educational content and facilities. China and Japan highlight the development of a local disaster prevention curriculum and studies on the role of science and technology in disaster reduction, respectively. Likewise, the Republic of Korea developed school content, namely fire science and management science, and constructed safety experience facilities.

The selected countries highlighted the key challenges in implementing disaster education in their countries. The study revealed that human capital is a key challenge to implementing disaster education in their countries. China has emphasized societal issues, namely regionalism and family. India cites the limited domain experts and sections of society that need to be accessed; Indonesia mentions that the low commitment among disaster professionals impedes the development of the DRR curriculum in the region; Japan underscores the limited number of enthusiastic teachers willing to teach the subject matter. The Republic of Korea emphasizes limited experts, training programs, and infrastructures that delay the implementation of DRR in the country.

Education and Training	China	India	Indonesia	Japan	Republic of Korea
1. Integration of DRR in the national curriculum	all levels	no primary curriculum	all levels	all levels	all levels
2. Integration of DRR in Teacher training	relevant trainings for in-service teachers	extensive teacher training for curriculum transition	training programs developed by non-government partners	distribution of reference materials during disaster risk reduction trainings for teachers	legislative mechanisms are being developed for disaster and safety education
3. Disaster education programs	development of local disaster prevention and reduction curriculum as well as school-based curriculum using modern technologies	preparation of supplementary text books and introduction of online courses for disaster management	preparation of educational materials and disaster related trainings	research studies on science and technology for disaster reduction	- school contents are under development covering subjects such as fire science and disaster management science. - construction of safety experience facilities
4. Challenges in the implementation of DRR education	- poor regionalism - close connection with family and society to be built	- Limited domain experts Investment in research and knowledge management - strategy to target sections of society who never attended formal schooling	low commitment of the region to develop curriculum	limited numbers of teachers who are enthusiastic to teach disaster reduction education	- limited number of educational institutes and experts - disaster and safety education programs are carried out separately - legislative mechanisms are to be developed for disaster and safety education and professional training

Fig. 12.2 Status of disaster education in selected countries from the Asia Pacific region. *Source* Accessed through the National Progress Report on the implementation of the Hyogo Framework for Action and compiled by the authors

12.4 Drivers of Disaster Education

The secondary data analysis highlights that the selected case studies from the Asia Pacific countries have revealed the following themes that serve as important drivers: disaster experience, government initiatives, infrastructures, educational curriculum, and teacher training programs that shape the evolution of disaster education programs in their home countries.

The disaster experience can help improve or establish disaster education programs. Disasters can trigger political and social responses that can push development. Governments have formulated programs, strategies, and policies to address the impact of disasters and protect their people. Government agencies are gearing up to construct infrastructures supporting research and training development. One of them is the establishment of disaster education programs and their integration into the educational system. This coincides as well with the goals of the HFA in the integration of disaster risk reduction into the curriculum.

It is essential to mention that disaster experiences provide key lessons and practices that influence the creation of educational curricula. The curriculum guides teachers to create better learning experiences for the students to learn the subject matter, particularly in the context of disaster education. Disaster education content aligned with the students' understanding level should be considered. The secondary data analysis has shown that tackling international or national disaster experiences and incorporating local disaster experiences can help students feel familiar with their immediate surroundings and enhance their awareness of the risks in their communities. Notably, the medium of instruction has shifted from traditional ways and integrated technologies to make learning more experiential, interesting, and innovative for the students.

The students are the core of disaster education. To train students with knowledge and skills that can assist them in understanding better ways to make solutions during a disaster is to train its frontline, the teachers. Teacher training programs, as cited in the selected case studies, have shown different levels of challenges to overcome and support needed to implement disaster education effectively. The availability of equipment and logistics relevant to teaching disaster education is a priority. It gives them the confidence to lead their students and increase motivation to learn about the subject matter. Establishing coordination not only with the academic community but with the private and public sectors within the school community can facilitate cooperation and foster good relationships in times of disaster. Significant to note is that the teacher's motivation plays an important role in disaster education. Some teachers are confident, and some are not when teaching disaster concepts. Some are enthusiastic, and some are indifferent. These psychological attributes of the teachers toward the teaching of disaster education should be considered in creating the pedagogy for disaster education.

12.5 Conclusion

The onset of the twenty-first century has brought a paradigm shift in what constitutes disasters. We are living in uncertainty, and new risks are starting to emerge. The ability to adapt to uncertainty and be resilient plays a crucial role that every society attempts to attain, and dealing with these adversities is an important skill to hone. Disaster education is a frontline to achieve this goal. The HFA has emphasized the value of disaster education in improving a country's resilience. The formulation of the Hyogo Framework for Action (HFA) has provided countries with a clear-cut template and guidelines on how disaster resilience can be incorporated, improved, and sustained in their existing policies and programs toward resilience.

This chapter traces the evolution of disaster education in selected countries in the Asia Pacific region, namely China, India, Indonesia, Japan, and the Republic of Korea. It highlights the government's initiatives to integrate disaster education into the educational system. It presents the education and training that occurs in their countries to enhance the teaching and learning of disaster education in their respective countries. Countries mentioned above are gearing up their capability to make disaster education available and accessible to all school levels. Thus, no child will be left behind in this learning journey. The implementation of disaster education considers what is suitable for the child's cognitive level as a key factor. Traditionally, children and youth are considered among the most vulnerable groups when a disaster occurs, but the risk landscape has changed. Research studies and practical experiences have a different take on the capability of the children and the youth.

They are no longer a spectator but proactive engaged to create solutions and take action during disasters. Empowering them is a key objective of disaster education. The knowledge and skills they can acquire are rooted in the educational content found in the educational curriculum. The educational curriculum is the blueprint for creating effective and efficient educational content. The concept of disaster education is integrated into the existing school subjects as an additional topic or a stand-alone subject, as exhibited in the cases of the selected countries.

The study also demonstrates teachers' gaps and challenges in implementing disaster education among students. The national government has provided policies and programs to promote disaster education. However, as demonstrated in the case studies from selected countries, large gaps between the policies and what is happening on the ground (i.e., the case of the teachers) impede the implementation. Notably, national governments have placed disaster education into the spotlight in the aftermath of disasters, creating a sense of urgency to create initiatives to address various issues. These initiatives are being directed to be put into practice by various government agencies, which brings into the importance of coordination and standardization of protocols and practices. Despite the efforts, well-coordinated and standardized dynamics can assist government agencies in gaining the people's trust and support, both essential during disasters. Also, this is an important competency to learn as a disaster education professional. However, training disaster professionals such as teachers is exacerbated by the limited institutions and human resources that can demonstrate sufficient knowledge and skills in dealing with disasters. Organizing training programs might be the best option but sustaining human resources to study disaster education boils down to building their motivation and enthusiasm by instilling the essence of disaster education. For instance, teachers are already burdened and busy dealing with their daily school tasks, subject matter, and classroom interaction with their students. Thus, creating a space to encourage them to learn while working with the community and government agencies for the safety of the students requires an openness to create possibilities that take time and not overnight solution-making. Also, the lack of infrastructure slows down the progress of disaster education. The limited funding and facilities impede disaster education professionals from implementing, innovating, and instructing disaster education among students. This calls for a direct and clear-cut direction on how public funding should be utilized.

Moreover, technological advancements have provided countries with mechanisms to innovate how teaching and learning experiences can be explored. This also extended to how disaster education can be implemented. It also provides a platform to create domestic and international partnerships to understand and learn from each other's disaster education experiences. It promotes a culture of information sharing and shared responsibility for how disaster education can be taught and explored. Technological advancements from the selected countries have different degrees of innovations and challenges, yet they are committed to utilizing this tool to promote disaster education.

Nevertheless, HFA offers countries a pathway to building their country's disaster resilience. The HFA emphasizes disaster education as a key component to achieve this. Disaster education empowers people to create solutions and actions during disasters. The government and the people should work together. Empowering disaster education professionals and dealing with disasters is not a linear path.

References

Amri A, Bird DK, Ronan K et al (2017) Disaster risk reduction education in Indonesia: challenges and recommendations for scaling up. Nat Hazards Earth Syst Sci 17:595–612. https://doi.org/10.5194/nhess-17-595-2017ß

Astuti NM, Werdhiana IK, Wahyono U (2021) Impacts of direct disaster experience on teachers' knowledge, attitudes and perceptions of disaster risk reduction curriculum implementation in Central Sulawesi, Indonesia. Int J Disaster Risk Reduct 53:101992. https://doi.org/10.1016/j.ijdrr.2020.101992

APCSS (2016) We are advancing comprehensive School Safety for Asia and The Pacific. Policy Brief. https://www.preventionweb.net/files/51084_policybriefadvancingcomprehensivesc.pdf

Cabinet Office (Japan) (2009) National progress report on implementing the Hyogo Framework for Action. https://www.preventionweb.net/files/9809_Japan.pdf

Cabinet Office (Japan) (2011) National progress report on the implementation of the Hyogo Framework for Action (2009–2011). Retrieved 02 July 2024 from https://www.preventionweb.net/files/19650_jpn_NationalHFAprogress_2009-11.pdf

Cabinet Office (Japan) (2013) National progress report on the implementation of the Hyogo Framework for Action (2011–2013). Retrieved 10 July 2024 from https://www.preventionweb.net/files/31426_jpn_NationalHFAprogress_2011-13.pdf

Cabinet Office (2023) White Paper on disaster management 2023. Retrieved 01 August 2024 from https://www.bousai.go.jp/en/documentation/white_paper/pdf/2023/SF1-1.pdf

Canlas I (2023) Three decades of disaster risk reduction education: A bibliometric study. Nat Hazards Res 3(2):326–335. https://doi.org/10.1016/j.nhres.2023.02.007

Center for Excellence in Disaster Management and Humanitarian Assistance (CFE-DM) (2018) Indonesia: Disaster management reference handbook. Retrieved 10 July 2024 from https://reliefweb.int/report/indonesia/indonesia-disaster-management-reference-handbook-june-2018

Chan C, Leung M, Pulmano N (2020) Guardians of the planet Asia Pacific children and youth voices on climate crisis and disaster risk reduction. UNESCAP. Retrieved 11 July 2024 from https://sdghelpdesk.unescap.org/sites/default/files/2020-10/Guardian%20of%20the%20planet.pdf

Chinoi T (2007) Disaster education in Japan. In: BRI & GRIPS. Disaster education, pp 50–66. https://www.preventionweb.net/files/3442_DisasterEducation.pdf

Das P, Malaviya R, Prof A, Prof A (2013) Role and capacity building of school teachers in disaster preparedness and prevention. https://www.gcoedu.in/pdf/evs2.pdf

Desilia NR, Lassa J, Oktari RS (2023) Integrating disaster education into school curriculum in Indonesia: A scoping review. Int J Disaster Manag 6:263–274. https://doi.org/10.24815/ijdm.v6i2.34867

Gupta B (2022) Role of education in mitigating disasters in India—thus saving lives. [Issue Brief] Center for land warfare studies. Retrieved 15 June 2024 from https://www.claws.in/static/IB-362_Role-of-Education-in-Mitigating-Disasters-in-India—thus-Saving-Lives.pdf

Kagawa, Fumiyo, Selby et al (2012) Disaster risk reduction in school curricula: Case studies from thirty countries

Kawasaki H, Yamasaki S, Yamakido M, Murata Y (2022) Introductory disaster training for aspiring teachers: A pilot study. https://doi.org/10.3390/su14063492

Korea, Republic of—government (2013) National progress report on the implementation of the Hyogo Framework for Action (2011–2013). https://www.preventionweb.net/files/30492_kor_NationalHFAprogress_2011-13.pdf

Ministry of Civil Affairs (2012) National progress report on implementing the Hyogo Framework for Action (2011–2013)—Interim. https://www.preventionweb.net/files/28446_chn_NationalHFAprogress_2011-13.pdf

Ministry of Home Affairs (2011) India National progress report on the implementation of the Hyogo Framework for Action (2009–2011). https://www.preventionweb.net/files/15602_ind_NationalHFAprogress_2009-11.pdf

Ministry of Home Affairs (2013) National progress report on the implementation of the Hyogo Framework for Action (2011–2013). https://www.preventionweb.net/files/40210_IND_NationalHFAprogress_2013-15.pdf

Ministry of Home Affairs (2015) National progress report on the implementation of the Hyogo Framework for Action (2013–2015) https://www.preventionweb.net/files/40210_IND_NationalHFAprogress_2013-15.pdf

Ministry of the Interior and Safety (2009) National progress report on the implementation of the Hyogo Framework for Action (2007–2009)—Interim. https://www.preventionweb.net/files/7587_KoreaRep[1].pdf

Ministry of the Interior and Safety (2022) Official statement the Republic of Korea midterm review of the implementation of the Sendai framework for disaster risk reduction 2015–2030. Retrieved 05 August 2024 from https://apmcdrr.undrr.org/sites/default/files/inline-files/The%20Republic%20of%20Korea_Statement_APMCDRR%202022.pdf?startDownload=true

National Atlas of Korea (n.d.) Natural disasters. Retrieved 10 August 2024 from http://nationalatlas.ngii.go.kr/pages/page_766.php

National Agency for Disaster Management (Badan Nasional Penanggulangan Bencana) (2009) National progress report on the implementation of the Hyogo Framework for Action. https://www.preventionweb.net/files/7486_Indonesia[1].pdf

National Agency for Disaster Management (Badan Nasional Penanggulangan Bencana) (2011) National progress report on the implementation of the Hyogo Framework for Action (2009–2011). https://www.preventionweb.net/files/15941_idn_NationalHFAprogress_2009-11.pdf

National Agency for Disaster Management (Badan Nasional Penanggulangan Bencana) (2013) National progress report on the implementation of the Hyogo Framework for Action (2011–2013). https://www.preventionweb.net/files/28912_idn_NationalHFAprogress_2011-13.pdf

National Agency for Disaster Management (Badan Nasional Penanggulangan Bencana) (2015) National progress report on the implementation of the Hyogo Framework for Action (2013–2015). https://www.preventionweb.net/files/41507_IDN_NationalHFAprogress_2013-15.pdf

National Disaster Management Division (2009) India National progress report on the implementation of the Hyogo Framework for Action. https://www.preventionweb.net/files/9808_India.pdf

National Disaster Management Authority (NDMA) (2021) Simplified Guideline for Earthquake Safety of Building from National Building Code of India 2016. Retrieved 07 March 2025 from https://ndma.gov.in/sites/default/files/PDF/Guidelines/Simplified_Guidelines_for_earthquake.pdf

National Disaster Management Authority (NDMA) (2023) Annual Report 2021–2022. Retrieved 03 August 2024 from https://ndma.gov.in/sites/default/files/PDF/Reports/NDMA_Annual_Report_2021-22.pdf

Park W, Lee H, Ko Y, Lee H (2023) "Safety" and "integration": Examining the introduction of disaster into the science curriculum in South Korea. J Curric Stud 55:580–597. https://doi.org/10.1080/00220272.2023.2239887

Ryoo S, Cheung C (2021) International students' perceptions of safety level and safety education needs in Korea. Int J Disaster Risk Reduct 55:102068. https://doi.org/10.1016/j.ijdrr.2021.102068

Sakurai A, Sato T (2016) Promoting education for disaster resilience and the Sendai framework for disaster risk reduction. J Disaster Res 3:402–412. Special Issue on the Third United Nations World Conference on Disaster Risk Reduction (WCDRR)

Shaw R (2020) Thirty years of science, technology, and academia in disaster risk reduction and emerging responsibilities. Int J Disaster Risk Sci 11. https://doi.org/10.1007/s13753-020-00264-z

Shaw R, Takeuchi Y, Ru Gwee Q, Shiwaku K (2011) Chapter 1 disaster education: An introduction. pp 1–22. In: Shaw R, Shiwaku K, Takeuchi Y (eds) Disaster education, pp 1–22. (Community, Environment and Disaster Risk Management, vol 7). https://doi.org/10.1108/S2040-7262(2011)0000007007

Song J, Yamauchi H, Oguchi T, Ogura T (2022) Application of web hazard maps to high school education for disaster risk reduction. Int J Disaster Risk Reduct 72. https://doi.org/10.1016/j.ijdrr.2022.102866

The Asia Foundation (2022) India's disaster risk reduction journey—opportunities for strengthening partnerships in the Indo-Pacific. The Asia Foundation: New Delhi. Retrieved 05 August 2024 from https://asiafoundation.org/wp-content/uploads/2022/10/Indias-Disaster-Risk-Reduction-Journey_TRIDEP_EN.pdf

The World Bank Group and the Asian Development Bank (2021) Climate risk country profile, China. Retrieved 01 August 2024 from https://www.adb.org/sites/default/files/publication/703641/climate-risk-country-profile-china.pdf

World Bank (2019) Strengthening the disaster resilience of Indonesian cities-a policy note. Retrieved 03 August 2024 from https://documents1.worldbank.org/curated/en/748581569515561529/pdf/Strengthening-the-Disaster-Resilience-of-Indonesian-Cities-A-Policy-Note.pdf

United Nations Office for Disaster Risk Reduction (2007) Hyogo framework for action 2005–2015: Building the resilience of nations and communities to disasters. https://www.undrr.org/publication/hyogo-framework-action-2005-2015-building-resilience-nations-and-communities-disasters

United Nations Office for Disaster Risk Reduction (2011) Hyogo framework for action 2005–2015 building the resilience of nations and communities to disasters. Mid-Term Review 2010–2011. Retrieved 10 June 2022 from https://www.unisdr.org/files/18197_midterm.pdf

United Nations, Economic and Social Commission for Asia and the Pacific (ESCAP) (2019) Summary of the Asia–Pacific disaster report 2019. Retrieved 15 July 2024 from https://www.unescap.org/sites/default/files/Summary%20of%20the%20Asia-Pacific%20Disaster%20Report%202019_English.pdf

United Nations, Economic and Social Commission for Asia and the Pacific (ESCAP) (2020) The disaster riskscape across South-East Asia: Key takeaways for stakeholders. ST/ESCAP/2885. Retrieved 10 July 2024 from https://www.unescap.org/sites/default/files/IDD-APDR-Subreport-SEA.pdf

United Nations, Economic and Social Commission for Asia and the Pacific (ESCAP) (2020) The disaster riskscape across East and North-East Asia: Key takeaways for stakeholders. ST/ESCAP/2882.Retrieved 10 July 2024 from https://www.unescap.org/sites/default/d8files/IDD-APDR-Subreport-ENEA.pdf

United Nations, Economic and Social Commission for Asia and the Pacific (ESCAP) (2022) Pathways to adaptation and resilience in South and South-West Asia Subregional Report. Retrieved 05 July 2024 from https://www.unescap.org/kp/2021/asia-pacific-disaster-report-2021

United Nations, Economic and Social Commission for Asia and the Pacific (ESCAP) (2023) Seizing the moment targeting transformative disaster risk resilience Asia—Pacific disaster report. Retrieved 10 July 2024 from https://un-ggim-ap.org/sites/default/files/media/docs/Asia-Pacific%20Disaster%20Report%202023-Full%20report.pdf

World Bank (2020) Learning from experience insights from China's Progress in disaster risk management. World Bank. Retrieved 01 August 2024 from https://www.gfdrr.org/sites/default/files/publication/Learning-from-Experience-Insights-from-China-s-Progress-in-Disaster-Risk-Management.pdf

Wahlström M (2009) Disaster risk and its reduction: Who is responsible?. Fletcher Forum World Aff:153–158. http://www.jstor.org/stable/45289481

Yildiz A, Shaw R (2024) Perspectives on future directions of risk-informed education: lessons from Japan and Türkiye. J Risk Res 1–12. https://doi.org/10.1080/13669877.2024.2368187

Zhong S, Cheng Q, Zhang S et al (2021) An impact assessment of disaster education on children's flood risk perceptions in China: Policy implications for adaptation to climate extremes. Sci Total Environ 757:143761. https://doi.org/10.1016/j.scitotenv.2020.143761

Zhu TT, Zhang YJ (2017) An investigation of disaster education in elementary and secondary schools: evidence from China. Nat Hazards 89:1009–1029. https://doi.org/10.1007/s11069-017-3004-2

Part II
Regional and Country Experiences

Chapter 13
Key Progress and Challenges in Tsunami Risk Reduction in the ASEAN Region

Laely Nurhidayah, Riyanti Djalante, Agus Suntoro, and Henny Warsilah

Abstract ASEAN as a regional cooperation plays a pivotal role in disaster management and emergency response. Two decades after the Indian Ocean tsunami which hit Indonesia the most, ASEAN has shown its immediate collective responses and has continued to evolve in the humanitarian landscape and natural disaster management ever since. There have been key progresses in strengthening cooperation and collective action to respond effectively to disaster, particularly following the adoption of the ASEAN Agreement on Disaster Management and Emergency Response (AADMER) in 2005 and entered into force in 2009, the establishment of the ASEAN Committee on Disaster Management in 2003 as well as. Strengthening the AHA Centre's role as the main coordinating body for disaster management, and realizing the vision of 'One ASEAN One Response is crucial. Despite this progress, however, ASEAN still faces many challenges to effectively respond to tsunami disasters. This chapter will examine opportunities to strengthen tsunami disaster risk management in ASEAN through 3 key factors, as outlined in the ASEAN Vision 2025 on Disaster Management, namely (i) institutions, (ii) finance and resource mobilization, and (ii) partnership and innovations at the regional. We find that current institutional arrangements recognize the need for multi-layer and cross-sectoral governance approach, but the implementation is still in infancy. The current financial and resource mobilization is not adequate to respond to the more complex, multi-faceted, and multi-characteristic nature of climate change-related disasters and the future humanitarian landscape. Moreover, there is still a gap in partnership in ASEAN in building its holistic partnerships from the international, regional, and local levels including public and private sectors, and the people. The paper recommends that ASEAN focus its future approaches for tsunami disaster risk reduction and other major disasters on partnerships, and to strengthen innovations at the regional and local levels.

L. Nurhidayah (✉) · A. Suntoro
Research Center for Law, National Research and Innovation Agency, Jakarta, Indonesia
e-mail: lael003@brin.go.id

R. Djalante
ASEAN Secretariat, Jakarta, Indonesia

H. Warsilah
Centre for Society and Culture, National Research and Innovation Agency, Jakarta, Indonesia

R. Shaw et al. (eds.), *Two Decades from the Indian Ocean Tsunami*, Disaster Risk Reduction, https://doi.org/10.1007/978-981-96-2669-4_13

Keywords Tsunami risk reduction · ASEAN · One ASEAN One response · Partnership · Financial mobilization

13.1 Introduction

The unprecedented Indian Ocean tsunami disaster in 2004 killed more than 200,000 people, the tsunami also has caused millions of survivors without food or shelter Gentner (2006). The coastal regions of four ASEAN members were hit by the tsunami wave with Indonesia suffering the most as it is located in the range of epicenter followed by Malaysia, Myanmar, and Thailand. Two decades after the Indian Ocean tsunami ASEAN has continued to evolve in the humanitarian landscape and natural disaster management. Key progress of ASEAN in responding to disaster is seen with the enactment of the ASEAN Agreement on Disaster Management and Emergency Response (AADMER) in 2009. AADMER is a legally binding regional agreement that was signed in 2005 and ratified on 24 December 2009 to address and guide regional cooperation in the field of disaster management and response. The objective of the agreement is to provide an effective mechanism to achieve a substantial reduction of disaster losses in the lives and social, economic, and environmental assets of the parties. Further, in 2016, ASEAN adopted the ASEAN Declaration on One ASEAN One Response. With this Declaration, ASEAN attempted to respond to Disaster as One in the Region and Outside the Region, which laid out the political commitment of ASEAN to achieve faster response, mobilize greater resources, and establish stronger disaster risk reduction. Moreover, in 2017, ASEAN adopted the ASEAN Vision 2025 on Disaster Management. With this vision, ASEAN has a view to develop partnerships with all partners and stakeholders, as such relationships are substantial for comprehensive, innovative, and robust disaster management and emergency response implementation in the region and beyond. Finally, the ASEAN Agreement on Disaster Management and Emergency Response (AADMER) Work Programme 2021–2025 emphasizes on partnership approach and multi-stakeholder cooperation. In the institutional framework, strengthening the AHA Centre's role as the main coordinating body for disaster management, and realizing the vision of 'One ASEAN One Response is crucial. AHA Centre in its efforts has built regional and national resilience relying on its Disaster Monitoring and Response System (DMRS). The DMRS system is considered a comprehensive early warning and decision-support technology for disaster risk reduction in the world. In addition, ASEAN Vision 2025 on Disaster Management which is considered a strategic policy document, identified the key areas that need to be strengthened in the implementation of AADMER include toward a people-centered, people-oriented, financially sustainable, and network approach by 2025. Despite these progress and commitments, ASEAN still faces many challenges to effectively respond to tsunami natural disasters. Reliance on individual states to manage and respond first to disasters, it makes difficult to collectively respond to disasters on time.

This chapter will examine the opportunity and challenges to strengthen tsunami disaster risk management in ASEAN in 3 key sectors, namely **institutions, finance, and resource mobilization, and partnership and innovations** at regional, and its implications to the ASEAN Member States. These key sectors are essential and considered the key guiding principles identified in the ASEAN Vision on Disaster Management 2025 to guide the direction of the future implementation of the AADMER. Following the introduction, the second section examines institutional framework for tsunami disaster risk reduction which highlights key progress of institutional framework for tsunami risk reduction. In the third section, financial resources and mobilization highlight possible alternatives of resources financing and mobilization in the region to respond to disaster such as tsunami, followed by review of partnership and innovation which highlighted the need for strengthening holistic partnership across the region. Finally, the chapter outlines ways forward and recommendations in the conclusion.

13.2 Institutional Framework for Tsunami Risk Reduction

13.2.1 ASEAN Cooperation in Disaster Risk Reduction

The ASEAN region is very vulnerable to natural disasters. It is reported that the number of disaster events per year in this region could increase from around 400 events in 2015 to 560 events per year in 2030 or is projected to increase by 40% (Mizan and Fathia 2022). In addition, ASEAN Disaster Information Network (ADINet) records show that disasters caused by weather and climate are the most common types of disasters in Southeast Asia (Mizan and Fathia 2022). Since its founding as reflected in the ASEAN charter, ASEAN has moved toward greater regional cooperation. The pinnacle of ASEAN cooperation in the field of disaster is the implementation of AADMER. The ASEAN Agreement on Disaster Management and Emergency Response (AADMER) is established as the policy backbone for the ASEAN Member States to increase their collective efforts in reducing disaster risks and responding to disasters in the region.

13.2.2 Progress of ASEAN Institutional Cooperation in for Disaster Preparedness and Emergency Response Including Tsunami Disaster Mitigation

Under AADMER the primary actors responsible for disaster management activities are nation states. The ASEAN Community already has cooperation and institutional framework to deal with natural disasters, in the form of (1). Institutional ASEAN Ministerial Meeting on Disaster Management (AMMDM), which has resulted in the

ASEAN Agreement on Disaster Management and Emergency Response (AADMER) with the ASEAN Committee on Disaster Management (ACDM). In addition, there is the ASEAN Working Group of Prevention and Mitigation which includes a Technical Working Group on Protection, Gender, and Inclusion, ASEAN Working Group of Preparedness and Response which includes a Technical Working Group on Civil-Military Coordination.

AADMER is a legally binding agreement for improving regional cooperation, coordination, technical assistance, and resource mobilization of disaster management in ASEAN. The Agreement was established in the wake of tsunami that hit the region and provides support ongoing and planned national initiatives of the ASEAN Member States, and strengthening national capacities and working with existing work programs in the region. Since its adoption and through various initiatives that have been implemented, ASEAN through AADMER has managed to increase both regional and national capacities for responding to disasters in Southeast Asia (ASEAN Vision).

One of the important components of the AADMER agreement is the establishment of the ASEAN Coordinating Center for Humanitarian Assistance on Disaster Management (AHA Center) in 2011. Under the AADMER, the AHA Centre is tasked to facilitate cooperation and coordination among the ASEAN Member States, and with relevant United Nations and International Organizations, in promoting collaboration in disaster management. The AHA Centre functions to implement operational strategies and procedures to enable rapid, joint, and effective response to disasters within ASEAN to respond to natural disasters. In the ASEAN region, the AHA Center will never act as a first responder and the national agency of the affected country will be in charge of operations on the ground (Angela 2015). To organize the mobility of resources, the Humanitarian Assistance on Disaster Management (AHA Center) was established, headquartered in Jakarta, Indonesia. The AHA Center also became a central point for mobilizing resources to disaster-affected areas. In carrying out its mandate, the AHA Center primarily works with the National Disaster Management Organizations (NDMOs) of ASEAN Member States, as well as partners with international institutions, the private sector, and civil society organizations (AHA Centre 2012). The commitment to resource mobility initiated by the AHA Center is an embodiment of the principle of "One ASEAN, One Response" through risk identification, early warning, and monitoring to reduce the loss of life and property (Sari 2022). Furthermore, AHA center coordinated various programs and projects including ASEAN-Emergency Response and Assessment Team/ERAT, Risk and Vulnerability Assessment/RVA, Disaster Monitoring and Response System/DMRS, Disaster Emergency Logistics System of ASEAN/DELSA, various ICT-based systems, and assets), One ASEAN One Response declaration, Technical Working Group (TWG) on civil-military.

Despite this progress of AHA center, there are still barriers faced by AHA Center in effectively conducting its task. Hughes argues that the AHA center's difficulties in coordinating DRM in the region are hampered by sovereignty and non-intervention which are the two guiding principles of the ASEAN Way. The preoccupation with state sovereignty and non-intervention makes ASEAN member countries reluctant to hand over even the slightest power to regional institutions, especially ASEAN. The country-centric nature of this way of doing business goes against the goals of AADMER and the vision of One ASEAN, One Response (Hughes 2015). At the national AMS level, gaps in the statutory and regulatory provisions governing Early Warning Systems (EWS) at national and regional levels were found to be a significant factor hindering the effective implementation of regional EWS (ASEAN Secretariat 2024). In addition, the lack of a coordinated and integrated approach between institutions in the AMS in handling and anticipating disaster risk is also a significant limitation in effectively dealing with the risk of tsunami disasters in the future.

A multi-layered and cross-sectoral governance approach driving the integration of the ASEAN on disaster management and emergency response is the core point and strategic element for institutionalization of AADMER in the region. Strengthening national and sub-national (city, provincial, and community) levels of disaster management and emergency response is crucial in order to develop more effective disaster management response. AHA center play crucial role of the importance of communication exchange between all stakeholders involved in disaster (ASEAN Vision 2025) (Fig. 13.1).

More connections by building on partnerships with all relevant stakeholders and working on resource mobilization beyond existing traditional approach will further enhance disaster management and emergency response at the regional level. This task will give the ASEAN Coordinating Centre for Humanitarian Assistance on Disaster Management (the AHA Centre) potential role to become a network coordinator of regional centers for excellence in disaster management leadership and management that can facilitate knowledge creation, policy analysis, and training for the present and next generation and entities working on disasters.

The successful implementation of AADMER will help boost the ASEAN into global leader which can provide strategic policy analysis and support to the development of the region and global disaster management and emergency response. Its encourages ASEAN as a pioneer in transforming the disaster management landscape in the Southeast Asian region and beyond and strengthens its leadership to maintain ASEAN Centrality through mutually inclusive strategic elements above (ASEAN Vision 2025). In addition, cooperation is conducted through the vast stakeholders including ASEAN Regional Forum (ARF) and ASEAN Defense Ministerial Meeting Plus (ADMM Plus), these organizations have a long-standing involvement as ASEAN dialogue partner countries. Practical cooperation in the ASEAN

Fig. 13.1 ASEAN VISION 2025. *Source* ASEAN Vision 2025

defense sector has grown significantly since its inception with various initiatives in the following areas include: The Use of ASEAN Military Assets and Capacities in HADR (SASOP Chap. 6); Logistics Support Framework (LSF) by the ASEAN Defense Sector; ASEAN Militaries Ready Group on HADR (AMRG on HADR); and Standard Operating Procedures on the Multinational Coordination Centre (MNCC SOP) (ADMM ASEAN 2024).

13.3 Financial and Resources Mobilization

Guided by AADMER, financial and resource mobilization is a significant element of disaster risk reduction that has been recognized in the agreement. It is stated in Article 9 stated that military and civilian assets, emergency stockpiles of disaster relief items, and disaster management expertise and technologies are set aside by each Member state voluntarily for regional standby arrangements for disaster relief and emergency response such as emergency response search and rescue directory. Article 24 AADMER further provides a financial arrangement mechanism that provides the ASEAN Disaster Management and Emergency Relief Fund. Furthermore, in the 7th Session of the Global Platform for Disaster Risk Reduction (GP2022) in Bali, the secretariat General of ASEAN highlighted several aspects include: building financial resiliency in facing multi-hazards through implementing the ASEAN Roadmap for Disaster Risk Financing and Insurance, operationalizing the Southeast Asia Disaster Risk Insurance Facility and strengthening ASEAN Disaster Management and Emergency Response Fund.

In the period 2012–2020, there were 2,302 disasters in the ASEAN region and there were three major disasters, namely the Indian Ocean Tsunami (which hit Indonesia, Malaysia, Thailand and Myanmar) in 2004, Cyclone Nargis in Myanmar (2008) and Cyclone Haiyan in the Philippines (2013). The combined losses from these disasters are estimated to reach USD 22.5 billion, with 278,000 fatalities. During that period, 36 responses were carried out for both emergency response missions and preparedness or assessment missions. All these activities were based on the mobility of resources owned by the ASEAN region and assistance from donor agencies (AHA Center 2023).

13.3.1 Resource Mobility

Resource mobility refers to the ability of resources to move from one place to another. This can include both tangible resources such as natural resources and people, as well as intangible resources such as information and ideas (Colombo and Labrecciosa 2024). The urgency of resource mobility is crucial in the prevention, handling, and recovery aspects of post-disaster in the ASEAN region. This consideration became a fundamental point in the ASEAN Meeting in Jakarta on June 6, 2005, which bound

ASEAN countries to cooperate in the regional and international context in the field of disaster emergency assistance, implementation of rehabilitation, and reconstruction, including disaster mitigation. Furthermore, the signing of the ASEAN Agreement on Disaster Management and Emergency Response (AADMER) was carried out at the 38th ASEAN Ministerial Meeting in Vientiane, Lao People's Democratic Republic, on July 26, 2005, which aims to provide an effective regional mechanism to reduce the impact of natural disasters (Tolentino Jr. 2007). The beginning of awareness of the importance of resource mobility in disasters has been the concern of the ASEAN expert group for the first time in 1971 on disaster management. The meeting confirmed that natural disasters and other calamities are problems faced by the ASEAN region that hamper the growth and development of ASEAN countries. The subsequent development was the signing of the ASEAN Declaration on Mutual Assistance in Natural Disaster Management (1976), which confirmed the commitment to help each other in the areas of mitigation, rescue, and recovery of victims of natural disasters (di Floristella 2016).

Some fundamental issues in the context of resource mobilization are, *first*, dependence on funds contributed by ASEAN countries. In less than a year (2025), the AADMER framework will be fully implemented, and there are challenges in terms of financing and mobilizing resources. Structurally, operations only rely on contributions from ASEAN countries of US$ 30 million per year managed by the AHA Center. Ultimately, ASEAN countries are highly dependent on partners in terms of resource mobilization because disaster management funds and emergency assistance from ASEAN, which are voluntary, cannot be relied on to mitigate, overcome, and recover post-disaster (ASEAN 2013). The AHA Center in 2020 received total funding of US$4.3 million, comprising US$900,000 from annual contributions from ASEAN countries, US$2.7 million from external partners, and US$36,800 from the ADMER Fund (Gong 2021a, b). The issue of limited funds and income disparities among countries in the ASEAN region have also become obstacles to increasing resource mobility to finance ADDMER. Brunei Darussalam and Singapore have high incomes, while Indonesia, Malaysia, and Thailand are middle-income countries, and Myanmar and Cambodia have low incomes (di Floristella 2016).

Second, there are challenges in consolidation for coordination between ASEAN countries in resource mobility. Although there is a new mechanism through the AHA Center, the differences in government systems and procedures in each ASEAN member country have an impact on the speed of response to disasters. As an illustration, for example, Indonesia has the National Disaster Management Agency (BNPB), whose legal basis is Law Number 24 of 2007 concerning Disaster Management in conjunction with Presidential Regulation Number 8 of 2008 concerning the National Disaster Management Agency. The organizational structure of disaster management involves the central government, provinces, and districts/cities following the division of authority (regional autonomy), which causes obstacles in decision-making and response. Likewise, in the aspect of distributing disaster assistance from abroad, requires a long bureaucracy from the BNPB, the Ministry of Finance, and the Ministry of Foreign Affairs (Affairs 2020). In comparison, the Philippines has the National Disaster Risk Reduction and Management Council as the main body for disaster management, which is focused on the central government. The system in the Philippines is faster in terms of decision-making, which has implications for simple coordination but is sometimes less flexible in implementation in the field. The cost of disaster management at the national level is met through an annual appropriation to the Calamity Fund. At the same time, local government units must reserve five percent of their budgets for this purpose (Bankoff and Hilhorst 2009).

Third, dependence on donors and international institutions. With the limited contribution funds of each ASEAN country, one of the breakthroughs made by the AHA Center is to maximize assistance from donor countries and other international institutions. As an illustration in the case of a natural disaster due to Typhoon Rai (Odette) that hit the Philippines on December 16, 2021, the AHA Center received assistance from the Japan-ASEAN Integration Fund (JAIF) and Direct Relief in the form of 541 housing repair kits, 275 family tents, 5000 family kits, 1000 rolls of tarpaulin, 5000 personal hygiene kits, and 1000 kitchen kits. All aid facilities and resource mobility were transported to areas severely affected by Typhoon Rai in several regions in the Philippines (Syifa 2021). With the assistance of the Mission of Japan to ASEAN (MoJ) and the JAIF Management Team (JMT), the AHA Center also operates the Disaster Emergency Logistics System for ASEAN (DELSA), a Satellite Warehouse in Chainat, Thailand; Malaysia and the Philippines to enhance the speed and scale of ASEAN's response (JAIF 2023). The warehouse in Chainat focuses on distribution to Thailand, Myanmar, Cambodia, Lao PDR, and Vietnam. Meanwhile, the Subang, Malaysia warehouse focuses on aid for Malaysia, Indonesia, and parts of the Philippines. While the one in the Philippines is for that country and serves other countries' requests if needed (Fig. 13.2).

13.3.2 Disaster Risk Fund

In less than a year (2025), the AADMER framework will be fully implemented, and there are challenges in terms of financing and mobilizing resources. Structurally, operations only rely on contributions from ASEAN countries of US$ 30 million per year managed by the AHA Center. Ultimately, ASEAN countries are highly dependent on partners in terms of resource mobilization because disaster management funds and emergency assistance from ASEAN, which are voluntary, cannot be relied on to mitigate, overcome, and recover post-disaster (ASEAN 2013). The AHA Center in 2020 received total funding of US$4.3 million, comprising US$900,000 from annual contributions from ASEAN countries, US$2.7 million from external partners, and US$36,800 from the ADMER Fund (Gong 2021a, b). The issue of limited funds and income disparities among countries in the ASEAN region have also become obstacles to increasing resource mobility to finance ADDMER. Brunei Darussalam and Singapore have high incomes, while Indonesia, Malaysia, and Thailand are middle-income countries, and Myanmar and Cambodia have low incomes (di Floristella 2016).

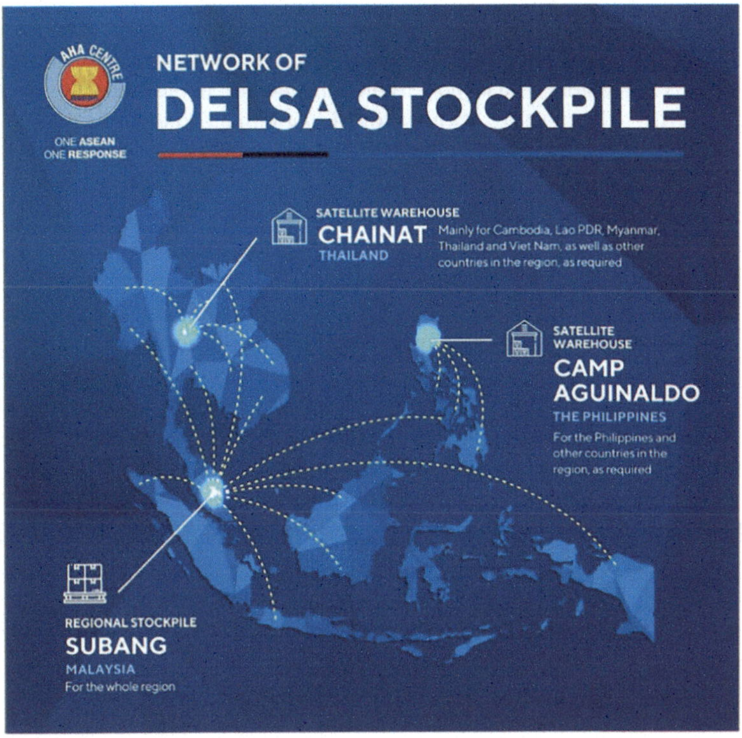

Fig. 13.2 Distribution of resource mobility for disasters in the ASEAN region

The second problem faced by ASEAN countries is dependence on donor agencies or other international institutions. With the limited contribution funds of each ASEAN country, one of the breakthroughs made by the AHA Center is to maximize assistance from donor countries and other international institutions. As an illustration in the case of a natural disaster due to Typhoon Rai (Odette) that hit the Philippines on December 16, 2021, the AHA Center received assistance from the Japan-ASEAN Integration Fund (JAIF) and Direct Relief in the form of 541 housing repair kits, 275 family tents, 5000 family kits, 1000 rolls of tarpaulin, 5000 personal hygiene kits, and 1000 kitchen kits. All of the aid facilities and resource mobility were transported to areas severely affected by Typhoon Rai in several regions in the Philippines (Syifa 2021). With the assistance of the Mission of Japan to ASEAN (MoJ) and the JAIF Management Team (JMT), the AHA Center also operates the Disaster Emergency Logistics.

System for ASEAN (DELSA), a Satellite Warehouse in Chainat, Thailand; Malaysia, and the Philippines to enhance the speed and scale of ASEAN's response (JAIF 2023). The funding initiative for risk mitigation and disaster management resulted from the ASEAN + 3 Finance Ministers and Central Bank Governors Meeting (The Meeting) held on 2 May 2023, attended by Japanese Finance Minister

Shunichi Suzuki, President of the Asian Development Bank (ADB), Director of the ASEAN + 3 Macroeconomic Research Office (AMRO), Deputy Secretary-General of the ASEAN Secretariat, and Deputy Managing Director of the International Monetary Fund (IMF). The meeting acknowledged that climate change, including more frequent and severe natural disasters, will affect the economic growth of the ASEAN region. Therefore, as a realization of the 2030 agenda for Sustainable Development, collaborative work is needed between ASEAN countries and various global institutions. The meeting promoted alternative emergency funding as a regional safety net that would benefit ASEAN countries. Tasked AMRO to prepare a proposal as a road map for emergency financing facilities and structures to be agreed upon by ASEAN countries. The meeting also endorsed the Action Plan 2023–2025 of the ASEAN + 3 Disaster Risk Financing Initiative, which sets out key actions to explore insurance solutions and other financial products, facilitate knowledge exchange, and develop technical solutions, including regional data repositories and monitoring tools, including the setting up of the ASEAN Disaster Risk Financing and Insurance/ ADRFI (ASEAN Secretariat 2020).

ASEAN countries also as members of the South-East Asia Disaster Risk Insurance Facility (SEADRIF), which focuses on providing disaster risk financing facilities and solutions through insurance for ASEAN countries, have prepared several follow-ups to the agreement, for example, Laos and Myanmar, which are developing a flood insurance scheme; while Indonesia and the Philippines have formed a Technical Working Group related to the protection of public assets or State Property. The programs developed are divided into two types, namely technical assistance and financing solutions. Cooperation in developing disaster risk financing policies is also carried out in the APEC Finance Minister Process forum. This cooperation is in the form of a Working Group on Disaster Risk Financing and Insurance/DRFI to assist APEC member economies in formulating the DRFI Strategy (Agency 2024).

Warner et al. (2009) present the experience of developed countries that illustrate the positive collaboration between the insurance industry and public institutions in disaster risk reduction. There are at least four arguments, namely (a) increasing risk awareness and education so that the government and insurance partners can partner together and share information and data; (b) accurate risk pricing mechanisms so that the government can make decisions for disaster risk reduction; (c) the creation of a legislative framework through legislation and monitoring efforts by the government so that insurance companies are more intense in carrying out risk reduction activities; (d) insurance companies can invest directly in risk reduction measures to avoid large compensation claims; and (e) risk reduction as a prerequisite for insurance, thus requiring policyholders to undertake certain disaster risk reduction measures (Warner et al. 2009). Overseas Development Institute (ODI) said that risk financing and disaster insurance are formulations as anticipation of vulnerable countries and communities to increase resilience from the impact of disasters. The disaster insurance mechanism will increase contingency funds, investment in preventing disasters, and risk. This method is a very important tool for dealing with residual risks from low-frequency and high-severity events (Fig. 13.3) (Weingärtner et al. 2017).

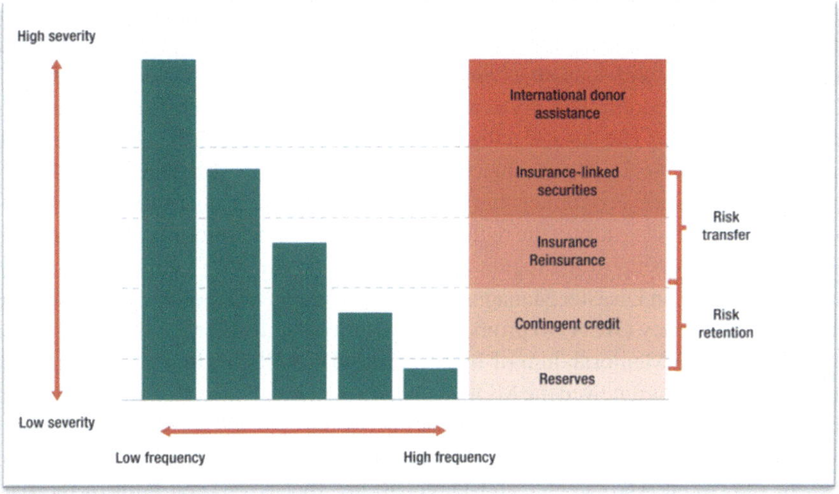

Fig. 13.3 Disaster risk financing options (*Source* Ghesquiere and Mahul 2010)

Specifically, Indonesia has formed a Disaster Pooling Fund (PFB), which is the main strategy for disaster management. PFB is a scheme for collecting, accumulating, and distributing special disaster funds to protect the APBN against disaster pressures by investing funds and transferring risks through insurance. PFB will also increase the government's capacity to fund mitigation and preparedness activities in the context of investment to reduce the risk of loss of life and property due to disasters. PFB is designed to be able to provide funds for pre-disaster, emergency response, and post-disaster financing. FDB initially had liabilities of up to 500 million US$ and became the main buyer of disaster insurance for all state and local government building assets, thus transferring risk through a disaster insurance scheme (Desdiani et al. 2021).

13.4 ASEAN Partnership and Innovation

13.4.1 Partnership

ASEAN required partnership and innovation in disaster management in order to execute AADMER effectively. The strategic element of partnerships and priorities for innovative ways is developing and establishing to work together with non-traditional partners for disaster management and emergency response. The strengths of partnership are showing by cooperation with entities at regional, national, and local levels in the public, private, and people sectors. In particular, this strategic partnership recognizes the importance of drawing on the local knowledge and capacity of civil society organizations.

Innovative, comprehensive and holistic responses to those affected by disaster need active collaborative effort by all parties including private parties to provide improvement that will lead to changes in humanitarian landscape in the future. ASEAN can also strengthen cooperation with the existing collaborative partners and donors include: The Asia–Pacific Economic Cooperation (APEC) Disaster Risk Reduction Framework, whose objective is to contribute to adaptive and disaster-resilient Asia–Pacific economies that can support inclusive and sustainable development in the face of disasters in the region (APEC 2024). In addition, as a part of efforts to promote partnership in the region the ASEAN Coordinating Centre for Humanitarian Assistance on Disaster Management (AHA Center) and the Japan International Cooperation Agency (JICA) to promote cooperation on disaster management in the region signed the Memorandum of Intent (MoI) (JICA 2023). In addition, in 2024, the AHA Center has renewed the MOU with Direct Relief. Direct Relief is a nonprofit humanitarian organization providing donations of the appropriate medical resources to poor people affected by disasters (Direct Relief 2024). These collaboration and activities between AHA center and Direct Relief that have been built in the past include: providing medical supplies donation at ASEAN disaster supply warehouses in Malaysia and the Philippines, joint participation in emergency response exercises and policy dialogues in the region, and funding in information sharing, data analysis, mapping, and collaboration during large-scale emergency response operations in Indonesia, Philippines, Cambodia, Myanmar, and Laos initiatives led by AHA-Center (Direct Relief 2024). To increase the capacities of DRM of ASEAN, the AHA center also collaborates with ASEAN Development Bank (ADB) by introducing high-level technology for regional cooperation on transboundary risks, and (iii) advancing capacity building and knowledge exchange for improving interregional cooperation (ADB 2024).

ASEAN began to extend its approach to disaster management and risk reduction by calling for more private-sector engagement in disaster risk financing and reduction AADMER's implementation through non-state actors in civil society and the private sector has implications for ASEAN more broadly, particularly it is in line with the ASEAN Charter's aim of making ASEAN a more people-oriented and participatory community. More generally, it also has implications for the emerging international law of disasters. To strengthen vulnerable community and engaging to more local resilience the ASEAN has a partnership with the International Federation of Red Cross and Red Crescent Societies (IFRC). In addition, A "whole of society approach" to disaster resilience in ASEAN has been introduced in the inaugural ASEAN Disaster Resilience Forum in Singapore, in 2023. This forum has highlighted the pioneering innovative disaster risk reduction and management strategies and approaches. This includes public–private partnerships with industries and finding the right private sector partners (or "connoisseurs") who can help better understand how the business community works in disaster management.

13.4.2 Technology and Innovation

There are opportunities and challenges in technology and innovation in the region regarding tsunami risk reduction. High level of digitalization in the ASEAN region is beneficial for communication risk information and early warnings (ASEAN 2024). There are existing regional cooperation on early warning system especially with ASEAN Specialised Meterological Center, AHA Center, and Disaster Monitoring and Response system (DMRS). Disaster Monitoring and Response System (DMRS) of ASEAN is one of the disaster monitoring tools utilized by the AHA Centre. The tool was designed in partnership with the Pacific Disaster Center (PDC), an applied science and information centre based in Hawaii, with the support of the Government of the United States of America. However, there are areas that need improvement from this EWS include indirect communication and information sharing mechanisms between countries, to improve risk analysis and EWS for cross-border hazards (ASEAN 2024). The further barriers for this technologies include: cultural and behavioral norms at local level and highlight the need to inform and strengthen effective risk communication for local EWS operators. For example, in Palu earthquake and tsunami based on research findings from UNDRR and UNESCO-IOC stated that it is important to building on community and indigenous knowledge of disaster and not becoming too dependent on technology. (UNDRR and UNESCO-IOC 2019). Further, ASEAN needs to consider the value of space technology for Humanitarian Assistant and Disaster Relief (HADR) with the climate change exacerbates the impacts of disasters AMSs in the need to coordinate efforts becomes more evident (Landicho 2024). With this pace technology state can make damage assessments to assess severity and prioritize response efforts.

13.5 Ways Forward

ASEAN through AADMER seeks to build humanitarian capacity which is often referred to as 'nationally led, regionally supported and international if necessary' so that it can lead responses to natural disasters (Cook 2021). At the ASEAN regional level there has been much progress in disaster risk management initiatives and action plans including tsunami risk reduction. Commitment to disaster risk reduction initiatives at the ASEAN regional level is very high. This is realized through agreements and declarations as well as action plans and an ASEAN approach to disasters with a commitment to cross-sectoral collaboration and an intra-regional collaboration network with relevant centers or institutions in ASEAN and beyond, in developing an early warning system and strategic foresight for disasters, disaster prevention, mitigation, preparedness, and increasing capability and capacity; provide rapid and effective responses to disasters, and undertake resilient recovery through building back better. However, challenges and limitations in reducing disaster risk in ASEAN still exist. There is a need to address and improve especially on three institutional aspects, financial and resource mobilization as well as partnerships and innovation.

At the institutional level in ASEAN, there is a need to strengthen the institutional capacity of the AHA center as a focal point for coordinating disaster management at the regional level, especially to facilitate increased skills, knowledge, and expertise in delivering and facilitating learning in the region. Additionally, align with localize approach, the AHA Center needs to assist in strengthening and networking at the local level, not just at the regional level and global. Align with localize approach, AHA center has facilitated a coordination of construction at local level through the establishment of an ASEAN Resilient village in Central Sulawesi after the tsunami with support from ASEAN Member States. In addition, in the future, the AHA Center as an ASEAN body can act as a center for disaster knowledge and information. Indonesia has a disaster preparedness school, and Malaysia also has one. This disaster preparedness school should be developed at the AHA Center and be able to become a center for cross-country knowledge. ASEAN leaders have launched "One ASEAN, One Response" in disaster management. In addition, there is already an agreement on disaster management and emergency response agreed by ASEAN countries to strengthen the AHA Center.

From the point of view of **financial and resource mobilization**, there are challenges in terms of financing and resource mobilization. The governments of ASEAN countries usually do not have sufficient funds to deal with major disasters. Therefore, in the future, there is a need to develop national disaster risk financing and insurance plans in the ASEAN region to improve the management of disaster-related fiscal burdens and the instability of inter-annual disaster spending needs. ASEAN also needs to develop a Trust Fund for Tsunami, Disaster and Climate Preparedness. These trust funds can be used to promote innovative pilot initiatives, increase the success of early warning systems, and facilitate regional cooperation, directly benefiting ASEAN member countries.

In relation to partnership and innovation, ASEAN needs to strengthen disaster cooperation and partnerships from regional to national and local community levels. The innovative partnership includes Penta Helix model which is based on five stakeholder types: businesses, public administration, local residents, the knowledge sectors and capital. ASEAN Village Network (AVN) as an innovative strategy probably can be developed in building tsunami awareness village through partnership and collaborative network facilitated by AHA Center. Technology for tsunami early warning system is already developed. However, the barriers still exist at local levels. Therefore, increasing traditional knowledge and awareness of local people to knowledge of disaster is really important. In addition, advancing space technology could help during and post-disaster relief.

13.6 Conclusion and Recommendation

ASEAN disaster management and humanitarian landscape have been robustly evolving since the tsunami disaster struck the ASEAN region in 2004. This is shown by the adoption of a legally binding AADMER and other declarations, initiatives, and action plans to implement and synergize with AADMER. Despite the limitations, ASEAN as a regional cooperation plays a pivotal role in disaster management and emergency response in the region. Recognizing the vulnerability of the ASEAN region to disaster, the momentum of responses to tsunami events and disasters provides key lessons learned and progress toward strengthening cooperation and collective action to respond effectively to disasters in the region to achieve a Resilient, Inclusive, People-Oriented, and People-Centered ASEAN.

Three aspects need to be strengthened in the ASEAN region in improving prevention, preparedness, and mitigating the disaster including institutional, finance, and resource mobilization and partnership and innovations. **Firstly**, Institutionally AHA centers play a crucial role in facilitating coordination amongst ASEAN Member States in disaster risk reduction. It calls for strengthening the capacity of the AHA Center and it calls for a re-conceptualization of how the region operates across different sectors and between different layers of governance to produce a robust synergy of whole-of-government layers and localized approach oriented toward disaster risk reduction. **Secondly**, Financial and Resources Mobilization is a crucial part of disaster risk reduction. It is critical for the Post-2015 AADMER Work Program to explore sustainable and innovative ways to transform the way it funds and mobilizes resources for disaster management and emergency response. Through collaboration with the private, public, and people sectors in new and creative ways the region will be well-placed to have attained financial sustainability and mobilize the necessary resources by 2025. **Thirdly**, ASEAN should look into forming stronger traditional and new non-traditional partnerships including public–private partnerships and civil society and local community engagement as such relationships are imperative for comprehensive, innovative, and robust disaster management and emergency response implementation in the region and beyond.

Finally, two decades after the tsunami and the Indian Ocean ASEAN should utilize this time for a momentum to reflect and move forward toward strengthening disaster management in the region, building more robust cooperation amongst ASEAN member countries. The tsunami acted as a wake-up call and understand how vulnerable people are to hazards. Learning from the tsunami that occurred in the Southeast Asia region shows that we need to build a more sustainable and responsible plan for our coastal urban and enhance understanding of risk amongst local government and coastal communities. In addition, aligning with Sendai Framework for Disaster Risk Reduction and Sustainable Development Goals (SDGs) strengthening institutional framework and partnership in Disaster Risk Reduction is very crucial for improving effective DRM (Djalante et al. 2012). Emerging pressing issues that are likely to challenge future resilience-building activities include the integration of DRR and CCA and urban risk governance, to build long-term resilience in the ASEAN region and globally.

References

ADB (2024) Strengthening cooperation on disaster risk management within the Association of Southeast Asian Nations. Manila, Philippines. https://www.adb.org/projects/55162-001/main

ADMM ASEAN (2024) About the ASEAN defence ministers' meeting. admm.asean.org. https://admm.asean.org/index.php/about-admm/about-admm.html

Affairs, Minister of Foreign (2020) Regulation of the Minister of Foreign Affairs Number 20 of 2020 Concerning the Formation of a Team and Work Procedures of the Ministry of Foreign Affairs for Handling Foreigners and International Assistance in Emergency Disasters in Indonesia. https://peraturan.bpk.go.id/Details/163045/permenlu-no-20-tahun-2020

Agency, Fiscal Policy (2024) SEADRIF. https://Fiskal.Kemenkeu.Go.Id/. https://fiskal.kemenkeu.go.id/strategi-drfi/seadrif.

AHA Centre (2012) About the AHA centre. Ahacentre.Org. https://ahacentre.org/about-us/

AHA Center (2021a) Disaster by numbers 2012–2020, data based on ADINET record

AHA Centre (2021b) Press Release, ASEAN stands in solidarity with the typhoon-affected people in the Philippines. Ahacentre.Org. https://ahacentre.org/press-release/typhoon-rai/

AHA Centre (2022) Rise, recover, reunite. Ahacenter.Org. Jakarta

APEC (2024) APEC disaster risk reduction framework. https://www.apec.org/docs/default-source/groups/epwg/2024/apecdisasterriskreductionframework_endorsed.pdf?sfvrsn=8d9ce067_2

ASEAN (2013) ASEAN Vision 2025 on disaster management. J Chem Inf Model 53(9):1689–1699

ASEAN Disaster Risk Communication Framework (2024) https://asean.org/wp-content/uploads/2024/12/ASEAN-Disaster-Risk-Communication-Framework_ACDM-endorsed_17-Sept-2024.pdf

ASEAN (2025) ASEAN vision 2025 on disaster management. https://asean.org/wp-content/uploads/2021/01/fa-220416_DM2025_email.pdf

ASEAN Secretariat (2020) Joint statement of the 6th ASEAN Finance Ministers, and, Central Bank Governors' Meeting (AFMGM). Video Conference Meeting, no. May: 1–6

ASEAN Secretariat (2024) Strengthening ASEAN multi-hazard end to end early warning system for naturak disasters an assessment of current capacity. https://asean.org/wp-content/uploads/2024/04/ASEAN_E2E_EWS_Report_Final.pdf

ASEAN Framework on Anticipatory Action in Disaster Management Jakarta, ASEAN Secretariat

Bankoff G, Hilhorst D (2009) The politics of risk in the Philippines: Comparing state and NGO perceptions of disaster management. Disasters 33(4):686–704. https://doi.org/10.1111/j.1467-7717.2009.01104.x

Chainikom P (2018) ASEAN: The invisible hand in disaster management1. Max Planck Encycl Public Int Law 2(2):14–23. https://papers.ssrn.com/abstract=3304535

Cook ADB (2021) Humanitarian diplomacy in ASEAN. Asian J Com Pol 6(3). https://doi.org/10.1177/20578911211045668

Cook ABD, Gong L, Nanthini S, Keith Paolo Catibog Landicho (2023) ASEAN Strategic Policy Dialogue in Disaster Management (SPDDM). https://reliefweb.int/report/world/asean-strategic-policy-dialogue-disaster-management-spddm-2023?gad_source=1&gclid=CjwKCAjw65-zBhBkEiwAjrqRMH162CLXIDO

Desdiani NA, Afifi FA, Cesarina A, Sabrina S, Husna M, Violeta RM, Adinegoro A, Halimatussadiah A (2021) Climate and environmental financing at regional level: Amplifying and seizing the opportunities, vol 067. LPEM Working Paper. Depok, Indonesia. https://www.lpem.org/wp-content/uploads/2021/12/WP-LPEM-067_Climate_and_Environmental_Financing_at_Regional_Level.pdf

Direct Relief (2024) Direct relief renews commitment to ASEAN Region, Building on strength of past partnership, directrelief.org. https://www.directrelief.org/2024/05/direct-relief-renews-commitment-to-asean-region-building-on-strength-of-past-partnership/

Djalante R, Thomalla F, Sinapoy MS, Carnegie M (2012) Building resilience to natural hazard in Indonesia: Progress challenges in implementing the hyogo framework for action. Nat Hazards:779–803

Feybe M (2016) BNPB: ASEAN countries are at very high risk of disaster. News One. Accessed from https://www.beritasatu.com/nasional/392135-bnpb-asean-country-at-risk-of-disaster-which-has-a-high-budget

Pennisi di Floristella AM (2016) Dealing with natural disasters: Risk society and ASEAN: A new approach to disaster management. Pacific Rev 29(2):283–305. https://doi.org/10.1080/09512748.2015.1013498

Gong, Lina (2021a) China's disaster diplomacy in Southeast Asia. China Rev 21(4):77–107

Gong SN, Lina (2021b) Financing ASEAN disaster management and resilience. https://Eastasiaforum.Org/. https://eastasiaforum.org/2021/11/04/financing-asean-disaster-management-and-resilience/

Gentner HH (2006) ASEAN: Cooperative disaster relief after the tsunami. Sudostasien Aktuell: J Curr Southeast Asian Aff, 24(4):3–9. https://nbn-resolving.org/urn:nbn:de:0168-ssoar-339208

Ghesquiere F, Mahul O (2010) Financial protection of the state against natural disasters: A primer (September 1, 2010). World Bank Policy Research Working Paper No. 5429, Available at SSRN: https://ssrn.com/abstract=1680722

Henny W (2010) Mapping social movements: civil society in Nanggroe Aceh Darussalam (NAD) to demand social, cultural, economic and political rights. LIPI Press, Jakarta

Hughes MSD (2015) ASEAN's role in regional natural disaster response. Thesis. https://www.rcrc-resilience-southeastasia.org/wp-content/uploads/2018/01/2015-Chulalongkorn-University-ASEANs-Role-in-Regional-Natural-Disasters.pdf

JAIF (2023) MoJ, JMT visit the disaster emergency logistics system for ASEAN (DELSA) satellite warehouse in Thailand. https://Jaif.Asean.Org/. https://jaif.asean.org/jmt-news/moj-jmt-visit-the-disaster-emergency-logistics-system-for-asean-delsa-satellite-warehouse-in-thailand/

JICA (2023) JICA Signs Memorandum of Intent with The ASEAN Coordinating Centre for Humanitarian Assistance on disaster management. https://www.jica.go.jp/english/overseas/indonesia/information/press/2023/20231017.html

Lai AY, He JA, Tan TB, Phua KH (2009) A proposed ASEAN disaster response, training and logistic centre enhancing regional governance in disaster management. Trans Stud Rev 16(2):299–315. https://doi.org/10.1007/s11300-009-0068-z

Landhicho K (2024) Space technology for humanitarian assistance and disaster relief in ASEAN, IDSS Paper, 1

Mizan BFB, Fathia L (2022) Common Resilience Challenges in ASEAN and Potential Areas of Collaboration in Disaster Management, ASEAN Socio Cultural Community Policy Brief. https://asean.org/wp-content/uploads/2023/03/20230321_Policy-Brief_Disaster-Management-2.pdf

Pennisi di Floristella A (2015) Building the ASEAN Center for Humanitarian Assistance and Emergency Response: is ASEAN learning from the experience of the European Civil Protection Mechanism? (KFG Working Paper Series, 62). Berlin: Freie Universität Berlin, FB Politik- und Sozialwissenschaften, Otto-Suhr-Institut für Politikwissenschaft KollegForschergruppe "The Transformative Power of Europe". https://nbn-resolving.org/urn:nbn:de:0168-ssoar-413134

Sari S (2022) ASEAN mechanism in disaster management: A case study of disaster management in Indonesia 2010–2020. J Glob & Strat 16(1):75–108. https://doi.org/10.20473/jgs.16.1.2022.75-108

Syifa M (2021) "Press Release, ASEAN Stands in Solidarity with the Typhoon-Affected People in the Philippines." Ahacentre.Org, 2021. https://ahacentre.org/pressrelease/typhoon-rai/

Tolentino Jr AS (2007) The challenges of tsunami disaster response planning and management. Int Rev Environ Strat 7(1):147–154. http://search.ebscohost.com/login.aspx?direct=true&db=aph&AN=24779403&custlogin.asp?custid=s7433796&site=ehost-live

UNDRR (2015) Sendai framework for disaster risk reduction 2015–2030. United Nations Office for Disaster Risk Reduction. https://www.undrr.org/publication/sendai-framework-disaster-risk-reduction-2015-2030

UNDRR (2019) Limitations and challenges of early warning systems: A case study of the 2018 Palu-Donggala tsunami. United Nations Office for Disaster Risk Reduction (UNDRR), Regional Office for Asia and the Pacific, and the Intergovernmental Oceanographic Commission of United Nations Educational, Scientific and Cultural Organization. IOC Technical Series N° 150

Warner K, Ranger N, Surminski S, Arnold M, Linnerooth-Bayer J, Michel-Kerjan E, Kovacs P, Herweijer C (2009) Adaptation to climate change: Linking disaster risk reduction and insurance. Unisdr, 18. http://www.microinsuranceconference2005.com/dms/MRS/Documents/MIC_Agriculture_Bibliography/9654_linkingdrrinsurance.pdf%5Cn. http://www.unisdr.org/we/inform/publications/9654

Weingärtner L, Simonet C, Caravani A (2017) Disaster risk insurance and the triple dividend of resilience. Overseas Development Institute (ODI)

Chapter 14
Urban Planning in Tsunami–Prone Areas: A Comparative Study on Integration of Urban Planning Strategies in Tsunami Preparedness Planning

U. T. G. Perera, Richard Haigh, Dilanthi Amaratunga, and H. Rahayu

Abstract Coastal cities are often subjected to extreme natural hazards such as sea level rise, coastal storms, and heavy rainfall. Tsunamis, despite their low frequency, continue to be the most devastating disaster for coastal communities due to their unpredictable nature and the high impact of a single hit. Indonesia, located on the Ring of Fire and at the intersection of three tectonic plates (Indo-Australian, Eurasian, and Pacific), is one of the world's most seismically and tsunami-prone countries. According to data that have been compiled, 105 tsunamis occurred in Indonesia alone, between 1600 and 1999. The 2004 Indian Ocean tsunami devastated Indonesia, prompting the country to revise its urban planning frameworks to include more tsunami preparedness measures. Nevertheless, Indonesia's urban planning guidelines lack the preparedness measures to implement in tsunami-prone cities. This further highlights the necessity for further sustainable and resilient urban planning. This study synthesizes and critically analyses the current level of integration of strategies for tsunami preparedness in Indonesian urban planning and related policies. This study adopts a multiple case-study approach to examine urban planning strategies for tsunami risk reduction in Indonesia. Padang City and Pangandaran Regency were selected as case studies for the study. Three qualitative research methods were employed, including analysing up-to-date local and regional planning documents, semi-structured interviews with experts, and field observation in both regions. The data analysis of documents and expert interviews employed theme

U. T. G. Perera (✉)
Department of Geography, Faculty of Arts, University of Colombo, Colombo, Sri Lanka
e-mail: Thisara.Uduwarage@hud.ac.uk

R. Haigh · D. Amaratunga
Global Disaster Resilience Centre, University of Huddersfield, Queensgate, Huddersfield HD1 3DH, UK

H. Rahayu
The School of Architecture, Planning and Policy Development, ITB, Bandung, Indonesia

R. Shaw et al. (eds.), *Two Decades from the Indian Ocean Tsunami*, Disaster Risk Reduction, https://doi.org/10.1007/978-981-96-2669-4_14

identifications, including spatial planning, community and stakeholder participation, evacuation planning, and soft and hard engineering. According to the findings, in both Padang City and Pangandaran Regency, community involvement is remarkably high, and several types of communities (Business community in Padang and tourist community in Pangandaran) are involved in DRR activities. Spatial planning in tsunami-prone areas has also considered the protection and restoration of natural resources, with the development of a coastal protection system with vegetation and conservation of sandbanks and coral reefs. However, the field visits highlighted that the maintenance of preparedness infrastructure needs to be improved, and tsunami evacuation shelters, signage, and soft and hard engineering structures should be well maintained with the support of disaster-related agencies and the local community. This detailed examination of current urban planning approaches in Indonesia provides valuable insights into the range of overall disaster risk reduction strategies available. Identified challenges associated with their implementation in policy and practice will be helpful for urban planners in better preparing the country for similar future hazards, Tsunamis in particular. A clear maintenance strategy and a set of urban policies are urgently required in Indonesia at the national and local levels to reduce its vulnerability to tsunamis.

Keywords Indonesia · Mitigation · Preparedness · Tsunami · Urban planning

14.1 Introduction

14.1.1 Coastal Cities and Disaster Risk

'In 2018, the UN estimated that 55% of the world's population was already living in cities and that this figure will rise to 68% by 2050 (United Nations 2017). Although precise figures vary due to differing definitions of urban areas, there is widespread agreement that the proportion of the population living in cities, as well as the number and size of cities, will continue to grow in the decades ahead. This urban expansion will result in a concentration of economic activity as well as contribute to the development of an increasingly complex, interconnected urban system.

Cities located in coastal areas tend to be more exposed to natural hazards than other regions (Gu 2019). Rapid-onset hazards such as major cyclones with high winds, waves, and surges, or tsunamis are examples of specific threats. Previous surge and flood events have already taken a heavy toll in terms of lives lost and economic damage. Every year, more than 10 million people, the majority of whom live in Asia, are affected by flooding caused by storm surges (Rentschler et al. 2022). Recent major coastal floods include the 2022 flood in Pakistan, which killed 1,200 people and displaced 33 million, and the record-breaking floods on June 2022 in Bangladesh, where 7.2 million people have been affected. Moreover, the ASEAN Disaster Information Network (ADINet) has reported more than 200 flood events in

Indonesia during the year 2022, including in Tapin and Samarinda Regencies, which affected more than 14,000 people and over 1,950 households.

Furthermore, (Fuchs 2010) notes that coastal cities face an increasing risk of future catastrophic loss as a result of climate change and rapid urbanization in low elevation coastal zones, especially in Asia. Global warming is causing sea levels to rise due to thermal expansion of ocean water and melting glaciers and ice sheets, as well as an increase in the intensity of tropical cyclones, bringing higher winds, heavier precipitation, and stronger storm surges, coastal flooding, and tsunamis as well. Such concerns were reaffirmed by the IPCC in a Special Report during the Sixth Assessment Cycle, which noted that human communities in close connection with coastal environments and small islands are particularly exposed to ocean and cryosphere change, such as sea level rise, extreme sea level, and extreme weather events (Oppenheimer et al. 2019).

14.1.2 Tsunami Events and Its Impact as a Disaster

Although earthquakes are usually the cause of tsunamis, other events that can set them off include volcanic eruptions, underwater landslides, and onshore landslides that send large amounts of debris into the ocean. Tsunamis can cause catastrophic property damage and fatalities in coastal cities, despite being less frequent than most other natural disasters. The International Tsunami Information Centre (2023) reported that over the last twenty years, deadly tsunamis have occurred in Peru (2001), Chile (2007, 2010), Japan (2011), Indonesia (2004, 2005, 2006, 2010, 2018), and Samoa—American Samoa - Tonga (2009, 2022).

In terms of economic losses, The Great East Japan Earthquake and Tsunami (The 2011 Tohoku Earthquake) was one of the most destructive natural hazards in recorded history. It caused the sea water to reach more than 5 kms inland in some coastal locations, reaching a height of 19.5 m above sea level. The tsunami overwhelmed protective seawalls, and the death toll eventually reached 15,641, with 5,007 people reported missing (Mori et al. 2011). The direct economic losses were estimated by the government of Japan as 16.9 trillion yen (US$211 billion) or 3.5 per cent of GDP, which was the costliest natural disaster on record (Kajitani et al. 2013). The most recent deadly tsunami was generated due to the volcanic eruption on Hunga Tonga—Hunga Ha'apai (Tonga) on January 15, 2022, and produced the strongest atmospheric blast ever recorded on modern instruments. This tsunami event affected the entire Pacific Ocean and was recorded in parts of the Atlantic Ocean and Mediterranean Sea too. Although such devastation has been rare, recent research suggests that even minor sea-level rise will increase the risk of destructive tsunamis for coastal communities (Li et al. 2018). Increased urbanization in coastal areas will also expose more people and assets to the threat posed by tsunamis.

14.1.3 Tsunami Experiences in Indonesia

Located on the Ring of Fire and at the meeting points of three tectonic plates—Indo-Australian, Eurasian, and Pacific—Indonesia is one of the most seismic and tsunami-prone countries in the world. Based on data compiled during the period, 105 tsunamis have occurred in Indonesia between 1600 and 1999. Ninety-five events (90%) of them were caused by earthquakes in a shallow region at subduction and plate boundaries, nine (8%) by volcanic eruption, and one (1%) by a landslide (Hamzah et al. 2000). Studies have identified six tsunami-prone zones in Indonesia. These include Irian Jaya, Makassar, Molucca, West and East Sunda, and Banda. In Banda and Molucca, the proportion of earthquakes with tsunamis to all earthquakes exceeds 50%. West Sunda is the worst area for tsunami-related fatalities due to the severe damage caused by the 1883 Karakatau volcanic eruption. Banda and Molucca have a very high potential for both frequent and destructive tsunamis. At Banda, the frequency of tsunamis has been particularly high since 1960. This was clearly evident in the 2004 Indian Ocean Tsunami, where 128,645 people in Aceh lost their lives, 37,063 were missing and 532,898 were displaced (Ozer and Longueville 2011).

In late 2018, Indonesia was hit by two further destructive tsunamis, which as reported by IOC (2019), challenged the traditional understanding of tsunami hazard, warning, and response mechanisms. The first event was the Palu and Donggala tsunami of 28 September 2018, which killed 4,340 people and followed a magnitude 7.5 earthquake in Central Sulawesi. The second was the Sunda Strait tsunami of 22 December 2018, which killed 426 people following an eruption and partial collapse of the Anak Krakatau volcano (Disasters Emergency Committee 2018).

These are examples of 'atypical' tsunami events, as they originate from other sources besides thrust earthquakes occurring on the subduction interfaces (Srinivasa Kumar and Manneela 2021). Cascade effects and multi-hazard threats are also linked to such events. According to Goda et al. (2019), the 2018 Sulawesi earthquake's cascading effects were extremely complex and varied greatly in space, which led to the creation of secondary hazards like liquefaction, submarine landslides, tsunamis, and massive mudflows in coastal areas. Imamura et al. (2019) adds that such cascading effects may accelerate over time, resulting in unexpected secondary events with significant consequences across Indonesia's coastal regions. Due to short response times, technological limitations, and currently available scientific knowledge, previous disasters highlight the complexities that tsunami early warning and preparedness efforts must overcome. Recently, recommendations from the Lessons Learnt Symposium (UNESCO IOC 2020) emphasized the importance of addressing a lack of proper evacuation plans and related infrastructure during a tsunami emergency, as well as poorly implemented and ineffective spatial planning and policy issues. Kumar and Manneela also note that "most of the current Tsunami Early Warning Systems are based on seismic origin earthquakes occurring at subduction zones. However, the other 'atypical' sources such as events related to submarine landslides and volcano eruptions are yet to be addressed globally."

The urban environment can be strained and exposed to these cascading effects by weak infrastructure, poorly planned urban environments, and ineffective urban governance. Rather than the danger, they have to do with how vulnerable coastal areas are. As Rus (2018) notes, safety precautions for the residents of these cities are essential when confronted with such hazards because these cities must be maintained and upheld. Because of this, research has shown that urban planning is essential to enhancing the resilience of urban coastal communities against rising risks and the likelihood of disasters (Perera et al. 2022).

14.1.4 Urban Planning and Disaster Risk

Preparation of plans for, as well as the regulation and management of, towns, cities, and metropolitan regions, falls under the purview of urban planning. It seeks to organize socio-spatial relations at various levels of government and governance, and it is concerned with the social, economic, and environmental consequences of delineating spatial boundaries and influencing spatial resource distributions (Huxley and Inch 2020). Also, it focuses on attempting to improve the plans, functions, and management of cities and areas, which have a clear role to play in disaster mitigation (Taheri Tafti 2017). With its grounding, at least in most developed countries, in legislative, agency, governmental, and professional fundamentals, it has established legitimacy as a base for action (Alan and Jorge 2013).

International and national policies emphasize the importance of disaster risk reduction and resilience enhancement in urban and regional planning. According to global agreements such as Sendai Framework for Disaster Risk Reduction (2015), and the New Urban Agenda (2016), urban and regional planning should be used to aid in disaster risk reduction, and local communities should be involved in these planning processes (UNISDR 2005, 2015; United Nations 2017).

Many urban planning instruments can be used to control land use to mitigate tsunami impact. These include relocation, the establishment of non-construction zones, land use restrictions, and tsunami-safe construction standards, as well as the protection of natural buffers such as forests, compacted dunes, and wetlands (Perera et al. 2022; Dissanayake et al. 2022). Urban planning can also enhance urban resilience through instituting statutory and non-statutory policies on evacuation, mandatory insurance, emergency, and social organizations (Herrmann-Lunecke and Villagra, 2020). Numerous previous studies found that in many countries, disaster risk has not been adequately considered as part of urban or regional planning (see for example, Herrmann Lunecke 2015; Villagra et al. 2017). Nor has the regulation of urban growth been effective with respect to risk management or resilience (Perera et al. 2022).

This was also evident in Indonesia, where many government agencies design their own sectoral plans without adequate synchronization between spatial plans and development plans at the national and local levels (World Bank City Planning Labs and CAPSUS 2020). Tauhid also revealed the lack of an adequate urban planning mechanism for tsunami events despite the country being highly exposed to and having experienced several highly destructive tsunamis in recent history. This was further discussed by Suppasri et al. (2021) where large tsunamis tend to be associated with earthquakes, liquefaction, and landslides that multiply the scale of impact in coastal regions in the country.

Despite recognition of the potential for urban planning to be used to reduce disaster risk in tsunami-prone areas, there has been a lack of detailed studies at the local level that concern the concurrent hazard risk along with tsunamis in Indonesia. In addressing this gap, this study explores the urban planning response to tsunami and other hazard risks in Indonesia at the national and local scale. Specifically, this study examines the extent to which national and local governments have developed and implemented urban policies and risk zone planning to address tsunami and other concurrent hazard risk in coastal cities in Indonesia.

14.2 Materials and Methods

This study addresses two main research questions:

1. What are the existing strategies and approaches for reducing disaster risk through urban planning in tsunami-prone areas of Indonesia?
2. What are the main gaps and limitations in those strategies?

In urban planning studies, case study methodology has often been employed as an approach to examine complex and under-explored topics (for example, see Baró et al. 2016). More specifically, there are several examples where a case study has been used to explore the complex phenomenon of urban planning and tsunami hazard vulnerability using different perspectives (Brune et al. 2010; Okumura et al. 2021; Rittichainuwat 2006). However, most of these studies have carried out tsunami risk and vulnerability assessments, rather than seek to understand the extent to which tsunami risk has been addressed through national and local urban planning efforts.

Therefore, this paper adopts a multiple case-study approach to examine urban planning strategies for tsunami risk reduction in Indonesia. Multiple-case studies are generally regarded as more robust than single-case studies, providing the observation and analysis of a phenomenon in several settings. This strategy allows the comparison of different cases selected to obtain equivalent results (literal replication) and allows for the exploration and confirmation of the strategies identified in the initial cases. In this study, two of the most impacted and vulnerable Indonesian regions to tsunami, Pangandaran and Padang, have been selected as the case studies for this research.

Padang City, with 913,450 inhabitants by the year 2021 (Padang City Statistic 2022), is one of the largest cities in West Sumatra. As it is on low coastal plains, its vulnerability to tsunami arrival is high. After the huge tsunami event in 2004 in the Aceh Province (northern part of Sumatra Island), a series of large earthquakes occurred toward the southern part of the Sunda Trench. There were two major earthquakes in 2007 and 2009, which damaged half of the city. Padang City, located in the center of the west coast of Sumatra Island, then became the focus of international concern due to its exposure to an existing seismic gap created by these recent earthquakes. This gap could generate an earthquake with a magnitude of Mw8.8 to Mw8.9 (Brune et al. 2010).

Pangandaran Regency was highly impacted by the 2006 tsunami event, which hit the southern coast of the Pangandaran Regency, West Java, Indonesia. This tsunami disaster caused 600 deaths, with 65 missing, 9,299 receiving treatments, and more than 75,000 people being displaced (Mardiatno et al., 2020). A recent study by Supendi et al. also found the potential for south Java megathrust earthquakes and tsunamis off the southern coast of West Java and Southeast Sumatera and predicted the potential for a 34 m maximum tsunami wave height. Earlier, Chlieh et al. also found an elevated probability of a large earthquake in this region due to the long-period strain accumulation by plate coupling and Coulomb stress transfer after the 2004 and 2005 Sumatran.

To analyse and compare the urban planning response to tsunami risk in these two regions, this research uses three qualitative research methods:

(1) An analysis of up-to-date local and regional planning documents;
(2) The application of semi-structured interviews with experts during 2021; and
(3) Field observation in both regions.

The analysis method followed an approach previously practiced in academic literature for reviewing similar urban planning strategies in Sri Lanka (Perera et al. 2022). Data analysis of documents and expert interviews including theme identifications developed using that study's conceptual framework. These themes included spatial planning, community and stakeholder participation, evacuation planning, and soft and hard engineering.

Using the above themes, the first step in this research was to analyse the content of national level and local level planning instruments in Padang city and Pangandaran city (see Table 14.1). A content analysis and detailed comparison of tsunami mitigation policies proposed in the country during the review period reveals the extent to which local planning documents included regulations, risk zone planning, and evacuation strategies to improve tsunami resilience in coastal cities. Table 14.1 contains a detailed list of reviewed documents at the national, provincial, and city levels in Indonesia.

The second step entailed conducting semi-structured interviews with experts to collect deep data and a contextualized understanding of:

(1) Tsunami warnings and evacuations preparedness;
(2) Spatial planning, relocation, and other tsunami mitigation policies proposed and implemented during each reconstruction phase (Padang with earthquake impact and Pangandaran with tsunami impact); and
(3) Community involvement in the planning process.

During October 2021 in Pangandaran and between September and November 2021 in Padang City, the authors carried out a total of 21 semi-structured interviews, six in Padang and 15 in Pangandaran Regency (Table 14.2). These were with local

Table 14.1 List of reviewed documents

Scale of the document	Type of document	Pangandaran		Padang	
		West Java province	Pangandaran regency	West Sumatera province	Padang city
National level	National plan	West Java Spatial plan	Pangandara Regency spatial plan	West Sumatera spatial plan	Padang city spatial plan
	Mid-term development plan	West Java mid-term development plan	Pangandaran regency mid-term development plan	West Sumatera Mid-Term Development Plan	Padang city mid-term development plan
Regional level/Sub district level	Long term development plan	West Java long term development plan	Pangandara regency detailed spatial (Sub-district level)	West Sumatera long term development plan	Padang city long term development plan
			Pangandaran regency long term development plan		
Micro level	Zoning plans	Coastal area and small Islands zoning plan	–	Coastal area and small Islands zoning plan	–
	Strategic plan	–	Pangandaran regency strategic plan	–	Padang city strategic plan
	Performance accountability report	–	Pangandaran regency performance accountability report	–	Padang city performance accountability report

agencies related to disaster management, public works, development, and spatial planning, environment, transportation, tourism agency, marine and fisheries, social, and the task force established in response to COVID-19, which were identified by the Indonesian National Board for Disaster Management as the leading actors in both regions for disaster management efforts (INBD 2015).

In the third step of the research, field observations were carried out in Padang City and Pangandaran Regency. The routes were selected based on the tsunami inundation maps, starting with the most populous areas (Figs. 14.1 and 14.2).

Researchers from the Bandung Institute of Technology (ITB), with support from the respective local government, observed the vulnerability and capacity indicators

Table 14.2 Interview respondent expertise

No	Institution	Position
Pd1	Badan Perencanaan Pembangunan Daerah—Regional Development Planning Office	Head of Infrastructure and Regional Division
Pd2	Badan Penanggulangan Bencana Daerah—Local Disaster Management Office	Head of LDMO
Pd3	Dinas Sosial—Social Office	Secretary
Pd4	Dinas Lingkungan Hidup—Environmental Office	Head of Supervision and Control Section
Pd5	Dinas Pekerjaan Umum dan Penataan Ruang—Department of Public Works and Spatial Planning	Head of Infrastructure Division
Pd6	Komunitas Siaga Tsunami (KOGAMI)—Tsunami Ready Community	Member
Pg1	Badan Penanggulangan Bencana Daerah—Local Disaster Management Office	Head of Prevention and Preparedness Division
Pg2	Badan Perencanaan Pembangunan Daerah—Regional Development Planning Office	Head of Infrastructure Division
Pg3	Dinas Pekerjaan Umum dan Penataan Ruang—Department of Public Works and Spatial Planning	Spatial Plan Division Staff
Pg4	Dinas Lingkungan Hidup—Environmental Office	Head of Division
Pg5	Dinas Perhubungan—Transportation Office	Head of General Affairs Division
Pg6	Dinas Kelautan, perikanan, dan Ketahanan Pangan—Marine, Fishery and Food Security Office	Head of Agency
Pg7	Dinas Pariwisata dan Kebudayaan—Tourism and Cultural Office	Head of Tourism Destination Planning Division
Pg8	Satuan Tugas Penanggulangan dan Percepatan Penanganan COVID-19—Covid 19 Task Force	Member
Pg9	Perusahaan Daerah Air Minum—Municipal Waterworks	Director
Pg10	Perhimpunan Hotel dan Restoran Indonesia (PHRI)—Indonesian Hotels and Restaurants Association	Head of Organization
Pg11	Asosiasi Perusahaan Perjalanan Wisata Indonesia (ASITA)—Association of The Indonesian Tours and Travel Agencies	Head of Organization
Pg12	Himpunan Pramuwisata Indonesia (HPI)—Indonesian Tourist Guide Association	Member
Pg13	Forum Kewaspadaan Dini Masyarakat (FKDM)—Early Public Vigilance Forum	Head of Organization
Pg14	Taruna Siaga Bencana (Tagana)—Disaster Preparedness Group under Ministry of Social Affairs	Head of Organization
Pg15	Nelayan—Fisherman	

Fig. 14.1 Observation points and routes in Pangandaran regency

for all inundation areas. They also observed the evacuation route, evacuation shelters, obstacles, signings, and critical facilities along the route in both regions.

14.3 Results and Findings

14.3.1 Coverage of Tsunami and Other Hazard Threats in Urban Planning

The National Mid-Term Development Plan (2020–2024) of Indonesia has mapped disaster risk areas, both hydro-meteorological and geological hazards. The occurrence of disasters due to the use of space that was not in accordance with the spatial plan has increased (about 2,000 cases of floods, landslides, forest fires, and so on). It further highlighted that the strategies in dealing with tsunami-prone areas are done through several physical and non-physical programmes and regulations such as avoiing inundation areas, site Buildings or infrastructure away from hazard area, and early warning systems. There are considerations regarding non-natural hazards (man-made hazards), but not specifics to any type of hazard.

Fig. 14.2 Observation points and routes in Padang city

In the regional and micro level urban planning documents including the West Sumatera spatial plan, Padang city spatial plan, Pangandaran Regency spatial and short-term plans (RTRW and RPJMD) have accommodated the mapping of tsunami-prone areas, both in the spatial structure and patterns. The strategies in dealing with tsunami-prone areas are provided in the document through controlling the spatial development of the area, both physical and non-physical developments such as infrastructure, economics, environment, or social conditions, that could result in the change of the spatial plan planned. As a result, regulations regarding the coastal area developments and limitations (especially in the coast–front areas) have been made. The Coastal Area and Small Islands Zoning Plan document for West Java Province and West Sumatra Province has accommodated the potential of tsunami within the area, especially in Pangandaran Regency and Padang City. The efforts are strategized in the form of an obligation to implement disaster mitigation activities that could damage coastal areas and small islands.

Apart from the tsunamis, other hazard threats have also been considered in urban planning documents in both regions. This was further confirmed during the semi-structured interviews with the Local disaster management office (Pd2) and social office (Pd3) in Padang city. They confirmed heavy rains, floods, landslides, and coastal abrasion scenarios were considered during the region's spatial planning process. As stated by one of the officials in the Local disaster management office in Padang city (Pd2):

> In addition, Padang City is prone to climate change which causes heavy rains, falling trees, floods, and landslides. There is also coastal abrasion. Our planning agencies consider these threats in the planning process.

Same situation can be identified in Pangandaran Regency where eight natural hazards including tsunamis, landslides, and floods have been identified and considered during the planning process of the region. However, the interviews revealed that the multi-hazard contexts are not yet accommodated into the planning documents in both regions. As stated by an official in local disaster management unit in Pangandaran Regency (Pg1).

"These plans do not yet accommodate multi-hazard scenarios. For the Rencana Detil Tata Ruang (RDTR)—sub district detailed spatial plan, the mega thrust scenario is the tsunami scenario."

As part of the multi-hazard scenario, the current spatial plans in both regions have not addressed epidemic and pandemic threats. For example, those agencies involved with the COVID-19 task force had not been included in the spatial planning process for either region. However, the current spatial plan is scheduled to be revised in the upcoming years. This was confirmed by the head of the Local Disaster Management Office in Padang (Pd2). The Health Department's involvement in the spatial plan is to be reevaluated due to COVID-19:

> The Covid-19 program initiative is still limited to providing social assistance such as food supply and capital for Kamicro and small home enterprise. For Covid-19 in BPBD so far, it is only related to health protocols and funeral protocols.

The same situation was identified in the Pangandaran Regency. Pangandaran Regency does not yet have regulations related to efforts to suppress the growth rate of COVID-19 in spatial planning. As stated by the local disaster management office (Pg2):

> The pandemic has not been included in the previous spatial plan. Only tsunami is considered in the spatial plan. In the detailed spatial plan (RDTR), the earthquake and tsunami are included.

As further explained by the officials of social service office (Pd3): "the Covid-19 program initiative is still limited to providing social assistance such as food supply and capital for micro and small home enterprise."

However, the efforts of Pangandaran District Government in dealing with COVID-19 have succeeded in making Pangandaran Regency one of the regions in West Java Province with level 1 for positive cases of COVID-19.

It was evident in both regions that pandemic mitigation and preparedness have not been considered in detail as part of the urban planning process. This is due to the unavailability of comprehensive studies and directions related to mainstreaming COVID-19 into spatial planning plans. However, based on the results of the interviews with the Padang City Public Works and Spatial Planning Service, it was identified that it is possible to incorporate COVID-19 into spatial planning. The current urban planning documents were not revised during the COVID-19. However, the COVID-19 considerations can be integrated into the existing documents based on the conditions and severity of the pandemic.

14.3.2 Spatial Planning

As a reactive approach to past earthquakes in 2007 & 2009, building codes and structural requirements are being drafted as a part of a national strategy plan to ensure resilience against earthquakes in both regions. Infrastructure retrofitting has also started especially for critical infrastructure such as hospitals, hotels, schools, and government offices. In Padang, more than 20% of the area is classified as green open space. This was a compulsory requirement of the regulation of the Ministry of Spatial Plan where a minimum of 20% of the urban area should be open space or green space. Similarly protected areas have been identified in the plan where development is restricted within the zone. This was further confirmed through the interview process. As explained by an official from the Department of Public Works and Spatial Planning (Pd5):

> There is RDTR preparation that regulates protected and cultivated areas. The coastal border area, which is a tsunami-prone area, is included in the protected area where development is restricted.

Similarly, in Pangandaran Regency, both Pangandaran's spatial plan 1: 20,000–25,000 and the detailed spatial plan accommodate disaster-prone areas (including tsunami). These areas are mapped in 1:5.000 scale in the detailed spatial plan, and this affects the spatial structure plan, spatial pattern plan, and General Requirement Ordinance for zoning regulations (Ketentuan Umum Peraturan Zonasi), as guidance of which activities are allowed within the area. The local government of Pangandaran Regency is also involved in spatial planning forums where planning, monitoring, and evaluation are conducted through multi-stakeholder meetings (Musrenbang—Meeting for planning and Development) with community members, government officials, and civil society organizations. However, interviews revealed that such meetings were not held during COVID-19, and have only been reactivated in 2022, at the end of the pandemic.

Even though disaster resilience is one of the notions in establishing land use regulations in Pangandaran Regency, these documents are mainly focused on tsunami events, whereas other hazards, including multi-hazard scenarios, have not been addressed. As stated by an official in the local disaster management office (Pg1): "These documents do not yet accommodate multi-hazard scenarios. For the RDTR—sub district detailed spatial plan, the megathrust scenario is the tsunami scenario."

During the field observations in both Padang city and Pangandaran Regency, it was revealed that the government office was relocated away from the coastline. However, the coastal boundary area, which is also the local protection area, continues to allow construction of hotels, restaurants, and other tourism-related establishments in both regions.

It was revealed through the reviewed sources that tsunami risk has been extensively discussed in both regions. Land use regulations and infrastructure retrofitting have started to practice in both regions to address government regulations. Field visits in both regions identified building zones in both regions where zoning regulations were taking place. In the sources that have been reviewed, none of the multi-hazard scenarios have gotten much attention. This was further revealed through the expert interviews where the main priority was identified as tsunami risk mitigation, rather than other hazards in the region.

14.3.3 Community and Key Stakeholder Participation

In Padang, disaster education for local communities has been improved through capacity-building programmes and curriculum development in educational institutes. A more integrated, coordinated, and correlated use of land, coast, and sea utilization (interdepartmental, inter-stakeholder, inter-sectoral) approach has been implemented in the region. This was evident through the interviews in the Padang region where community associations are directly engaged in the preparation of the RDTR:

"The preparation of the RDTR, including during the pandemic in 2021, was done by accommodating aspirations/concepts/ideas from the urban villages. In the preparation of the RDTR in 2021, 104 urban villages participated. Several associations were also invited. Apart from spatial planning-related agencies, there is also a health office"—Pd1 (Bappeda).

Apart from the local communities, other representatives from sub-districts including Community-based organizations (KOGAM, KSB- Community resilience group, SSB—School resilience group) and private sector organizations (hotels, shopping malls, and banks) also involved in this planning process. As stated by an official from the Local disaster management office (PD 2), institutional collaboration was one of the main concerns for relevant disaster-related agencies in the region:

"There are coffee mornings, which is a forum to discuss disaster issues, including disaster risk reduction efforts attended by disaster-related agencies and forums, such as Kogami, RAPI, ORARI, PMI, SAR, Classy FM Radio, print media, Kelurahan, KSB, and others"—Pd2 (BPBD).

As per document reviews on Pangandaran Regency, the local community was involved in spatial planning for disaster resilience. Educating the citizens to increase awareness of potential disasters was a main concern in disaster management master plan. However, community involvement in spatial planning is limited only to the implementation (socializing planned programmes) rather than to the planning process (aspiration collecting). It is also revealed that the involvement of representatives from sub-districts, including community-based organizations and private sector organizations in planning is a key aspect of Pangandaran Regency too. As highlighted through the interviews, MUSRENBANG (Meeting for planning and Development) is a main activity for both disaster risk reduction and urban planning in both regions, and the involvement of disaster-related agencies and local communities in urban planning process was a critical factor in both Padang and Pangandaran regions.

Evacuation Planning

In Padang, Evacuation routes are planned, including temporary evacuation sites/ shelters and the route to get there. There are 5 tsunami vertical evacuation shelters:

a. TES Mesjid Muhajirin, Pasir Putih Raya, Bungo Pasang, Kec. Koto Tangah
b. TES Mesjid Darussalam, Parupuk Tabing, Kec. Koto Tangah
c. TES Mesjid Nurul Haq, Jalan Komplek Jondul 4 Parupuk Tabing Koto Tangah, and
d. TES Ulak Karang, Ulak Karang, Padang Utara
e. Air Manis (New Government official location).

which were constructed by the National Government such as the Ministry of Public Works (PUPR) and National Disaster Management Agency (BNPB). However, the field observation revealed that two of these were well-maintained and three of them were abandoned. Apart from these evacuation shelters, several schools, hotels, and Government offices have been checked and recommended as vertical evacuation centres for the region.

During the field observations, multiple tsunami sirens were also found in Padang City. Most of the sirens are optimizing speakers found in mosques and mash Allah. Field observations revealed that several of the sirens are no longer functioning, but the siren in the regional disaster management agency (BPBD) of West Sumatera Province are able to fulfil the combined role of those sirens in spreading the news during a tsunami event.

In Pangandaran, a tsunami vertical evacuation shelter has been constructed by the national government: the Ministry of Public Works (PUPR) and the National Disaster Management Agency (BNPB). However, field observation identified that this tsunami vertical evacuation shelter was not adequately maintained. Since there have not been recent tsunamis in the region, it is used as a parking area for tourist buses. Interviewee (Pd 2) suggested that several hotels, schools, and government offices in the area were already identified and recommended as vertical evacuation centres in the region.

Apart from the above-mentioned buildings, a hilly area, an elevated bridge, and a sea world aquarium are also designated for vertical evacuation in the region. The sea world aquarium (PIAMARRI) is owned by the National Ministry of Marine and Fisheries as a research centre and tourist attraction. This aquarium was constructed as a tsunami vertical evacuation shelter for officials, workers, fishermen, and the community surrounding. Field observation revealed that the building was adequately maintained to date. Interviews also revealed that disaster mitigation infrastructure facilities such as evacuation routes have been identified throughout the region. Tsunami warning sirens were also found in Telkom Regional Office and Bojongsalawe Beach where the noise is said to be heard within a 2 km radius.

In comparison, Pangandaran Regency has identified particular locations in the region as evacuation sites, such as hotels, schools, and government offices. Further, the local government maintained them effectively so that they can be functional during an emergency. Even though Padang city also identified and assigned tsunami vertical evacuation shelters in the region, maintenance of these locations was not adequately carried out in the region.

14.3.4 Soft and Hard Engineering

Padang city is more into hard engineering structures for coastal protection such as groynes and low sea walls. These have been established for abrasion rather than tsunami events. Green coastal protection including green open spaces is planned to be expanded through city forests, city parks, seeding areas, water conservation areas (water infiltration), and coastal woodland (mangrove). However, several green coastal protection applications also exist in the city including mangroves and pine forests. Mangrove protection is given due consideration in Port Bungus and Plantation of pine trees as a buffer zone on the coast is initiated by the Ministry of Marine and Fishery in collaboration with the local community. Interviews revealed that the local community was paid for maintaining the pine trees along the coastal belt.

In Pangandaran, planning documents (spatial plans, detailed spatial plans, and zoning plans for coastal areas and small islands (Provincial level)) have integrated resilience as a main concern in tsunami-prone areas where there are efforts through plans such as green open space management and mangrove seeding, involving the communities. There is no designated area where hotels are allowed to build beside the road between the road and the coast.

It was evident through the interviews that mangroves and coastal forests were frequently used in the region's coastal protection activities. Also, a coastal lagoon in the region also integrated as a water retention and tsunami reduction reserve. As stated by a member of the Indonesian Tourist Guide Association (Pg12):

> Yes, there are mangroves and natural coral reefs. These Coral reefs are a part of National conservation area.

In comparison, both cities have both soft and hard engineering techniques for coastal protection. However, Padang city has practiced more grey protection strategies whereas Pangandaran Regency has implemented more green protection strategies. Sea walls have been established to protect the coast from abrasion but not from tsunamis. Mangroves, coastal lagoons, and forests have been used in both regions as tsunami-reducing strategies. The collaboration of the local community to maintain these establishments can be identified as positive development initiatives in both regions.

14.4 Discussion

This study discussed specific urban planning approaches under four high-level categories; spatial planning, community participation, evacuation planning, and soft and hard engineering. Some of them are phase-specific strategies that cover one phase in the disaster management cycle (prevention, preparedness, response, and recovery), while others cross multiple phases, though the specific urban planning approaches within each strategy tend to differ. Examples of various approaches are provided above within each strategy.

Importance of spatial planning and community participation in planning processes is widely recognized in the literature across all three phases, i.e., before, responding, and recovering from disaster. The majority of the document's planning strategies for areas vulnerable to tsunamis involve managing the area's spatial development, including both tangible and intangible developments like the environment, infrastructure, economy, and social conditions. However, inadequate implementation of building and planning regulations by developers, and lack of monitoring by the local authorities were also highlighted as main concerns in both regions. This also relates to the issues of governance, which is highlighted by Perera et al. (2022) in tsunami-prone areas in Sri Lanka. This includes the inter-organisational, intra-organisational, and infrastructure challenges faced by the country to introduce better urban planning strategies to tackle disaster risk. Murray (2017) also highlighted the same issue in most of the low- and middle-income countries where overlapping bureaucracies, limited decentralisation, fragmented roles, and responsibilities were the main constraints for urban disaster risk governance.

In addition to the concerns over the implementation of planning regulations in Indonesia, existing approaches have reached out to members of the community. The local government gathers inspiration from communities on the potential new programmes and regulations that are needed and feedback on current programmes and regulations. Musrenbang has played a vital role in this aspect where all the key stakeholders provide their insights to strengthen the urban planning strategies in the region. Principles of disaster resilience such as protecting persons and communities, their livelihoods, health, and heritage have been applied in the spatial planning of both regions. This includes the consideration on the needs of low-income communities, building requirements, building techniques, sustainable urban design, preservation of cultural heritage, and community involvement in the planning process. Several studies conducted in Chile and Sri Lanka also emphasized the inclusion of different layers of the community in the planning process, especially focusing on participation in decision making as well as for implementation (Baudoin et al. 2016; De Zoysa et al. 2023; Herrmann-Lunecke and Villagra 2020; Lim et al. 2016). The integration of different stakeholders' perspectives is also a particular concern in the planning process, and both interviews and reviewed documents have highlighted the importance of institutional collaboration for urban planning in both regions.

In Indonesia, the lack (or not sufficient) of health- and population-related planning strategies in the pre-disaster phase is notable; for example, the COVID-19 pandemic was yet to be considered as part of the spatial planning process. COVID-19 is considered a biological disaster, however, COVID 19 taskforce has not been considered in local or national level spatial planning process in Indonesia. This is also the case in multi-hazard scenarios in Indonesia, where tsunami events were the main priority whereas other hazards were not adequately considered during the urban planning process. A common feature among reviewed sources and current practices in Indonesia is the primary focus on single-hazard events. However, The COVID-19 pandemic has brought to the forefront of importance and challenges associated with multi-hazard threats. It can be especially challenging where they are concurrent but independently occurring hazards that interact. The nature of COVID-19, which has disrupted society for several years, has clearly exposed the potential for this type of situation. Several studies have highlighted the occurrence of other hazard events while tackling the pandemic threat (Mavroulis et al. 2022; Turay 2022; Kamalrathne et al. 2023). It will be important to understand how these circumstances will impact wider hazard vulnerability, and to what extent urban planning can reduce or exacerbate risks associated with multi hazards in such a situation.

In tsunami-prone areas, a combination of soft and hard engineering approaches can be effective in reducing disaster risks. Although the potential for hard engineering structures is well established, maintenance of hard engineering structures was comparatively low in both regions. However, Indonesia appears to have embraced several soft engineering measures, such as nature-based solutions, which are increasingly being promoted for their potential benefits in biodiversity conservation, safety, and cost-effectiveness. These Programmes are also implemented through mangrove planting and mangrove tourism (West Sumatra Province), disaster risk management, and rehabilitation of coastal areas and small islands (West Java Province), wherein both programs the communities are involved in executing the plan.

14.5 Conclusion

This study on urban planning in the Padang and Pangandaran regions, as well as a more detailed examination of current urban planning approaches in Indonesia, provide several useful insights into the range of disaster risk reduction strategies available, as well as several of the many challenges associated with their implementation them in practice.

Firstly, tsunamis are one of the coastal hazards that, while infrequent, are extremely destructive. Severe impacts on coastal cities in Indonesia during previous tsunamis prompted researchers and policymakers to study urban planning strategies that can reduce disaster risks in tsunami-prone areas. Natural hazard potential, particularly tsunami and other coastal hazards, are already considered in the planning processes in both study locations. The pandemic, on the other hand, is a new phenomenon that was not anticipated in the planning documents. The policies that have been implemented appear as a response to a pandemic, not as a form of preparedness in dealing with a pandemic. However, these considerations have not been implemented in the urban planning process, either at the national, provincial, or regional scale. This could be further explained due to the timescales associated with the review process of planning documents, where revisions are made every five-year period. However, the review can be carried out more than once in 5 years, if there are conditions including large-scale natural hazards, changes to the country's territorial boundaries determined by law, Changes to regional boundaries determined by law, etc. COVID-19, as it is not included in the category of natural hazards, planning process needs adjustments so that it can be mainstreamed in urban planning process in both regions.

Regulations are updated and new programmes have been carried out to strengthen infrastructures' ability to deal with potential disasters and make sure the community understands and can save themselves from them. These are done through the provision of evacuation infrastructures and capacity building programmes. There has also been multi-stakeholder involvement in DRR initiatives. Especially, for both Padang city and Pangandaran Regency, community involvement is remarkably high where diverse types of communities (Business community in Padang and Tourist community in Pangandaran) are involved in DRR activities.

Spatial planning in tsunami-prone areas has also been considered including the protection and restoration of natural resources, with the development of a coastal protection system with vegetation and conservation of sandbanks and coral reefs. However, maintenance of preparedness infrastructure needs to be improved where Tsunami evacuation shelters, Signage and soft and hard engineering structures should be well maintained by the support of disaster-related agencies as well as the local community.

References

Alan M, Jorge L (2013) Urban planning for disaster risk reduction : Establishing 2nd wave criteria. State Aust Cities:1–13

Baró F, Palomo I, Zulian G, Vizcaino P, Haase D, Gómez-Baggethun E (2016) Mapping ecosystem service capacity, flow and demand for landscape and urban planning: a case study in the Barcelona metropolitan region. Land Use Policy 57:405–417

Baudoin M-A, Henly-Shepard S, Fernando N, Sitati A, Zommers Z (2016) From top-down to "community-centric" approaches to early warning systems: exploring pathways to improve disaster risk reduction through community participation. Int J Disaster Risk Sci 7(2):163–174. https://doi.org/10.1007/s13753-016-0085-6

Brune S, Ladage S, Babeyko AY, Müller C, Kopp H, Sobolev SV (2010) Submarine landslides at the eastern Sunda margin: observations and tsunami impact assessment. Nat Hazards 54:547–562

De Zoysa CJ, Abeysinghe AASE, Perera UTG, Siriwardana CSA, Dissanayake PBR, Haigh R, Amaratunga D (2023) Critical analysis of tsunami preparedness at local level for sustainable urban planning in Sri Lanka, pp 327–339. https://doi.org/10.1007/978-981-19-2886-4_24

Dissanayake R, De Zoysa C, Abeysinghe S, Haigh R, Amaratunga D, Perera T (2022) Briefing Paper: Integrating pandemic, tsunami, and other multi-hazard preparedness into urban planning in Sri Lanka: Ministry of Rural Roads and other Infrastructure, Sri Lanka

Disasters Emergency Committee (2018) 2018 Indonesia tsunami appeal six-month report, pp 1–20. www.dec.org.uk

Fuchs RJ (2010) Cities at risk: Asia's coastal cities in an age of climate change. Asia Pacific Issues 96:1–12

Goda K, Mori N, Yasuda T, Prasetyo A, Muhammad A, Tsujio D (2019) Cascading geological hazards and risks of the 2018 Sulawesi Indonesia earthquake and sensitivity analysis of tsunami inundation simulations. Front Earth Sci 7:261. https://doi.org/10.3389/feart.2019.00261

Gu D (2019) Exposure and vulnerability to natural disasters for world's cities. United Nations Department of Economic and Social Affairs, Technical Paper No. 4., 1–43. www.unpopulation.org

Hamzah L, Puspito NT, Imamura F (2000) Tsunami catalog and zones in Indonesia. J Nat Dis Sci 22(1):25–43. https://doi.org/10.2328/jnds.22.25

Herrmann Lunecke MG (2015) Urban planning and tsunami impact mitigation in Chile after February 27, 2010. Nat Hazards 79(3):1591–1620

Herrmann-Lunecke MG, Villagra P (2020) Community resilience and urban planning in tsunami-prone settlements in Chile. Disasters 44(1):103–124. https://doi.org/10.1111/disa.12369

Huxley M, Inch A (2020) Urban planning. In: International encyclopedia of human geography. Elsevier, pp 87–92. https://doi.org/10.1016/B978-0-08-102295-5.10228-8

Imamura F, Takakura H, Matsuzawa T, Ito K (2019) A platform for multidisciplinary research in disaster science through experiences from the 2011 Tohoku earthquake and Tsunami. J Disaster Res 14(9):1318–1322

Kajitani Y, Chang SE, Tatano H (2013) Economic impacts of the 2011 Tohoku-oki earthquake and tsunami. Earthq Spectra, 29(SUPPL.1):457–478. https://doi.org/10.1193/1.4000108

Li L, Switzer AD, Wang Y, Chan CH, Qiu Q, Weiss R (2018) A modest 0.5-m rise in sea level will double the tsunami hazard in Macau. Sci Adv, 4(8):1–12. https://doi.org/10.1126/sciadv.aat1180

Lim MBB, Lim HR, Piantanakulchai M, Uy FA, Weerasinghe WK, Shigemura T, Ndah AB, Odihi JO, Liao KH, Djalante R, Holley C, Thomalla F, Zawawi EMA, Yusof NS, Ismail Z, Hettige S, Haigh R, Setiati NJ, Arlikatti S, Dong J (2016) Community resilience and urban planning in tsunami-prone settlements in Chile. Nat Hazards 7(1):103–124. https://doi.org/10.1111/disa.12369

Mavroulis S, Mavrouli M, Kourou A, Thoma T, Lekkas E (2022) Multi-hazard emergency response for geological hazards amid the evolving COVID-19 Pandemic: Good practices and lessons learned from earthquake disaster management in Greece. Sustain (Switz), 14(14). https://doi.org/10.3390/su14148486

Mori N, Takahashi T, Yasuda T, Yanagisawa H (2011) Survey of 2011 Tohoku earthquake tsunami inundation and run-up. Geophys Res Lett 38(18):6–11. https://doi.org/10.1029/2011GL049210

New Urban Agenda (2016) The New Urban Agenda, UN Habitat. https://habitat3.org/the-newurban-agenda/

Okumura Y, Matsuoka H, Arakawa H, Tokanai F, Suzuki A, Irizuki T, Kajita H, Hara M (2021) The influence and impact of tsunamis on the microorganism assembly of Nagatsura-Ura Lagoon, Miyagi, northeastern Japan. Fish Sci 87:121–130

Oppenheimer M, Glavovic BC, Hinkel J, van de Wal R, Magnan AK, Abd-Elgawad A, Cai R, Cifuentes-Jara M, DeConto RM, Ghosh T, Hay J, Isla F, Marzeion B, Meyssignac B, Sebesvari Z (2019) Sea level rise and implications for low-lying islands, coasts and communities. In: Pörtner H-O, Roberts DC, Masson-Delmotte V, Zhai P, Tignor M, Poloczanska E, Mintenbeck K, Alegría A, Nicolai M, Okem A, Petzold J, Rama B, Weyer NM (eds) IPCC special report on the ocean and cryosphere in a changing climate

Ozer P, de Longueville F (2011) The tsunami in South-East Asia—a retrospective analysis of the management of an apocalyptic natural disaster. Cybergeo. https://doi.org/10.4000/cybergeo.24607

Perera UTG, Zoysa C De, Abeysinghe AASE, Haigh R, Amaratunga D (2022) A study of urban planning in Tsunami-Prone Areas of Sri Lanka, pp 562–592

Rentschler J, Salhab M, Jafino BA (2022) Flood exposure and poverty in 188 countries. Nat Commun 13(1):1–11. https://doi.org/10.1038/s41467-022-30727-4

Rittichainuwat BN (2006) Tsunami recovery: a case study of Thailand's tourism. Cornell Hotel Restaur Adm Q 47(4):390–404

Srinivasa Kumar T, Manneela S (2021) A review of the progress, challenges and future trends in tsunami early warning systems. J Geol Soc India 97(12):1533–1544. https://doi.org/10.1007/s12594-021-1910-0

Suppasri A, Kitamura M, Tsukuda H, Boret SP, Pescaroli G, Onoda Y, Imamura F, Alexander D, Leelawat N, Syamsidik (2021) Perceptions of the COVID-19 pandemic in Japan with respect to cultural, information, disaster and social issues. Prog Disaster Sci 10:100158

Taheri Tafti M (2017) Reconstruction of informality: Can formal reconstruction re-create informality? In: Urban planning for disaster recovery. https://doi.org/10.1016/B978-0-12-804276-2.00008-6

Turay B (2022) Flood hazard management in a multiple hazard context: a systematic review of flood hazard management during the COVID-19 pandemic in Africa. Discov Water, 2 (1). https://doi.org/10.1007/s43832-022-00014-w

UNESCO IOC (2020) Capacity assessment of tsunami preparedness in the Indian Ocean: status report of 2018 and Supplement 1. https://unesdoc.unesco.org/ark:/48223/pf0000373680. Accessed 7 Mar 2025

UNISDR (2005) Hyogo framework for action 2005–2015. Encycl Earth Sci Ser 2005:508–516. https://doi.org/10.1007/978-1-4020-4399-4_180

UNISDR (2015) Sendai framework for disaster risk reduction 2015–2030. CMAJ 144(2):169–173

United Nations (2017) New urban agenda. In: Habitat III secretariat

Villagra P, Herrmann MG, Quintana C, Sepúlveda RD (2017) Community resilience to tsunamis along the Southeastern Pacific: a multivariate approach incorporating physical, environmental, and social indicators. Nat Hazards 88:1087–1111. https://doi.org/10.1007/s11069-017-2908-1

World Bank City Planning Labs & CAPSUS (2020) Urban planning tools as agents of change: Collaborative spatial data for sustainable urban development in Indonesia. https://www.worldbank.org/en/data/statistical-capacity-building/data-innovation-fund/urban-planning-tools

Chapter 15
Disaster Risk Reduction in Indonesia: 20 Years After Aceh Tsunami

Ariyaningsih, Riyan Benny Sukmara, Ardiansyah Fauzi, Nurrohman Wijaya, Kamaluddin Latief, Ray-Shyan Wu, and Rajib Shaw

Abstract The 2004 tsunami had the greatest impact on Indonesia, particularly the province of Aceh. Aside from the loss of human lives, the tsunami also caused damage to infrastructures, public facilities, the fishery sector, and the loss of the mangrove forests. As a result, the economy of Aceh experienced a 15% decline in 2005. The Aceh tsunami, which took place two decades ago, aided as a starting point for the establishment of the National Disaster Management Agency (BNBP). However, the government has faced numerous challenges in its efforts to recover from the tsunami and develop resilience in areas vulnerable to disasters. This chapter tries to identify the state, gaps, and challenges in the reduction of disaster risks in Indonesia after the 2004 Aceh tsunami.

Keywords Aceh tsunami · Post-tsunami · Post-recovery · Indonesia

Ariyaningsih (✉) · R. Shaw
Graduate School of Media and Governance, Keio University, Minato, Japan
e-mail: ayya19@keio.jp

Ariyaningsih · R. B. Sukmara
Department of Civil Engineering and Planning, Institut Teknologi Kalimantan, Balikpapan, Indonesia

A. Fauzi
Department of Mathematics, Physics and Electrical Engineering, Northumbria University, Newcastle upon Tyne, UK

R. B. Sukmara · R.-S. Wu
Department of Civil Engineering, National Central University, Taoyuan City, Taiwan

N. Wijaya
Department of Urban and Regional Planning, Bandung Institute of Technology, Bandung, Indonesia

K. Latief
College of Public Health, Program in Global Health and Health Security, Taipei Medical University, Taipei, Taiwan

R. Shaw et al. (eds.), *Two Decades from the Indian Ocean Tsunami*, Disaster Risk Reduction, https://doi.org/10.1007/978-981-96-2669-4_15

15.1 Introduction

Frequent occurrences of unpredictable risk events sometimes referred to as natural hazards, have become common in several countries. Earthquakes, tsunamis, hurricanes, landslides, and volcanic eruptions are all examples of natural hazards (Amezquita-Sanchez et al. 2017; Gouramanis and Morales-Ramirez 2020). Due to its proximity to the convergence of three significant tectonic plates—the Eurasian, Indo-Australian, and Pacific plates—Indonesia is vulnerable to regular natural hazards. In addition, Indonesia is vulnerable to several natural hazards, including tsunamis. A tsunami is a series of oceanic waves that move due to a massive seismic event or volcanic activity occurring at the boundary between two tectonic plates (Ramalanjaona 2011). Furthermore, it can be triggered by various natural hazards such as earthquakes in subduction zones, landslides, volcanic eruptions, and other similar events (Steinritz et al. 2021).

The 2004 tsunami, which occurred at 00:59 GMT, had a magnitude of 9.1 on the Richter scale, creating it the seventh-largest earthquake globally. The event was triggered by a seismic disturbance of significant magnitude occurring at a depth of 30 kms below the ocean surface, specifically in the vicinity of the western coast of Sumatra (Rose Enid Teresa et al. 2020). Furthermore, the 2004 Indian Ocean tsunami devastated the western coastal communities of Aceh, Indonesia. The tsunami caused the most severe damage in Banda Aceh and Aceh Besar District, resulting in a minimum of 96,000 fatalities in those areas alone. Overall, Aceh experienced a total of 130,000 mortalities. In addition, the 2004 tsunami destroyed 120,000 buildings and significant damage to 70,000 houses. It also completely destroyed 3,000 government structures, such as schools, hospitals, and 14 ports (Suppasri et al. 2015). Furthermore, it was projected that the tsunami caused over ninety percent damage to coral reefs. The tsunami also impacted Mangrove trees, which influenced the fishing industry. Due to the paralysis and decreased production of numerous critical sectors of the Aceh economy, the administration predicted that the economy would deteriorate by as much as 15% in 2005.

Particularly in Banda Aceh City, Aceh Province, the earthquake and tsunami seriously impacted the development that had been going on in Aceh. In addition, the city's growth suffered from a series of disasters, which had a greater impact on the coastal areas (Syamsidik et al. 2023). Over two decades, the area was extensively reconstructed and redeveloped. The Indonesian government legally supervised and supervised several of the procedures. From 2005 to 2011, BRR NAD-Nias was responsible for managing the process of restoring and rebuilding Banda Aceh City after the destructive earthquake and tsunami. During this period, the process of reconstructing housing, schools, hospitals, roads, clean water, and other vital facilities was a huge endeavour. The Aceh government collaborated with the Asian Development Bank (ADB) to rebuild the city's transport infrastructure in the aftermath of the tsunami and earthquake (Nugroho et al. 2023).

Since the destructive Aceh Tsunami in 2004, two global frameworks for managing disasters have been established: the Hyogo Framework for Actions and the Sendai

Framework for Disaster Risk Reduction (SFDRR). The city's current structural mitigation measures, implemented to prevent losses like the 2004 tsunami, are still inadequate in efficiently reducing wave energy (Syamsidik et al. 2023). The post-2004 Aceh tsunami recovery has been an extensive and gradually implemented process. Following the 2004 Aceh disaster, there has been a prolonged and progressive phase of growth in the economy. In response to the 20th anniversary of the 2004 Aceh disaster, many local and international non-governmental organizations (NGOs), along with federal, provincial, and city administrations, have implemented several programmes to decrease the risk of future tsunamis. However, the efforts to mitigate the impact of tsunamis have not been fully implemented in Aceh. It is challenging for local governments to meet the financial needs of certain mitigation initiatives, particularly those with a structural component (Syamsidik et al. 2021).

This chapter tries to examine the transformations that occurred in Aceh and Indonesia following the 2004 tsunami and identify challenges to the recovery process during the previous two decades. To achieve this goal, the chapter discusses the subchapter on the Evolution of disaster management institutions in Indonesia following the Aceh Tsunami. The purpose is to analyse policy changes or institutions' establishment because of the 2004 tsunami. The following sub-chapter examines the Impact of Tsunami Aceh and Its Recovery Progress to learn about the Indonesian government's recovery efforts over the last two decades. The next subchapter talks about the Actor of Disaster Risk Reduction for Post Disaster. In addition, we also show the Progress of Sendai Framework's Implementation for Post Recovery Aceh Tsunami. After taking stock of existing conditions and progress in the previous section, we identify key Challenges for Disaster Risk Reduction for Post-Tsunami and provide recommendations to address these challenges.

15.2 Evolution of Disaster Management Institutions in Indonesia After Aceh Tsunami

15.2.1 Review of Disaster Trend in Indonesia

Indonesia's unique geographic position, high population density, and rapid urbanization along its coastlines have heightened the country's exposure and risk to natural hazards. Approximately 70% of inhabitants reside on the coast, and an additional seven million directly derive their source of income from the sea. This significant coastal population and economic dependence on marine resources heighten the country's vulnerability to coastal hazards, including tsunamis, coastal flooding, and storm surges (BPS Statistics Indonesia 2020; Pratami).

Beginning with the aftermath Indian Ocean earthquake and tsunami in 2024 and extending through 2023, the frequency of disaster occurrences has continued to increase, peaking in 2020. During this period, the average number of disasters per year was 2,224. Since 2020, fluctuations in disaster figures have persisted,

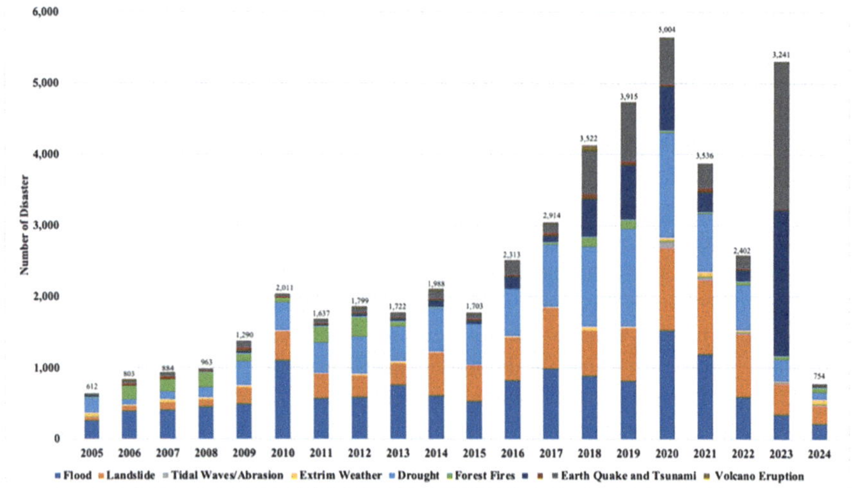

Fig. 15.1 Number of disasters per year from 2005–2024. (*Source* The data has been processed based on information from the National Disaster Management Agency (BNPB). The data for the year 2024 is only up to June 30, 2024, based on the latest data extraction date, National Agency for Disaster Management)

with the latest figure exceeding 3,241 in 2023. Floods, droughts, landslides, and combined earthquake-tsunami events represent the most common types of disasters, with average annual occurrences of 704, 600, 489, and 293, respectively. The most notable increase has been in the incidence of earthquakes and tsunamis, which reached its apex in 2023 with 2,066 cases recorded that year (Fig. 15.1).

The second criterion for assessing the disasters's impact is to quantify the total of casualties. More than 237,578 people have been killed by all combined disasters in Indonesia. Figure 15.2 shows the trend of death number from 2005 until June 2024. The earthquakes and tsunami are the deadliest disaster. The seismic event in year 2006, Yogyakarta and the earthquake-triggered tsunami of Mount Krakatau, Sunda Strait are identified as the foremost and second-most catastrophic incidents in Indonesia, correspondingly.

Figure 15.3 illustrates the total of people affected by disasters yearly. The data shows an upward trend in the number of individuals impacted from 2015 to 2018, with the highest level recorded in the latter year. This significant increase was largely caused by the impact of the eruption that occurred on December 22, 2018, on Mount Krakatau. The volcanic event resulted in 431 fatalities, 7,200 injuries, 15 missing persons, and the displacement of 46,646 individuals. The district of Pandeglang experienced the highest number of deaths due to this particular disaster (Suryaratri et al. 2020).

Figure 15.4 depicts the overall destruction resulting from disasters between 2005 and June 2024. The data indicates that the residential sector has been the group most impacted by these events, with notable spikes in damage recorded in 2006, 2009,

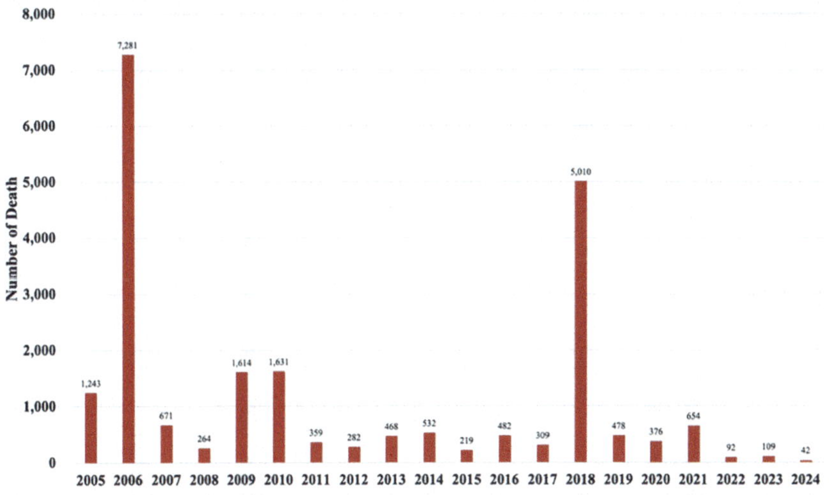

Fig. 15.2 Number of disasters per year from 2005 to 2024. (*Source* The data has been processed based on information from the National Disaster Management Agency (BNPB). The data for the year 2024 is only up to June 30, 2024, based on the latest data extraction date (National Agency for Disaster Management))

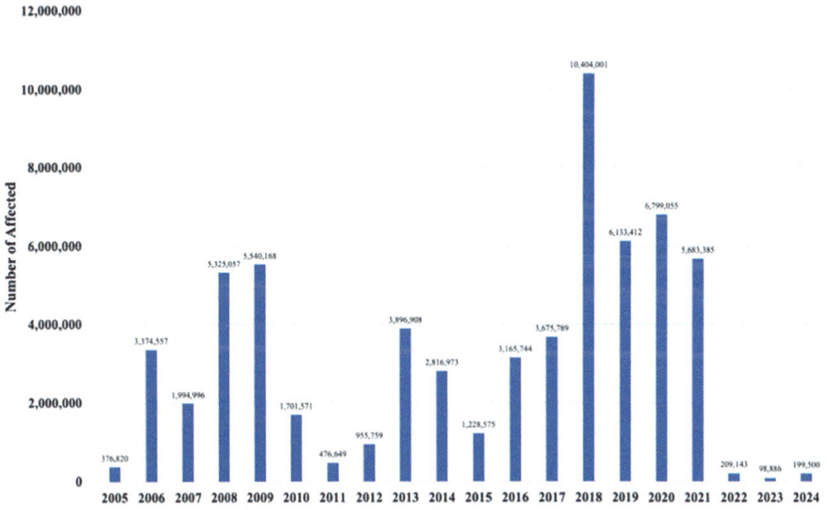

Fig. 15.3 Total of individuals affected by disaster per year from 2005 to 2024. (*Source* The data has been processed based on information from the National Disaster Management Agency (BNPB). The data for the year 2024 is only up to June 30, 2024, based on the latest data extraction date, (National Agency for Disaster Management))

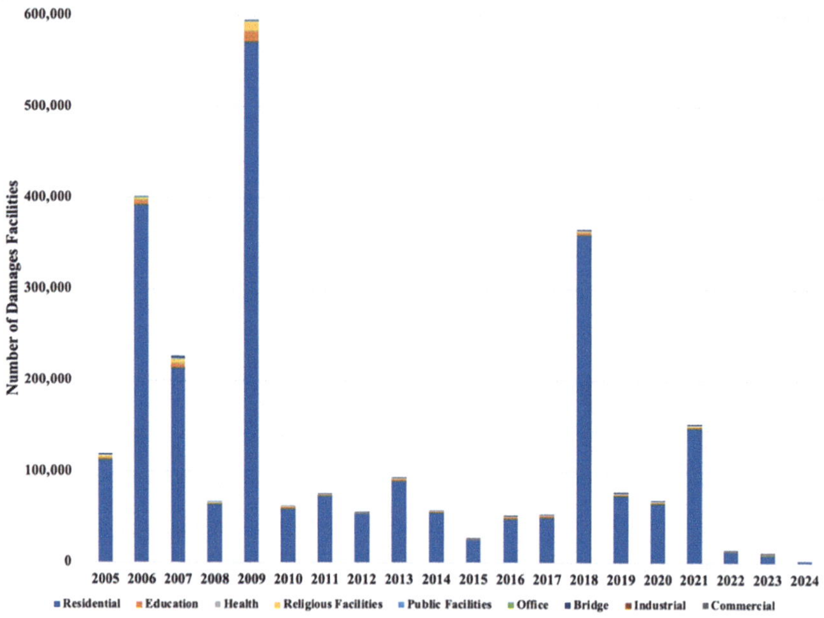

Fig. 15.4 Number of damages facilities by disaster per year from 2005 to 2024. (*Source* The data has been processed based on information from the National Disaster Management Agency (BNPB). The data for the year 2024 is only up to June 30, 2024, based on the latest data extraction date, (National Agency for Disaster Management))

and 2018. The most severe impact occurred in 2009, which can be directly attributed to the devastating earthquake that struck West Sumatra. This seismic event not only destroyed housing units but also caused significant damage to educational, religious, and healthcare facilities.

These findings underscore the significant and persistent threat natural hazards pose to Indonesia. The country's susceptibility is heightened by its distinctive geographic position along the Circum-Pacific Belt which subjects it to a significant risk of seismic and volcanic activity. Furthermore, the rising frequency and severity of hydrometeorological disasters, including floods and droughts, may be associated with the climate change impact.

15.2.2 Review of Disaster Governance in Indonesia and Development of BNPB After Tsunami Aceh 2004

The devastating earthquake and tsunami that impacted Aceh Province and its neighboring regions in 2004 drew significant attention from the global community and the Indonesian Government towards disaster risk management. Since 2004, policies have been reformed by establishing a more structured and comprehensive institutional and legal framework for disaster risk reduction (DRR). The period since 2014 has been marked by increased consideration of environmental hazards, city-related threats, enhancing the capacities of governmental and community entities, and a focus on community-based approaches (Djalante et al. 2017).

The Indonesian government's approach to disaster management has undergone a significant transformation in recent years. Previously, statutory foundation for the nation's disaster risk reduction initiatives was the National Coordinating Board for Disaster Management (*Badan Koordinasi Nasional Penanganan Bencana*), established through Presidential Regulation 83/2005. However, this centralized entity has since been replaced by the National Disaster Management Agency (*BNPB*), that was according to the legal provisions of Law of Law No. 24/2007 and Presidential Regulation 8/2008. The National Agency for Disaster Management, known as BNPB, is the Indonesian authority for disaster-related matters. It was founded in 2008 to succeed the National Disaster Management Coordinating Board (National Agency for Disaster Management 2007).

At the subnational level, a similar institutional shift has occurred. The government previously relied on the Provincial Disaster Management Coordination Agency (Satkorlak PB) to organise disaster response and recovery efforts at the provincial level. The Provincial Disaster Management Agency (BPBD-Propinsi) has now supplanted this entity, with the legal basis provided by supporting regional regulations. An analogous transition has also taken place at the district and regency levels. Formerly, the the body for Crisis Response and Displacement (Satlak PBP), established through a Governor's Decree, was the primary local-level disaster management body. This unit has since been transformed into the Local Disaster Management Agency (BPBD-Kabupaten/Kota), with the legal foundation of a Local Regulation *(Peraturan Daerah).*

These institutional and legal changes reflect Indonesia's efforts to strengthen its disaster management framework, transitioning from a centralized, top-down approach to a more decentralized and regionally empowered system. This shift aligns with the broader decentralization of governance in the country and the recognition of the need for enhanced local-level preparedness and response capabilities. The evolving legal and organizational landscape of catastrophe mitigation in Indonesia underscores the state's commitment to building a more thorough and resilient disaster risk reduction and management infrastructure.

The emergency management framework has experienced modifications at the national, regional, and village/community tiers. The authority of the National Disaster Management Agency (BNPB) and its coordination functions at the national, provincial, and village levels have been carried out as part of the efforts to address disasters in the era of decentralization. To effectively mitigate the impacts of these disasters, policymakers and disaster management authorities must develop and implement comprehensive risk reduction strategies. These strategies should prioritize fortifying early warning mechanisms, increasing infrastructure robustness, and augmenting community readiness and response capabilities. Furthermore, the integration sustainable development planning and DRR (disaster risk reduction) is essential for building long-term resilience and reducing the socioeconomic consequences of these events (Fig. 15.5).

The National Disaster Management Agency (BNPB) is responsible for coordinating the systematic, coordinated, and comprehensive implementation of disaster management activities. This demonstrates the Government's strong commitment to addressing disasters in Indonesia. From an organizational viewpoint, the National Disaster Management Agency (BNPB) continues to be a crucial actor in enhancing coordination and cooperative governance among involved parties. Due to limited

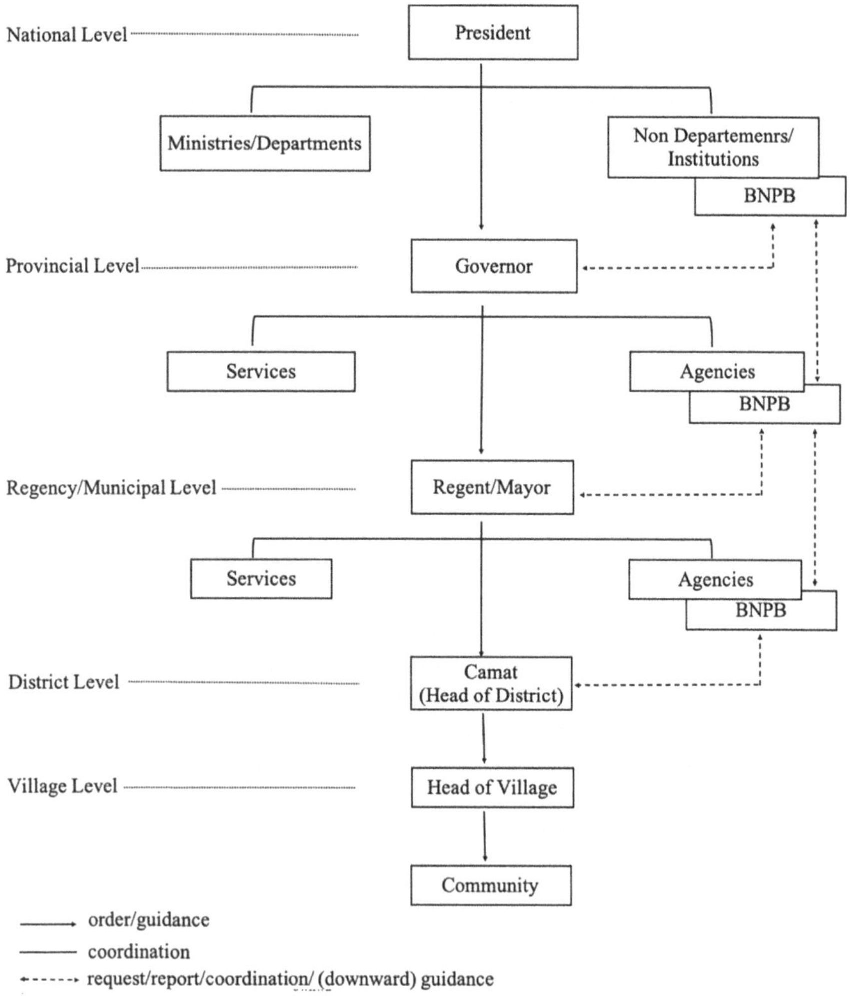

Fig. 15.5 Disaster management system. *Source* (Japan International Cooperation Agency 2015)

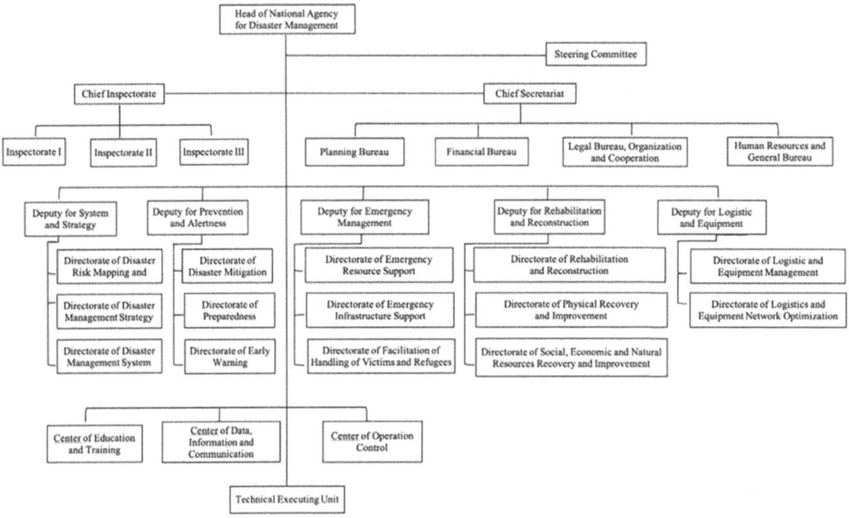

Fig. 15.6 Organizational structure of BNPB 2024. *Source* (National Agency for Disaster Management 2024)

active community engagement, a lack of integration among different government agencies, local government, community organizations, the private sector, and the BNPB's ministerial-level status and enabling legislation, the BNPB has to do comprehensive disaster risk governance. The BNPB's efforts to foster collaboration and coordination among these diverse stakeholders remain a work in progress. In order for Indonesia to effectively mitigate and manage disaster risk, it is imperative to bolster the BNPB's capacity to collaborate and integrate the efforts of all crucial stakeholders from various social and governmental sectors (Fig. 15.6).

15.2.3 Policy Related Disaster Risk Reduction from National Level to Local

The decentralization process, initiated in Indonesia in the year 2000, has fundamentally transformed the country's political and governance landscape. This transition has shifted power and authority from the centralized national level to greater regional and provincial autonomy. This structural change has also been reflected in the domain of disaster risk reduction (DRR), where the responsibility and decision-making authority have increasingly devolved to the local and provincial tiers of government. Following the catastrophic seismic event and tsunami that occurred on December 26, 2004, the Indonesian Government implemented the Disaster Management Act No. 24/2007 on April 26, 2007. This legislative action sought to minimize disaster risk and incorporate disaster risk reduction principles into the nation's wider development strategies (National Disaster Management Authority 2007).

The enactment of this new disaster management legislation clearly indicates a transition in emphasis from Disaster Response (DR) to Disaster Risk Reduction (DRR). Subsequently, the National Disaster Management Agency (*BNPB*) was formed through a Presidential Regulation in 2008, reporting directly to the President. Furthermore, the Minister of Home Affairs issued Decree 46/2008 (The Minister of Home Affairs 2008), mandating the creation of Regional Disaster Management Agencies (*BPBD*) in every provinces by 2009. More recently, Indonesia has developed a comprehensive Master Plan for Disaster Management (RIPB) 2020–2044, which serves as the national guideline for Disaster Management (PB) and provides a reference for the the national and local governments, along with other essential stakeholders, in the planning and execution of disaster management programs. Over time, critiques have emerged regarding the progress of Disaster Risk Reduction (DRR) in Indonesia, suggesting a national-level focus and the need for more efforts to enhance regional organizations, boost community engagement, and implement a more unified approach in development planning Although not maximized, these issues have been addressed through the Indonesia's National Medium-Term Development Plan (RPJM), which serves as the overarching framework. The RPJM emphasizes empowering subnational entities, adopting a decentralized and community-centric approach to enhance responsiveness and adaptability, increasing the community's role in DRR initiatives, and integrating DRR with broader development planning to mainstream disaster resilience across sectors and governance levels (Cabinet Secretariat of The Republic of Indonesia 2020).

The existence of village development planning forums (*musrenbangdes*) and the Village Fund (*Dana Desa*) also offer promising avenues for supporting mitigation and preparedness efforts at the local level. The development of community resilience, with a commitment to building disaster-resilient villages and cities, safe school facilities, and disaster risk-adaptive infrastructure, is expected to be a key focus area. Additionally, achieving a pentahelix cooperation between the Government, academia, the community, the business sector, and the mass media represents another crucial challenge in strengthening the country's disaster management capabilities.

15.3 A Review of Impact of Tsunami Aceh and Its Recovery Progress

15.3.1 Impact of Aceh Tsunami

As mentioned in the introduction part, an undersea earthquake with magnitude of 9.1 occurred on December 26, 2004, at 7:59 AM local time (00:59 GMT), triggering a massive tsunami in the Indian Ocean, with significant impact in Aceh (Vidyattama et al. 2021) and become the largest in the world in the past 40 years (Satake 2014).

The Aceh Tsunami had impacted several African and Asian countries such as India, Maldives, Thailand, Sri Lanka and Indonesia (Suppasri et al. 2015). According to the report by USGS (U.S. Geological Survey), the tsunami wave heights reached approximately 15 m (50 ft) to 30 m (98 ft) in certain locations in the northern region of Sumatra Island, specifically Aceh and North Sumatra Province (Associated Press 2014; USGS 2005). Moreover, a team from Japanese coastal engineering was measured the maximum tsunami runup height in Rhiting located in the south-western region of Banda Aceh to be 49 m (Shibayama et al. 2005). This catastrophic tsunami resulted in extensive damages and significant losses. The waves, propagated at a velocity of approximately 800 km/h, hit Banda Aceh. Waves with heights of about 12 m affected the cities of Moelaboh, Calang, and Lamno, flooding areas up to 10 km from the coastaline (Athukorala and Resosudarmo 2005). As a result, Indonesia suffered the greatest number of casualties, with over 160.000 fatalities and approximately 500.000 people forced to be displaced (Vidyattama et al. 2021). Additionally, over 3000 civil servants lost their lives and another 2.275 were reported missing (Fan 2013). Furthermore, this disaster significantly impacted the environment, particularly in coastal areas. Several coastal areas in northwestern part of Sumatra Island were severely damaged and also destructing of infrastructures in those areas.

The coastal area of Aceh experienced significant erosion, extending approximately 500 m, with exception of area with rocky headlands. This hydrodynamic phenomenon severely damaged beaches, dunes, and wetlands (Faiez and Fan 2023). Ulee Lheue beach in Banda Aceh lost 15% of its total sediment due to the tsunami and experienced land subsidence due to the 2004 earthquake (Meilianda et al. 2019). In Meulaboh, the Tsunami eradicated the beach ridge, altering the river's morphology, causing it flow directly into the ocean and inundating the swamps (Liew et al. 2010). Additionally, the major tsunami in Indian Ocean caused severe damage to coral reefs. According to a report by Reef Check Foundation, extensive stands of toppled and broken coral lined the southern coast of Weh Island, spanning a 7-km stretch and 50 m wide. These patch reefs primarily composed of Blue Coral (Heliopora coerulea) at depths ranging from 2 to 6 m (Foster et al. 2006).

The Aceh Tsunami was not only impacted to the environment but, it was also severely impacted various other sectors, including infrastructure, social and economic sector. In tterm of infrastructure, the tsunami impacted both private and public infrastructure, such as Harbour, Mangroves, fish pond (*tambak ikan*) (Faiez and Fan 2023), settlements (Mirza 2015), road (Ohsumi and Gejo 2005), and buildings (Syamsidik et al. 2023). According to an assessment report of Government of Indonesia (GoI), the total losses and damage to infrastructure were estimated at approximately 8.2 trillion IDR (501,900,483 USD) as shown in Fig. 15.7. This encompasses the impact on a total 316 km national road, 1621 km local road, 437 bridges, 19 seaport, and 4 airports. Furthermore, the economic impact was significant, since large number of businesses being disrupted, fisheries and agriculture were severely damaged, and the livelihood of countless people were destroyed. The total damage and losses in economic sector (productive sector) are estimated to be around 1182.1 million USD, comprising 224.8 million USD for agriculture and livestock,

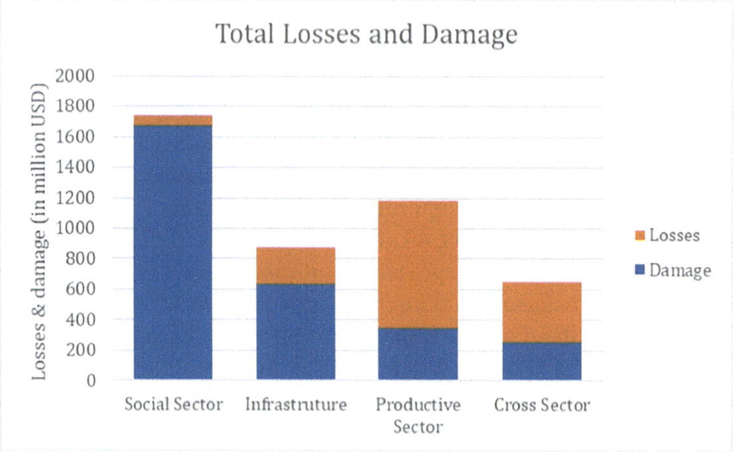

Fig. 15.7 Total loss and damage Indian Ocean tsunami (Indonesia). (*Source* Adopted from MoNDP, 2005)

510.9 million USD for fisheries, and 446.6 million USD for enterprise (MoNDP 2005).

Figure 15.7 also shows that the social sector also incurred substantial damages, with losses amounting to 776 million IDR (47,530 USD) for worship places, 13.4 trillion IDR (820,762,060 USD) for housing sector, 1.2 billion ID (73,502 USD) for education sector, and 854 million IDR (52,307 USD) for Health sector (MoNDP 2005). This comprehensive damage highlights the widespread ad multifaceted devastation caused by the tsunami, necessitating extensive recovery and rebuilding effort across multiple sectors.

Regarding the impact on communities, specifically with regards to mental well-being, it imperative to address the psychological issues faced by survivors. Unfortunately, psychological evaluation related to the impact of major disaster are still rare, especially for developing countries (Bland et al. 2005). Undoubtedly, evaluating this aspect is essential for the psychological well-being of individuals who have experienced a disaster. Post-disaster traumatic situation can greatly impede their capacity to return to their regular life (Davidson and McFarlane 2006).

Previous study conducted by Rousseau et al. in (2004) examined psychological issues in vulnerable people impacted by the tsunami on the northern coastal area of Aceh. The study implemented The Hopkins Symptom checklist as screening instrument to asses severe emotional distress and depression. The result indicated a substantial prevalence of severe psychological distress and depressive symptom among susceptible people residing in tsunami-affected communities in Aceh. Furthermore, the number of tsunami-related deaths in households has identified at high risk of mental health issues in disaster situation. Additionally, the research also shown that younger persons were more likely have severe distress than older individuals (Souza et al. 2007). and it also confirmed by other previous study state that adolescents and young adults form another at-risk population related to stresses due to natural haza previous research have confirmed this finding, showing that adoloscents and young adults are also a vulnerable group when it comes to stress caused by natural hazards. (Rousseau et al. 2004).

According to the research finding by Wiguna et al. (2010), indicates that after a year following tsunami, 24.6% of children (aged between 4 and 10 years old) and 35.6% of children (aged between 11 and 18 years old) were found to be suffering from Port-traumatic Stress Disorder (PTSD) (Wiguna et al. 2010). Furthermore, research indicates that women are particularly susceptible to psychological distress of any kind as a result of disasters (Frankenberg et al. 2008). Married woman, in particular, receive more social assistance compared to widows or single women (Musa et al. 2014).

15.3.2 Recovery of Aceh Tsunami 2004

The 2004 Aceh's Indian Ocean Tsunami entered its twenty years of commemoration in 2024. It had significant short- and long-term impact on both socioeconomic and environment aspect of the region. The tsunami caused widespread destruction and profoundly impacted the lives and livelihood of millions (MacDonald 2005). Besides the direct impact on communities and the environment, the disaster also substantially altered organization and structure (Stephan et al. 2017). The magnitude of the catastrophe required substantial efforts in reconstruction and recovery procedures, underscoring the crucial requirement for comprehensive disaster management techniques to tackle both direct and long-term difficulties. These efforts involve reconstructing infrastructure, restoring economic activities, and addressing the psychological well-being of affected populations.

The Aceh Province formally known as Nangroe Aceh Darussalam, suffered significant damage from tsunami disaster and obtain the highest portion of recovery aid, amounting to around 5.8 billion USD (Fengler 2007). This assistance encompassed approximately 1500 project managed by government, donor, and non-government organization (NGO) agencies. To carry out the work, the Government of Indonesia (GoI) established a special official agency called Aceh-Nias Rehabilitation and Reconstruction Agency (Badan Pelaksana Rehabilitasi dan Rekonstruksi Nangroe Aceh Darussalam dan Nias, BRR NAD-Nias). B RR NAD-Nias implemented the initial rehabilitation programme in nine districts in collaboration with multiple donor agencies and NGOs from March 2005 to April 2009. The attention of the rehabilitation program was on resettlement and infrastructure rehabilitation, including the construction of seven tsunami-survival buildings-four in Banda Aceh, and one each in Aceh Barat, Aceh Jaya, and Biereun (Syamsidik et al. 2021).

The four-year rehabilitation process led by BRR NAD-Nias, with the participation of donors and NGOs, achieved substantial progress in embodying the "Build Back Better" spririt advocated by Indonesian President Susilo Bambang Yudhoyono. This advancement enabled communities to resume productive activities and strengthened both local and regional economies. These endeavours were crucial in the transition process from emergency relief to long-term recovery (Thorburn 2009). In addition, the concept of community engagement during the recovery process, particularly in rebuilding their homes, was implemented by United Nation Human Settlements Programs—Aceh-Nias Settlement Support Program (UN-HABITAT ANSSP). This initiative positively impacts knowledge sustainability, including coordination strategies, environmental awareness, and proper construction techniques (Manaf 2011). After the 2004 Indian Ocean Tsunami, there are numerous coastal tides gauges have been installed in Indian Ocean regions and available real time monitoring by accessing www.ioc-sealevelmonitoring.org (Satake 2014).

Government agencies and NGOs at national and international levels too the lead in the first restoration of coastal areas throughout the rehabilitation and reconstruction phase. The key government actors involve in the restoration of coastal forest were Watershed Management Agency (BP DAS) under Ministry of Forestry, and the foresty service at both regency and provincial levels. Wetland International, Ofam International, Care International Indonesia, and the Beach Care Program-Japan Surf, which are international NGOs, played significant role in the rehabilitation of coastal areas in Aceh after the tsunami. Following the formal tsunami rehabilitation and reconstruction process in Aceh concluded in April 2009, the Ministry of Marine Affairs and Fisheries took over most of the efforts to restore and maintain coastal forest for tsunami disaster mitigation, particularly in Peukan Bada and Baitussalam sub-district in Aceh Besar) (Nugroho et al. 2023). A prominent NGO in terms of mangrove rehabilitation in recent years is Yayasan Gajah Sumatera (YAGASU) which has a mangrove planting program. Currently, with the support of the private sector, the Coastal Carbon Corridor Development program aims to rehabilitate approximately 720 km of coast from Aceh Besar to the west coast of North Sumatra Province. The program involves at least 176 community groups (Oktari et al. 2018).

Nonetheless, the number of evacuation facilities built has not increased significantly. Between 2005 and 2009, at least seven tsunami evacuation facilities were built in Banda Aceh (four units), Bireuen (one unit), Aceh Jaya (one unit), and West Aceh (one unit). From 2009 to 2019, only one more tsunami evacuation facility was built in Aceh Besar's Peukan Bada sub-district. Meulaboh City, West Aceh, has alternative tsunami evacuation structures utilizing the use of individual buildings. Several rows of stores in the city have been modified to serve as alternate evacuation sites in the event of a tsunami. Typically, the government continues to prefer constructing tsunami evacuation shelters that serve a particular purpose. There have been numerous suggestions to optimize public buildings with routine tasks so that they can be used as tsunami evacuation structures by adapting their facilities and practicality (BNPB 2022).

According to tsunami experts, a preferred type of building is a mosque. If mosque roofs or mosque architecture are altered, they could serve as alternative locations for tsunami evacuation. The current number of designated evacuation buildings is insufficient to accommodate all residents who need to leave. During emergencies, residents may not effectively utilize evacuation structures, as demonstrated in the earthquake that occurred on April 11, 2012 (McCaughey et al. 2017).

15.4 Stakeholders of Disaster Risk Reduction for Post Disaster in Aceh

Prior initiating a comprehensive restoration effort, the Indonesian Government developed a "Master Plan for the Rehabilitation and Reconstruction of Aceh and Nias, North Sumatra.". In line with the information provided in the previous sub-chapter on the recovery of Aceh Tsunami 2004, it is interesting to note that in April 2005, the government established the Agency for the Rehabilitation and Reconstruction of Aceh and Nias (BRR). This ad-hoc agency was assigned significant responsibilities in carrying out and coordinating rehabilitation and reconstruction initiatives that were financed by various donors and non-governmental organizations (NGOs), such as German Agro Aid, Red Cross, Oxfam, Care, German Agro Aid, and Muslim Aid. In total, BRR has managed and executed approximately 12,000 projects. During 4-year period, activities were divided into four stages of reconstruction development, including housing, public infrastructure, institutional and social, and economy. The agency's overall cost for reconstruction reached 6.7 billion USD, which was funded by the Indonesian government, bilateral and multilateral donors, as well as national and international NGOs. Most of the donors and NGOs focused on saving thousands of people by providing clean water, food, medical necessities, and shelter (Brennan 2005; Kennedy et al. 2008; Rofi et al. 2006; Steinberg 2007). Some others focused on the psychological side through trauma healing (ACFID 2006). The Aceh reconstruction using a village-level method that involved active participation from local communities. Communities were given the authority to determine whether their region would undergo land consolidation techniques for reconstruction or remain in its original condition. (Suppasri et al. 2015).

In 2024, two decades after the Aceh disaster, the Aceh Disaster Management Agency Work Unit is establishing the Work Plan (Renja). The team was established in accordance with Minister of Home Affairs Regulation No. 86/2017. In 2024, the Work Plan in Aceh has also been directed to support the achievement of national development goals in disaster management implemented by the Indonesian National Disaster Agency. The primary objective of post-disaster recovery interventions implemented by the government, donor agencies, and NGOs is to effectively tackle core issues, including restoring sustainable livelihoods and enhancing economic activities. The post-tsunami recovery and reconstruction efforts achieved substantial progress across various sectors, including infrastructure, housing, environment, livelihood, agriculture, health, local economy, education, and disaster management. Several non-governmental organizations (NGOs) shifted their primary focus from providing immediate assistance to engaging in reconstruction activities. This change was prompted by an exceptional surge of grant funds from both the general public and their respective governments. From 2005 to 2008, the collaboration between foreign donors (United Nations, Asian Development Bank, World Bank, German Development Bank), NGOs, and the BRR was able to build 61,000 temporary and more than 124,000 new permanent housing units (Steinberg 2007).

The government and NGOs devised a range of income restoration programs and methods to aid in the rehabilitation of individuals' livelihoods following the disaster. The primary objective of most livelihood programs is to restore and enhance the means of earning a living in order to provide social security to communities that have been affected. These programs also aim to construct community infrastructure and implement local economic initiatives or interventions for the benefit of the recipients. Various institutions ranging from the central government through the Ministry of Labor and local government, local, national and international NGOs and other institutions offered various forms of support and technical assistance by establishing employment service centres in order to restore livelihoods for the people of Aceh during the recovery period. The service centers offered a range of skills training programs, including language and computer training, debris clearance work (CFW), textile production and embroidery, business proposal writing, seafood processing labor, and physical and building construction. They also provided job placement services (UNDP 2013). NGOs also offer employment opportunities through cash-for-work programs, small-scale aquaculture initiatives, the restoration of aquaculture ponds, and various livelihood programs for women and youth, including snack production and handicrafts (Scoones 2009). Community-based cash-for-work (CFW) programs are highly favoured among communities due to their provision of a temporary income ranging from 50,000 IDR to 100,000 IDR, typically 2–3 months.

After 20 years of the 2004 Indian Ocean tsunami, several mitigation efforts have been implemented by the central government, local governments, as well as local/international NGO initiatives. However, not all tsunami mitigation efforts (structural and non-structural) have been implemented in Aceh. Several types of mitigation efforts, especially structural ones, require large funding sources that are difficult to meet from the local government budget. There is limited physical protection such as a sea wall or breakwater built as part of tsunami disaster mitigation efforts in Aceh. All types of sea walls or coastal protection construction in Aceh were built to control the coastal sediment transport. However, several types of non-structural mitigation efforts are often on the agenda of local government and community activities even though the consistency of implementation in several types of activities is not very good. The enactment of Law No. 24/2007 concerning Disaster Management and the subsequent restructuring of disaster management agencies signified a critical shift in Indonesia's disaster management paradigm, thereby enhancing the country's preparedness for future disaster events.

Indonesia has placed significant emphasis on the establishment of a tsunami early warning system following the occurrence of the Indian Ocean tsunami in 2004. The system was developed in a partnership between the Indonesian and German governments and has undergone testing for both near-field and far-field tsunamis. The main goal of the Indonesian Tsunami Early Warning System (Ina-TEWS) is to promptly assess tsunamis following an earthquake within a time frame of under 5 min. A significant challenge for the system is issuing timely warnings and determining the appropriate time to cancel them. During the 2010 Mentawai earthquake, the warning was terminated less than one hour after the earthquake, despite the tsunami continuing for up to two hours later (Suppasri et al. 2015). A significant drawback of the current system is the regular maintenance required by the buoys. Between 2012 and 2018, 22 buoys in Indonesian waters were reported as damaged or missing. This highlights the urgent need for Indonesia to develop a reliable backup system to work in synergy with Ina-TEWS, enhancing the effectiveness of early tsunami warnings (Adityawan et al. 2023). Therefore, a non-buoy-based warning system such as using a maritime wireless connection (Adityawan et al. 2023), ship-borne (Hossen et al. 2021), or airborne GNSS data (Mulia et al. 2020), can be an alternative to enhance the performance of tsunami warning systems in Indonesian waters. In accordance with the problem, another critical issue is the human response to tsunami warnings. During the 2012 Indian Ocean earthquake, the tsunami sirens in Banda Aceh were delayed, sounding more than an hour after the earthquake occurred. Although vertical evacuation buildings were available, many residents lacked confidence in these structures' ability to withstand the earthquake and potential tsunami. Consequently, the majority of Banda Aceh's population opted for horizontal evacuation as the preferred strategy (Suppasri et al. 2015). Therefore, an evacuation plan developed from deterministic or stochastic earthquake scenarios (e.g. Muhammad et al. 2017) is necessary to provide clear guidance when an earthquake followed by a tsunami occurs.

Evacuation procedures can be conducted when an official tsunami hazard map is available. The fragility functions are an essential tool to estimate the damage probabilities of houses/structures and death rate in terms of the hydrodynamic features of tsunami inundation. With a case study in Banda Aceh, this method is further useful in developing tsunami hazard maps as they provide a relationship between local vulnerability and tsunami hazard. The potential damage to structures or casualties due to the variety of tsunami scenarios can be estimated by multiplying the number of tsunami exposures (e.g. houses and populations) by the fragility functions according to the local features of tsunami obtained by tsunami inundation models (Koshimura et al. 2009). A more sophisticated way to develop a tsunami hazard map is through probabilistic hazard analysis. Since it is derived from thousands of earthquake scenarios, the influence of the spatiotemporal earthquake rupture models is also included. The approach has been applied to the western part of Sumatra considering the future earthquake in the Mentawai region (Muhammad et al. 2022).

Since 2004, studies have been mainly focused on fault-generated tsunamis such as the 2004 Indian Ocean tsunami, 2006 Pangandaran tsunami, and 2010 Mentawai tsunami. However, in the last decade, two landslide tsunamis occurred in the same year, that is the 2018 Palu tsunami (Carvajal et al. 2019; Ho et al. 2021; Takagi et al. 2019) and the Sunda Strait tsunami (Heidarzadeh et al. 2020; Muhari et al. 2019). This raised the awareness that further studies need to be conducted since there is a difference between fault- and landslide-generated tsunamis. Most of the fault zones which has potential tsunamis have been identified, but not for the landslide tsunami. A recent study found a potential tsunami landslide in the eastern part of Kalimantan, even though with a relatively low probability of occurrence (Brackenridge et al. 2020). Similar studies should be performed in other regions, especially in the estuary area that is close to the active fault.

15.5 Progress of Sendai Framework's Implementation for Post Recovery Aceh Tsunami

15.5.1 Current Practice of Sendai Framework in Indonesia

Adopted by the United Nations in 2015, the Sendai Framework for Disaster Risk Reduction (SFDRR) 2015–2030 provides a global blueprint for mitigating disaster risks and enhancing resilience. The framework emphasizes the need of having a comprehensive understanding of disaster risk, enhancing governance in disaster risk management, investing in risk reduction initiatives, and enhancing capacities for disaster preparedness and response. As one of the most disaster-prone countries globally, Indonesia contends with various natural hazards, including earthquakes, tsunamis, landslides, flooding, and droughts. As an archipelagic nation with diverse topography—including coastal, mountainous, and mainland areas—the country has actively integrated the SFDRR's principles and priorities. In 2007 and 2008, significant efforts were made to develop the National Disaster Management Law and establish the National Disaster Management Agency (BNBP), with the goal of reducing the effects of disasters on the country's population and infrastructure.

The Government of Indonesia (GoI) is actively involved in worldwide discussions on disaster risk reduction (DRR), including participation in the Global Platform for Disaster Risk Reduction (GPDRR). Such conversations are held on a regular basis internationally. This active engagement allows Indonesia to share its experiences and best practices with other countries, contributing to a global understanding of DRR. A notable example is the 7th GPDRR Forum in Bali, Indonesia, with the theme "From Risk to Resilience: Toward Sustainable Development for All in a COVID-19 Transformed World." At this forum, Indonesia emphasized its dedication to reducing disaster risk by fostering collaboration among government, society, businesses, media, and academia. The forum also underscored the importance of tackling major disaster threats, including climate change and the COVID-19 pandemic.

The three primary groups affected by disasters are the government, businesses, and communities. In line with the Sendai Framework's second priority of improving disaster risk governance, Indonesia has launched initiatives such as the Disaster Alert Cadet Program and Disaster Resilient Villages (Desa Tangguh Bencana/Destana). These initiatives aim to be expanded to a citywide scale to support resilient city programs based on spatial planning. National institutions have enhanced data collection, information dissemination, and storage capacity. BNPB is the leading agency overseeing all aspects of disaster management, including prevention, mitigation, preparedness, emergency response, and recovery. It applies the principles of the

Sendai Framework to develop the National Disaster Risk Reduction Strategy (RIPB) for 2020–2044. A notable objective of this strategy is to enhance the distribution of digital information to remote areas for early warning of natural hazards. Furthermore, the third priority of the Sendai Framework, which emphasizes investment in DRR, is being implemented through a pooling fund scheme that supports the restoration of the government's fiscal capacity.

The Government of Indonesia (GoI) is focused on strengthening its disaster risk governance by refining its institutional framework and promoting improved stakeholder coordination. Moreover, it has invested in DRR initiatives, including early warning systems, evacuation planning, and the development of disaster-resistant infrastructure raining and capacity-building programs for emergency responders and the public are the example of efforts to improve disaster preparedness and response. Indonesia's commitment to implementing the Sendai Framework is crucial in reducing the losses and damages caused by disasters to its population and infrastructure. By integrating the framework's principles into its DRR strategies, Indonesia seeks to enhance its resilience, mitigate disaster risks, and support the attainment of the United Nations' Sustainable Development Goals (SDGs) (Uchiyama et al. 2021).

Indonesia has utilized the Sendai Framework as a guiding principle in developing Presidential Decree No. 87 of 2020, which pertains to the RIPB 2020–2044. The implementation roadmap for the RIPB is structured into five-year achievement phases (Rachel and Hidayat 2024). The RIPB Implementation Roadmap has adopted four main priorities in the Sendai Framework, namely (1) increasing disaster risk assessment; (2) increasing institutional synergy and resilience in disaster risk governance; (3) strengthening investment in disaster risk management; and (4) increasing readiness for an effective response. In the current conditions, Indonesia has improved its disaster risk assessment and mapping using advanced technologies like GIS and remote sensing to identify hazard-prone areas, especially in the analysis for formulating spatial plans. Also, the GoI under BNPB has collected and disseminated disaster risk data, making it accessible to public and private stakeholders. Regarding disaster risk governance, the GoI has strengthened and mainstreamed its legal and policy framework for DRR. Laws and regulations have been updated to align with the Sendai Framework (for example, RIPB), ensuring a coordinated approach across all government levels, from provinces to cities and regencies. At the national level, BNPB, with support from Local Disaster Management Agencies (BPBD), plays a crucial role in coordinating DRR efforts. It collaborates closely with local governments, NGOs, and international partners. The GoI has prioritized investments in DRR to enhance resilience, focusing on infrastructure projects like earthquake-resistant buildings, flood control systems, early warning systems, and community-based DRR programs (Lassa et al. 2018). For instance, local community engagement in DRR through socialization, training, capacity building, and participatory planning processes. Programs like Disaster Resilient Villages strengthen communities to manage and reduce their risks. At the national level, BNPB, with support from Local Disaster Management Agencies (BPBD), plays a crucial role in coordinating DRR efforts. It collaborates closely with local governments, NGOs, and international partners. The GoI has prioritized investments in DRR to enhance

Fig. 15.8 The span of the Sendai framework practices in Indonesia. (*Source* Authors)

resilience, focusing on infrastructure projects like earthquake-resistant buildings, flood control systems, early warning systems, and community-based DRR programs (Lassa et al. 2018). The span of the Sendai Framework Practices in Indonesia can be seen in Fig. 15.8.

15.5.2 Integration Among Sendai Framework, Disaster Law of Indonesia, Climate Change Adaptation Regulation

In integrating the principles of the SFDRR, national disaster management laws, and climate change adaptation (CCA) regulations, Indonesia has made significant initiatives. This integration creates a cohesive approach to DRR and climate resilience (Wijaya et al. 2017). This effort is reflected in the country's legal and policy framework, including Law Number 24 of 2007 on Disaster Management, which offers a thorough framework for disaster management, encompassing prevention, mitigation, preparedness, emergency response, and recovery. Other vital components are the RIPB, which sets out strategies and priorities in line with SFDRR objectives; the National Action Plan for CCA (Rencana Aksi Nasional-Adaptasi Perubahan Iklim/RAN-API), which focuses on addressing climate change impacts and incorporating CCA into development planning; and Presidential Regulation No. 16/2015, which established the National Council on Climate Change to coordinate national climate policies and align them with broader development goals.

Besides the legal framework, the efforts for integration can be seen from the institutional perspective (Uchiyama et al. 2021). For example, in 2008, the GoI established BNPB, the central agency responsible for coordinating DRR activities and collaborates with the Ministry of Environment and Forestry, which leads CCA efforts. The cross-sectoral coordination has also been adopted through regular meetings and joint initiatives between BNPB and other relevant ministries and agencies to harmonize disaster management and CCA efforts. As mentioned in the legal aspect, the policy and planning have been integrated among the issue of disaster, for instance, the alignment of RIPB and RAN-API to ensure cohesive strategies.

This effort involves incorporating climate risk assessments into DRR plans. Institutional integration occurs not only at the national level but also at the local level. Local governments are urged to include DRR and CCA measures in their regional development plans (Rencana Pembangunan Jangka Panjang-RPJP/Rencana Pembangunan Jangka Menengah-RPJM). National agencies provide guidelines and technical support to facilitate this integration.

Lastly, integrating financial and programs is also essential (Uchiyama et al. 2021), such as the dedicated budget allocations for DRR and CCA are integrated into national (APBN) and local budgets (APBD), ensuring that resources are available for implementing integrated strategies. Programs and projects often have components addressing both disaster risk and CCA, such as flood management projects that incorporate both structural measures and ecosystem-based approaches. For example, the case of Jakarta's Flood Management strategy combines infrastructure projects (sea walls and retention basins) with CCA measures (mangrove restoration and green spaces). This integrated approach aligns with both the Sendai Framework and climate adaptation goals. Another case in the coastal resilience programs through ecosystem-based adaptation, particularly in regions like Semarang and Demak, has integrated DRR and climate adaptation efforts to restore mangroves and other natural barriers and coastal erosion and flooding.

Despite progress, several challenges hinder Indonesia's effective DRR and climate adaptation efforts. Complex coordination among multiple agencies and sectors often leads to overlapping or implementation gaps (Peters 2018). Limited financial and human resources also pose significant barriers to integrating DRR and climate adaptation measures. Additionally, gaps in data availability and insufficient information on climate risks and disaster impacts can impede planning and decision-making processes. However, there are opportunities to overcome these challenges. Leveraging technology for improved risk assessment, early warning systems, and data management can enhance integration efforts. International support and collaboration, including technical assistance and funding from global partners, can help bridge gaps in capacity and resources. Furthermore, increasing community participation in DRR and climate adaptation initiatives can lead to more sustainable and locally tailored solutions.

In conclusion, Indonesia has made significant strides in integrating the Sendai Framework, national disaster management laws, and CCA regulations. While challenges persist, the country's commitment to cohesive and comprehensive strategies provides a promising path toward building resilient and sustainable communities. Indonesia can effectively reduce the impacts of climate-related disasters by improving coordination, optimizing resource allocation, and encouraging community involvement.

15.6 Key Challenges for Disaster Risk Reduction for Post-Tsunami

The disaster management in Aceh has referred to the national disaster management system for its implementation and action. As mentioned that changes have occurred, from a responsive perspective (emergency response) to disaster risk reduction. This is evidenced by the Aceh Development Agenda which contains the handling of pre-disaster, disaster, post-disaster (emergency response) and post-disaster (rehabilitation and reconstruction) events. First, the challenge in providing Disaster Management services in Aceh is mostly due to the substantially high Disaster Risk Index, which stands at approximately 160. That number means Aceh Province is highly vulnerable to hazards from nature due to its condition. The Aceh Disaster Risk Index Book 2022 reveals that 12 districts/cities (52%) in Aceh continue to have a high disaster risk index value. These districts/cities include Aceh Besar, Nagan Raya, West Aceh, Southwest Aceh, Aceh Jaya, and Aceh Singkil. The areas included are East Aceh, North Aceh, South Aceh, Bireun, Pidie, and Simeuleu. Currently, 11 districts or cities have a moderate risk level (48%), including Aceh Tamiang, Langsa City, Pidie Jaya, Southeast Aceh, Sabang City, and Central Aceh. The areas mentioned include Bener Meriah, Gayo Lues, Subulussalam City, Lhokseumawe City, and Banda Aceh City. This indicates that every district and city in Aceh is vulnerable to risk of disaster.

Second, some disaster risk reduction methods used in developed countries cannot be straightforwardly implemented in Aceh. An important challenge is the significant disparity in financial resources and socio-cultural context. Nevertheless, Aceh has the potential to adopt and implement successful practices observed in developed country. An approach involves modifying existing structures or infrastructure to fulfill their original function and enhance their ability to withstand and mitigate the impact of tsunamis. The term used to describe this idea is well recognized as the co-benefits of structure/infrastructure (Samarasekara et al. 2017). Furthermore, the implementation of techniques in Indonesia concerning the early warning system, as indicated in Chap. 4, is crucial due to enhancing public comprehension of appropriate actions to take in response to a tsunami. It is essential to recognize the importance of providing public education and counseling on the correct procedures to follow in the aftermath of a tsunami warning (Supani and Andriani 2020). People need to understand and be able to notice warnings attentively, accurately identify safe evacuation areas, and promptly and effectively follow evacuation protocols. Effective communication between relevant officials and communities plays an important role in achieving this awareness (Suppasri et al. 2015).

Third, incorporating tsunami risk into spatial planning is the most important aspect of mitigating risks (Santos et al. 2020) as discussed in Chap. 5. Coastal areas that vulnerable to tsunamis should limit inappropriate development and regulate land use wisely (Butt 2014). It is essential to ensure that critical infrastructure, such as hospitals, schools, and evacuation centers, are located in areas that are safe from the risk of potential tsunami disasters (Ibrion et al. 2020). This involves the government's role in developing spatial policies that focus on disaster risk mitigation.

Fourth, it is also important to consider the use of more advanced technologies in tsunami risk reduction or mitigation. For instance, the application of sensor-based early warning systems and real-time monitoring can provide more accurate and rapid information, as mentioned in Sect. 15.4. In a global context, cooperation between countries is also very important. Tsunamis are natural hazards that do not recognize national borders, and often affect more than one country. Therefore, countries must be willing to help each other in emergency situations (United Nations 2020). These activities include sharing information, developing humanitarian assistance, and providing technical assistance to reduce the impact of potential risks. Furthermore, technological innovation has expanded the range of options available to enhance a disaster resilience and reduce risk. Robots and drones, along with advancements in disruptive technologies like as artificial intelligence (AI), the internet of things (IoT), and big data, are transforming various industries, including a disaster risk reduction and management. However, the cost of implementing digital technologies for disaster management can be expensive. Although the expenses associated with hardware, such as UAVs and sensors, are consistently decreasing, the costs related to operating, integrating, and evaluating the data they provide remain expensive. Finally, from the sub-chapters described above, it can be concluded that the challenges and recommendations to reducing disaster risk in Indonesia, especially for tsunamis, are as follows:

a. Natural factors, i.e. Climate Change

It cannot be ignored that climate change affects the stability of coastal ecosystems. Phenomena such as sea level rise alter coastal boundaries and affect seawater flow, which can alter tsunami dynamics (IPCC 2019). In this context, the resilience of coastal ecosystems is important. For example, healthy mangrove forests act as a 'living wall' that protects against tsunami impacts (Wetlands International 2017).

b. Urbanization and Spatial Planning Factors

Considering tsunami risk is crucial when considering spatial planning. Numerous coastal cities possess infrastructure that is inadequate for dealing with this natural hazards. Effective urban planning must consider the establishment of easier access to areas of evacuation and the continued economic activities in the aftermath of a disaster (UN-Habitat 2020). The existence of the disaster resilient village (Desa Tangguh Bencana) program is a good case and should be sustainable to reduce disaster risk. Besides, AI has the capacity to significantly enhance response and recovery times in the field of disaster management. Humanitarian groups are employing machine learning techniques to expedite the map creation process by detecting elements such as roads and residences in aerial photographs. Many individuals are considering artificial intelligence as a promising technology for the detection and, possibly, prediction of earthquakes. Integrating this technology with spatial planning can also help the fast recovery for the natural hazard.

c. Early Warning and Education System

In order to provide effective early warning, it is essential to not only possess the necessary technological skills, but also prioritize public education. People have to understand the right direction to follow after receiving a warning. This includes basic understanding such as reading evacuation maps, familiarizing oneself with the quickest route to higher ground, and detecting natural signals that indicate the threat of a tsunami (UNESCO 2016). As described in Sect. 15.3 (three), optimizing mosques as evacuation centers may be a cost-effective alternative when local governments have limited funding for constructing new evacuation sites. Using technology is more important to reducing risk. For example, the use of sensors to monitor hazards that could cause disasters has a long history. Technological developments in areas such as cloud computing, wireless broadband, sensors, and data processing have made possible the expansion of interconnected, more reliable, and real-time systems that make up the Internet of Things (IoT).

d. Multi-sectoral and International Collaboration

National or sectoral boundaries do not limit the potential for a tsunami to occur. The issue is worldwide and necessitates answers on a global scale. International cooperation in geological and oceanic research and monitoring is crucial. Collaboration among many sectors, including the commercial sector, governments, and NGOs, is equally important. They must join forces to develop efficient strategies (World Economic Forum 2021). A tsunami is a large ocean wave caused by an underwater earthquake, volcanic eruption, or other disturbance. Global cooperation is crucial in mitigating the risk of tsunamis. It is imperative for countries that face the same tsunami threats, such as Indonesia, to collaborate in the exchange of data, technology, and resources to enhance early warning systems and collective risk reduction efforts. This partnership also involves engagement in regional and international institutions that facilitate the reduction of tsunami risks (UNDRR 2015). Best partnerships from developed countries is crucial. Indonesia can take best practice, for instance, like in Japan. Japan is a leading country when it comes to the utilization of robots in scenarios of disaster. Tohoku University's Human–Robot Informatics Laboratory has developed a range of robots for disaster response, including a snake-like robot fitted with a camera. This robot is capable of maneuvering through confined spaces, overcoming obstacles, and tracking along walls. Current research endeavors to provide search-and-rescue dogs with compact and lightweight inertial measurement units, Global Positioning System (GPS) receivers, and cameras.

15.7 Conclusion

The tsunami that occurred in Aceh 20 years ago had a major effect on the lives of numerous people in Aceh as well as another nation. The last twenty years should be recognized as awakening our fundamental human nature and the realization that life is unpredictable, requiring constant preparedness to deal with unanticipated events. This chapter highlights that:

a. Indonesia experienced the most impact from the 2004 tsunami. Aceh province was the most affected provinces. There was massive damage and lost after tsunami 2004. Furthermore, that tsunami resulted in significant alterations to the coastal areas of Aceh, including changes to coastlines, land use, settlements, and infrastructure.
b. Since the 2004 Aceh Tsunami raised awareness of disaster risk, disaster management efforts have advanced significantly. The regulatory framework for disaster management has been improved by the enactment of Law No. 24/2007 on Disaster Management, as well as a number of additional laws and regulations. Disaster management planning has improved, with several areas developing disaster management, disaster recovery, and emergency plans.
c. Indonesia has actively engaged in implementing the Hyogo Framework for Action 2005–2015 internationally. Additionally, Indonesia has established several policies and disaster management planning documents that are aligned with the Sendai Framework for Disaster Risk Reduction.
d. There are several challenges to implementing proper disaster risk reduction especially for tsunami, for instance: natural factors, urbanization and spatial planning, early warning system, and collaboration.

Developing a culture that prioritizes disaster awareness and preparedness is an important aspect of human resource development and an understanding of science and technology. In order to enhance human resource capacity, it is essential also to enhance knowledge and understanding of disasters. This should be followed by the establishment of strong awareness and continuous education in preparedness. These efforts will ultimately encourage a culture and moral commitment to disaster preparedness and resilience across various development sectors. This approach should be accompanied by efforts to enhance human resource capacity in the fields of science and technological proficiency, which can serve as sustainable development assets.

References

ACFID (2006) NGO report on the Asian Tsunami (26 December 2004–31 December 2005)

Adityawan MB, Nurendyastuti AK, Purnama MR, Arifianto MS, Farid M, Kuntoro AA, Widyaningtias (2023) Development of a tsunami early warning system on the coast of Palu based on maritime wireless communication. Prog Disaster Sci 19:100290. https://doi.org/10.1016/j.pdisas.2023.100290

Amezquita-Sanchez J, Valtierra-Rodriguez M, Adeli H (2017) Current efforts for prediction and assessment of natural disasters: Earthquakes, tsunamis, volcanic eruptions, hurricanes, tornados, and floods. Sci Iran 24. https://doi.org/10.24200/sci.2017.4589

Associated Press (2014) Numbers that tell story of 2004 Tsunami disaster. Associated Press

Athukorala P, Resosudarmo BP (2005) The Indian Ocean tsunami: Economic impact, disaster management and lessons. Asian Econ Pap 4(1):1–39. https://doi.org/10.1162/asep.2005.4.1.1

Bland SH, Valoroso L, Stranges S, Strazzullo P, Farinaro E, Trevisan M (2005) Long-term follow-up of psychological distress following earthquake experiences among working Italian males: A cross-sectional analysis. J Nerv Ment Dis 193(6):420–423. https://doi.org/10.1097/01.nmd.0000165297.49452.41

BNPB (2022) Badan Nasional Penanggulangan Bencana. 17 Februari 2022, 13120. https://bnpb.go.id/berita/bnpb-verifikasi-5-402-kejadian-bencana-sepanjang-tahun-2021

BPS Statistics Indonesia (2020) Number and distribution of population. Retrieved from https://sensus.bps.go.id/main/index/sp2020

Butt S (2014) Disaster management law in Indonesia: From response to preparedness? In Butt S, Nasu H, Nottage L (eds) Asia-Pacific disaster management: Comparative and socio-legal perspectives. Springer, pp 1–303. https://doi.org/10.1007/978-3-642-39768-4_9

Brackenridge RE, Nicholson U, Sapiie B, Stow D, Tappin DR (2020) Indonesian throughflow as a preconditioning mechanism for submarine landslides in the Makassar Strait. Geol Soc, Lond, Spec Publ 500(1):195 LP–217. https://doi.org/10.1144/SP500-2019-171

Brennan RJ (2005) Rapid health assessment in Aceh Jaya District, Indonesia, following the December 26 tsunami. Emerg Med Australas 17(4):341–350. https://doi.org/10.1111/j.1742-6723.2005.00755.x

Cabinet Secretariat of The Republic of Indonesia (2020) President Jokowi Signs Regulation on 2020–2044 disaster management master plan. Retrieved from https://setkab.go.id/en/president-jokowi-signs-regulation-on-2020-2044-disaster-management-master-plan/

Carvajal M, Araya-Cornejo C, Sepúlveda I, Melnick D, Haase JS (2019) Nearly instantaneous tsunamis following the Mw 7.5 2018 Palu earthquake. Geophys Res Lett 46(10):5117–5126. https://doi.org/10.1029/2019GL082578

Davidson JRT, McFarlane AC (2006) The extent and impact of mental health problems after disaster. J Clin Psychiatry 67(SUPPL. 2):9–14

Djalante R, Garschagen M, Thomalla F, Shaw R (2017) Disaster risk reduction in Indonesia: Progress, challenges, and issues disaster risk reduction, methods, approaches and practices. Retrieved from https://doi.org/10.1007/978-3-319-54466-3

Faiez Z, Fan D (2023) Study of coastal morphological changes by tsunamis in Aceh (Indonesia) using satellite images. J Geosci, Eng, Environ, Technol 8(4):295–304. https://doi.org/10.25299/jgeet.2023.8.4.12757

Fan L (2013) Disaster as opportunity?—building back better in Aceh, Myanmar and Haiti (Issue November)

Fengler W (2007) Tracking financial flows after disasters—Reconstruction Expenditure Tracking Analysis Methodology (RETAM)

Foster R, Hagan A, Perera N, Gunawan CA, Silaban I, Yaha Y, Manuputty Y, Hazam I, Hodgson G (2006) Tsunami and earthquake damage to coral reefs of Aceh, Indonesia. Reef Check Foundation, Pacific Palisades, California, USA 33:3

Frankenberg E, Friedman J, Gillespie T, Ingwersen N, Pynoos R, Rifai IU, Sikoki B, Steinberg A, Sumantri C, Suriastini W, Thomas D (2008) Mental health in Sumatra after the tsunami. Am J Public Health 98(9):1671–1677. https://doi.org/10.2105/AJPH.2007.120915

Gouramanis C, Morales-Ramirez C (2020) Deep understanding of natural hazards based on the Sendai framework for disaster risk reduction in a higher education geography module in Singapore. Int Res Geogr Environ Educ 30:1–20. https://doi.org/10.1080/10382046.2020.1751391

Ho T-C, Satake K, Watada S, Hsieh M-C, Chuang RY, Aoki Y, Mulia IE, Gusman AR, Lu C-H (2021) Tsunami induced by the strike-slip fault of the 2018 Palu earthquake (Mw = 7.5), Sulawesi Island, Indonesia. Earth Space Sci 8(6):e2020EA001400. https://doi.org/10.1029/2020EA001400

Hossen MJ, Mulia IE, Mencin D, Sheehan AF (2021) Data assimilation for tsunami forecast with ship-borne GNSS data in the Cascadia Subduction Zone. Earth Space Sci 8(3). https://doi.org/10.1029/2020EA001390

Heidarzadeh M, Ishibe T, Sandanbata O, Muhari A, Wijanarto AB (2020) Numerical modeling of the subaerial landslide source of the 22 December 2018 Anak Krakatoa volcanic tsunami, Indonesia. Ocean Eng 195:106733. https://doi.org/10.1016/j.oceaneng.2019.106733

Ibrion M, Paltrinieri N, Nejad AR (2020) Learning from non-failure of Onagawa nuclear power station: An accident investigation over its life cycle. Results Eng 8(November):100185. https://doi.org/10.1016/j.rineng.2020.100185

IPCC (2019) Climate change and land: an IPCC special report on climate change, desertification, land degradation, sustainable land management, food security, and greenhouse gas fluxes in terrestrial ecosystems. Shukla PR, Skea J, Calvo Buendia E, Masson-Delmotte V, Pörtner H-O, Roberts DC, Zhai P, Slade R, Connors S, van Diemen R, Ferrat M, Haughey E, Luz S, Neogi S, Pathak M, Petzold J, Portugal Pereira J, Vyas P, Huntley E, Kissick K, Belkacemi M, Malley J (eds). In press

Japan International Cooperation Agency (2015) Country report Indonesia. Retrieved from https://openjicareport.jica.go.jp/pdf/1000023394.pdf

Kennedy J, Ashmore J, Babister E, Kelman I (2008) The meaning of 'Build back better': Evidence from post-tsunami Aceh and Sri Lanka. J Contingencies Cris Manag 16(1):24–36. https://doi.org/10.1111/j.1468-5973.2008.00529.x

Koshimura S, Oie T, Yanagisawa H, Imamura F (2009) Developing fragility functions for tsunami damage estimation using numerical model and post-tsunami data from Banda Aceh, Indonesia. Coast Eng J 51(3):243–273. https://doi.org/10.1142/S0578563409002004

Lassa JA, Boli Y, Nakmofa Y, Fanggidae S, Ofong A, Leonis H (2018) Twenty years of community-based disaster risk reduction experience from a dryland village in Indonesia. Jàmbá: J Disaster Risk Stud 10(1):1–10

Law of the Republic of Indonesia No. 24/2007 Concerning Disaster Management (2007) National Agency for Disaster Management (2024) Organizational structure of BNPB. Retrieved from https://bnpb.go.id/struktur-organisasi-bnpb

Liew SC, Gupta A, Wong PP, Kwoh LK (2010) Recovery from a large tsunami mapped over time: The Aceh coast, Sumatra. Geomorphology 114(4):520–529. https://doi.org/10.1016/j.geomorph.2009.08.010

MacDonald R (2005) How women were affected by the tsunami: A perspective from Oxfam. PLoS Med 2(6):0474–0475. https://doi.org/10.1371/journal.pmed.0020178

Manaf A (2011) Trusting the survivor to recosntruct their own shelter, challenges, constraint and advantages—Lesson learnd from UN HABITAT's post disaster recostruction project in Aceh, Indonesia. Jurnal Tata Loka 13(1):16–26. https://doi.org/10.14710/tataloka.13.1.16-26

McCaughey JW, Mundir I, Daly P, Mahdi S, Patt A (2017) Trust and distrust of tsunami vertical evacuation buildings: Extending protection motivation theory to examine choices under social influence. Int J Disaster Risk Reduct 24(April):462–473. https://doi.org/10.1016/j.ijdrr.2017.06.016

Meilianda E, Pradhan B, Syamsidik, Comfort LK, Alfian D, Juanda R, Syahreza S, Munadi K (2019) Assessment of post-tsunami disaster land use/land cover change and potential impact of future sea-level rise to low-lying coastal areas: A case study of Banda Aceh coast of Indonesia. Int J Disaster Risk Reduct 41(October 2018):101292. https://doi.org/10.1016/j.ijdrr.2019.101292

Mirza F (2015) Disaster mitigation approach of urban green structure concept in coastal settlement. DIMENSI—Journal Arch Built Environ 42(2):51–58. https://doi.org/10.9744/dimensi.42.2.51-58

MoNDP (2005) Indonesia: Preliminary damage and loss assessment. The December 26, 2004 natural disaster. Ministry of National Development Planning/National Development Planning Agency (Bappenas)

Muhammad A, Goda K, Alexander NA, Kongko W, Muhari A (2017) Tsunami evacuation plans for future megathrust earthquakes in Padang, Indonesia, considering stochastic earthquake scenarios. Nat Hazard 17(12):2245–2270. https://doi.org/10.5194/nhess-17-2245-2017

Muhammad A, Goda K, Werner MJ (2022) Time-dependent probabilistic tsunami hazard analysis for Western Sumatra, Indonesia, using space-time earthquake rupture modelling and stochastic source scenarios. Nat Hazards Earth Syst Sci

Muhari A, Heidarzadeh M, Susmoro H, Nugroho HD, Kriswati E, Supartoyo, Wijanarto AB, Imamura F, Arikawa T (2019) The December 2018 anak krakatau volcano tsunami as inferred from post-tsunami field surveys and spectral analysis. Pure Appl Geophys 176(12):5219–5233. https://doi.org/10.1007/s00024-019-02358-2

Mulia IE, Hirobe T, Inazu D, Endoh T, Niwa Y, Gusman AR, Tatehata H, Waseda T, Hibiya T (2020) Advanced tsunami detection and forecasting by radar on unconventional airborne observing platforms. Sci Rep 10(1):2412. https://doi.org/10.1038/s41598-020-59239-1

Musa R, Draman S, Jeffrey S, Jeffrey I, Abdullah N, Halim NAM, Wahab NA, Mukhtar NZM, Johari SNA, Rameli N, Midin M, Jaafar NRN, Das S, Sidi H (2014) Post tsunami psychological impact among survivors in Aceh and West Sumatra, Indonesia. Compr Psychiatry 55(SUPPL. 1). https://doi.org/10.1016/j.comppsych.2012.12.002

National Agency for Disaster Management (2024) Struktur Organisasi. Retrieved from https://bnpb.go.id/strukturorganisasi-bnpb

National Agency for Disaster Management. Profil Bencana di Indnesia. Retrieved from https://dibi.bnpb.go.id/home/index2

Nugroho SS, Pasaribu D, Fadiyah D, Ayu Andini L (2023) Tata kelola bencana non alam di Indonesia: Temuan Awal. Jiapi: J Ilmu Adm Dan Pemerintah Indones 4(1):13–26

Ohsumi T, Gejo T (2005) Field investigation on the damage of road and infrastructure caused by the tsunami in Meulaboh, Aceh, Indonesia. In: Proceedings of the JSCE earthquake engineering symposium, vol 28, pp 1–8. https://doi.org/10.11532/proee2005a.28.110

Oktari RS, Shiwaku K, Munadi K, Syamsidik, Shaw R (2018) Enhancing community resilience towards disaster: The contributing factors of school-community collaborative network in the tsunami affected area in Aceh. Int J Disaster Risk Reduct 29(April 2017):3–12. https://doi.org/10.1016/j.ijdrr.2017.07.009

Peters K (2018) Accelerating Sendai framework implementation in Asia. The Overseas Development Institute (ODI), London

Pratami D (2023) Riding the waves for a blue future in Indonesia. Retrieved from https://unsdg.un.org/latest/blog/riding-waves-blue-future-indonesia

Rachel I, Hidayat M (2024) Implementation of disaster risk reduction policy in Indonesia. Jurnal Ekonomi 13(01):2104–2120

Ramalanjaona G (2011) Impact of 2004 tsunami in the islands of Indian Ocean: Lessons learned. Emerg Med Int 2011:1–3. https://doi.org/10.1155/2011/920813

Rofi A, Doocy S, Robinson C (2006) Tsunami mortality and displacement in Aceh province, Indonesia. Disasters 30(3):340–350. https://doi.org/10.1111/j.0361/3666.2005.00324.x

Rose Enid Teresa A, Jerry Vasanth GDN, Vidya C (2020) A review on the potential effects of tsunami on built environment. Mater Today: Proc 33:711–715. https://doi.org/10.1016/j.matpr.2020.06.019

Rousseau C, Drapeau A, Platt R (2004) Family environment and emotional and behavioural symptoms in adolescent Cambodian Refugees: Influence of time, gender, and acculturation. Med Confl Surviv 20(2):151–165. https://doi.org/10.1080/1362369042000234735

Samarasekara RSM, Sasaki J, Esteban M, Matsuda H (2017) Assessment of the co-benefits of structures in coastal areas for tsunami mitigation and improving community resilience in Sri Lanka. Int J Disaster Risk Reduct 23:80–92. https://doi.org/10.1016/j.ijdrr.2017.04.011

Santos A, Sousa N, Kremers H, Bucho JL (2020) Building resilient urban communities: The case study of setubal municipality, Portugal. Geosci (Switz) 10(6):1–13. https://doi.org/10.3390/geosciences10060243

Satake K (2014) Advances in earthquake and tsunami sciences and disaster risk reduction since the 2004 Indian ocean tsunami. Geoscience Letts 1(1):1–13. https://doi.org/10.1186/s40562-014-0015-7

Scoones I (2009) Livelihoods perspectives and rural development. J Peasant Stud 36(1):171–196. https://doi.org/10.1080/03066150902820503

Shibayama T, Okayasu A, Sasaki J, Wijayaratna N, Suzuki T, Jayaratne R, Masimin, Ariff Z, Matsumaru R (2005) Disaster survey of Indian Ocean tsunami in South Coast of Sri Lanka and Aceh, Indonesia. Ann J Coast Eng JSCE 52:1401–1405

Souza R, Bernatsky S, Jong RR, de K (2007) Mental health status of vulnerable tsunami-affected communities: A survey in Aceh province, Indonesia. J Trauma Stress 20(3):251–262. https://doi.org/10.1002/jts

Steinberg F (2007) Housing reconstruction and rehabilitation in Aceh and Nias, Indonesia—Rebuilding lives. Habitat Int 31(1):150–166. https://doi.org/10.1016/j.habitatint.2006.11.002

Steinritz V, Pena-Castellnou S, Marliyani GI, Reicherter K (2021) GIS-based study of tsunami risk in the Special Region of Yogyakarta (Central Java, Indonesia). IOP Conf Ser: Earth Environ Sci 851(1). https://doi.org/10.1088/1755-1315/851/1/012007

Stephan C, Norf C, Fekete A (2017) How "Sustainable" are post-disaster measures? Lessons to be learned a decade after the 2004 tsunami in the Indian Ocean. Int J Disaster Risk Sci 8(1):33–45. https://doi.org/10.1007/s13753-017-0113-1

Supani A, Andriani Y (2020) Designing and applying flood early warning system based on waterfall and water level for special of Palembang City. J Phys: Conf Ser 1500(1). https://doi.org/10.1088/1742-6596/1500/1/012129

Suppasri A, Goto K, Muhari A, Ranasinghe P, Riyaz M, Affan M, Mas E, Yasuda M, Imamura F (2015) A Decade after the 2004 Indian Ocean tsunami: The progress in disaster preparedness and future challenges in Indonesia, Sri Lanka, Thailand and the Maldives. Pure Appl Geophys 172(12):3313–3341. https://doi.org/10.1007/s00024-015-1134-6

Suryaratri RD, Akbar Z, Ariyani M, Purwalatia AT, Wahyuni LD (2020) The impact of disaster awareness towards household disaster preparedness among families on the coast of Banten, Sumur District, Indonesia. IOP Conf Ser: Earth Environ Sci 448(1):012122. https://doi.org/10.1088/1755-1315/448/1/012122

Syamsidik, Al Farizi MD, Tursina, Yulianur A, Rusydy I, Suppasri A (2023) Assessing probability of building damages due to tsunami hazards coupled with characteristics of buildings in Banda Aceh, Indonesia: A way to increase understanding of tsunami risks. Int J Disaster Risk Reduct 90:1–24. https://doi.org/10.1016/j.ijdrr.2023.103652

Syamsidik, Oktari RS, Nugroho A, Fahmi M, Suppasri A, Munadi K, Amra R (2021) Fifteen years of the 2004 Indian Ocean Tsunami in Aceh-Indonesia: Mitigation, preparedness and challenges for a long-term disaster recovery process. Int J Disaster Risk Reduct, 54(Jan):102052. https://doi.org/10.1016/j.ijdrr.2021.102052

Takagi H, Pratama MB, Kurobe S, Esteban M, Aránguiz R, Ke B (2019) Analysis of generation and arrival time of landslide tsunami to Palu City due to the 2018 Sulawesi

The Minister of Home Affairs (2008) Permendagri No 46 Tahun 2008 tentang pedoman organisasi dan tata kerja BPBD earthquake. Landslides, 16(5):983–991. https://doi.org/10.1007/s10346-019-01166-y

Thorburn C (2009) Livelihood recovery in the wake of the tsunami in aceh. Bull Indones Econ Stud 45(1):85–105. https://doi.org/10.1080/00074910902836171

Uchiyama C, Ismail N, Stevenson LA (2021) Assessing contribution to the Sendai framework: Case study of climate adaptation and disaster risk reduction projects across sectors in Asia-Pacific (2015–2020). Prog Disaster Sci 12:100195

UN-Habitat (2020) World cities report. The value of sustainable urbanization. United nations human settlements programme. https://unhabitat.org/sites/default/files/2020/10/wcr_2020_report.pdf

UNDRR (2015) Global assessment report on disaster risk reduction, making development sustainable: the future of disaster risk management. UN, ISDR

United Nations (2020) Sustainable development goals. https://www.un.org/sustainabledevelopment/

United Nations Development Programme (UNDP) (2013) The rise of the South: human progress in a diverse world. Human development report. New York: UNDP

USGS (2005) The 26 December 2004 Indian Ocean tsunami: Initial findings from Sumatra. U.S. Geological Survey (USGS)

Vidyattama Y, Merdikawati N, Tadjoeddin MZ (2021) Aceh tsunami: Long-term economic recovery after the disaster. Int J Disaster Risk Reduct, 66(Sept). https://doi.org/10.1016/j.ijdrr.2021.102606

Wetlands International (2017) Wetlands annual report. https://rsr.akvo.org/media/db/org/474/document/WL-AnnualReport%202017.pdf

Wiguna T, Guerrero APS, Kaligis F, Khamelia M (2010) Psychiatric morbidity among children in North Aceh district (Indonesia) exposed to the 26 December 2004 tsunami. Asia Pac Psychiatry 2(3):151–155. https://doi.org/10.1111/j.1758-5872.2010.00079.x

Wijaya N, Bisri MBF, Aritenang AF, Mariany A (2017) Spatial planning, disaster risk reduction, and climate change adaptation integration in Indonesia: Progress, challenges, and approach. Disaster Risk Reduct Indones: Prog, Chall, Issues:235–252

World Economic Forum (2021) The global risks report 2021. 16th Edition, World Economic Forum, Cologny. http://www3.weforum.org/docs/WEF_The_Global_Risks_Report_2021.pdf

Chapter 16
Two Decades from the Indian Ocean Tsunami: Key Challenges and Advancements in Thailand

Indrajit Pal, Ratchaneekorn Thongthip, Subhajit Ghosh, Ayush Baskota, and Jeeten Kumar Bhagnani

Abstract Although 80% of tsunamis occur in the Pacific Ocean, it has always been an impending threat to the countries within the Indian Ocean region. Tsunami risks in coastal communities in Thailand are imminent due to inadequate coastal protection infrastructure, high population density, tourist-dependent economy and social inequalities, as witnessed during the large-scale damage and economic loss caused by the 2004 Indian Ocean Tsunami. In response to the 2004 tsunami, regional early warning systems have been developed to detect and enhance the effectiveness of response and evacuation. However, tsunami risk management entails not only technological advancement in tsunami detection but also a deeper understanding of local contexts and investment in community engagement, disaster preparedness education, evacuation facilities and procedures, land use zoning, livelihood and coastal infrastructure planning. Therefore, the objective of this work is to study the advancements made by Thailand over the past two decades in tsunami risk management and elaborate on success, challenges, and gaps at national, sub-national and particularly community levels. The findings of the study will be helpful in streamlining existing policies and interventions for risk reduction in the local context.

Keywords Tsunami · Indian Ocean · Thailand · Coastal communities · Risk

16.1 Introduction

In the active seismic ocean boundaries, tsunamis are a perpetual threat to lives, livelihood, and infrastructure along the coast. The United Nations estimates that global exposure to tsunami dangers 60,000 people and US$4 billion in assets (Pal et al. 2023) Due to poor infrastructure and high population density, developing countries

I. Pal (✉) · R. Thongthip · S. Ghosh · A. Baskota · J. K. Bhagnani
Disaster Preparedness, Mitigation and Management, Asian Institute of Technology, Bangkok, Thailand
e-mail: indrajit-pal@ait.ac.th

R. Shaw et al. (eds.), *Two Decades from the Indian Ocean Tsunami*, Disaster Risk Reduction, https://doi.org/10.1007/978-981-96-2669-4_16

and small Island nations are particularly vulnerable to tsunami but modern tsunami early warning system can help to reduce catastrophic fatalities (Bernard and Titov 2015). The key to building a resilient community is not only to provide risk information early but embedded to develop overall human and infrastructural capacity through policy and planning.

Prior to the 2004 Indian Ocean Tsunami, most of the United Nations Organizations coordinated efforts regarding tsunami early warnings were mostly concentrated in the Pacific Region. Nonetheless, the 2004 Indian Ocean tsunami, which claimed over 235,000 lives and left 1.6 million people stranded across 16 countries (Titov et al. 2005), altered the perspective of the international community to take the Indian Ocean region into account (Kanoglu et al. 2015; Synolakis and Kong 2006). The term Tsunami was mostly unheard of in many countries around the Indian Ocean region and inattention to Paleo Tsunami evidence and pre-curser research led to no preparedness related to tsunami, which resulted in unprecedented damage caused by this single event (Bernerd and Titov 2015).

The 2004 Indian Ocean Tsunami remains one of the 10 deadliest disasters in recorded history (Suppasri et al. 2015; Bernard 2012). December 26th, 2004 a magnitude of 9.3 earthquakes struck the Sumatra coast, the Northwest part of Indonesia (Koshimura and Takashima 2005), which resulted in a tsunami that rose 30 m in height (Synolakis and Kong 2006). Indonesia was most severely hit by the combined effects of the earthquake and tsunami. A reported wave above 10 m in height hit Banda Aceh city almost at a velocity of 800 km/Hour. Places like Meulaboh, and Calang which are situated western part of Banda Aceh city, faced almost 12–15 m high water and submerged almost 10 km inland. The economic losses alone in Indonesia stand at around US$ 4.5Bn due to the resultant effect of earthquake and tsunami combined (World Bank 2005). The tsunami wave flattened approximately 115,000 houses and additionally damaged almost 150,000 houses alone in Indonesia (Asian Development Bank 2005).

After Indonesia, Sri Lanka suffered a high number of casualties, and the reported estimated death toll reached up to 35,000 (Noyalhr 2005). The fishing industry in Sri Lanka suffered the worst due to the tsunami which destroyed 10 of the country's 12 major fishes harbour (Rathnayake et al. 2019).

India also suffered high casualties during the 2004 Indian Ocean tsunami disaster. Although the impact of the tsunami was felt on the east coast of India. The state of Tamil Nadu and the Andaman Nicobar Islands faced the brunt of it and suffered unusually high casualties. Besides having a high death toll (approximately 7000) the Island state of Andaman and Nicobar suffered massive infrastructure damage to the point where no basic civil infrastructure was functioning after the disaster (Patterson Edwards et al. 2005).

Records of high fatalities and wave height including Indonesia, Thailand, India, Sri Lanka were later surveyed and reported by many experts (UN 2005; Tsuji et al. 2005; Yasuda et al. 2005; Matsutomi et al. 2005; Satake et al. 2005; Tsuji et al. 2006; Fujima et al. 2005; NOAA 2005) and a collective summary of Asia shown in Table 16.1.

Table 16.1 Fatalities and missing and reported wave height in Asian countries

Country	Total of fatalities and missing	Wave height (m)
Indonesia	167,776	3–30
Sri Lanka	35,322	2.4–10
India	18,045	2.5–5.5
Maldives	108	0.6–3.4
Thailand	8,212	2.5–19.6
Myanmar	61	0.4–2.9
Malaysia	75	1.0–4.0

16.2 Impacts of the 2004 Indian Ocean Tsunami on Thailand Coast

The Indian Ocean Tsunami of 2004 also caused substantial damage in Thailand. The impact was particularly felt in the 6 provinces (Phang Nga, Krabi, Phuket, Ranong, Trang, Satul) along the Andaman coast. The impact of the tsunami was exacerbated due to the high tide and surged on top of the elevated water on the west coast of Thailand). The tsunami height fluctuated by a factor of three to four due to the variation in offshore bathymetry. The highest recorded height was 19.6 m at Ban Tung Dap near Prathong Island, 15 m at Khao Lak and around 5–6 m at Patong Beach (Srivichai et al. 2007).

In terms of human tragedy, the 2004 tsunami is probably the worst natural disaster (Nidhiprabha 2007) Thailand has faced in recorded history. Over 8000 persons lost their lives or went missing, along 4500 houses were damaged and 14 million Thai Bhat worth of economic losses were assessed (Department of Disaster Prevention and Mitigation 2005). Phang Nga particularly accounts for the largest number of deaths or missing (more than 65%) as well as property damage (about 50%) due to high waves (Table 16.2).

The region has faced severe impacts in multiple dimensions and incurred the largest economic losses in past ten years. The heigh economic losses incurred by Sri Lanka and Indonesia came from physical damage to property, but in contrast, Thailand's largest losses were felt in the productive sectors such as tourism and fisheries. The tourism sector was particularly affected because the tsunami struck the southern provinces, which consist of Thailand's key tourist spots for domestic and foreign tourists. In total tsunami affected 12,815 households comprising 58,500 residents. Damages to the business property were found most in Phang Nga, Phuket, and Krabi whereas damages to the fisheries sectors are higher in the other three provinces.

Most of the economic impact on livelihood came from the damages in the business property which are mostly Hotels and properties that provide services or support the tourism industry (approximately 85%). Fisheries were the second most affected sector (12%) while damages to the agriculture and livestock industry were negligible (Nidhiprabha 2007).

Table 16.2 Number of deaths and injuries (*Source* DDPM 2005)

Si no	Province	Death				Injured			Missing		
		Thai	Foreigner	Unidentified	Total	Thai	Foreigner	Total	Thai	Foreigner	Total
1	Phang Nga	1389	2114	722	4225	4344	1253	5597	1352	303	1655
2	Krabi	357	203	161	721	808	568	1376	314	230	544
3	Phuket	151	111	17	279	591	520	1111	245	363	608
4	Ranong	153	6		159	215	31	246	9		9
5	Trang	3	2		5	92	20	112	1		1
6	Satul	6			6	15		15			
Total		2059	2436	900	5395	6065	2392	8457	1921	896	2817

Regarding the transport infrastructure (road network and bridges), the damage was rather insignificant and estimated at less than 2 million USD. However extensive damage was found in piers which also affected the quick recovery of the fishing and tourism industry (DDPM 2005).

Initially, the primary concern was about how tsunami waves affected the coral reefs, a precious natural resource that is crucial for the ocean ecosystem, ecosystem services, fisheries and the region's tourism industry. A rapid impact assessment study after three weeks of tsunami revealed only 13% of the coral reef in the Andaman Sea was severely damaged, while 40% showed no significant impact (Brown 2005; DMRC 2005). However, the high velocity of the tsunami waves and excessive salt-water intrusion caused massive damage in the coastal mangrove forest (321.6 Ha) (ONEP 2006) and the freshwater swamp forest area, which needs extensive and longer rehabilitation to ensure long-term sustainability (FAO 2005).

16.3 Recovery from the Mega-Disaster

The initial obstacle in the relief effort was complicated as it required dealing with many dead, injured and missing foreigners as well as natives of the land. Although it is widely acknowledged the effective emergency responses deployed by the Royal Thai Government as well as the contribution of the local Thai people in rescue operations in the face of such devastation. Following the catastrophic event, the Thai government initially set up a fund of USD 2.5 million and many international organizations contributed significantly later in the recovery process (Asian Development Bank 2009). Distribution of aid and supplies was coordinated through the Department of Disaster Mitigation (DPMM) as well and the debris cleaning process was ensured by the local administration led by the army and the people who volunteered across Thailand. The government began working on the restoration of the environmental damage caused by tsunamis in conjunction with national and international organizations and NGOs which also included bringing new legislation to protect and rehabilitate natural resources in the affected areas (Gubić 2010). Although the scale of destruction in Thailand was relatively small compared to Sri Lanka or Indonesia successful relief efforts immediately after the tsunami without significant logistic failure attributed to a number of factors such as synergy generated within the Thai community in the provision of the different forms of assistance, close collaboration among government department and civil society as well as immediate influx of humanitarian support from the international community and Organizations (WCDR 2006).

During the reconstruction phase, the Thai government's approach was strongly centralized. A budget of 506 million Bhat was approved under the supervision of the Ministry of Social Development and Human Security to build 3,661 permanent houses. The government planned to use this reconstruction opportunity to implement new land use planning and building codes. As a part of this, a key strategy was to move people without legal land tenure to public land further from the sea. The new

Table 16.3 Compensation package for fishing vessels (DoF 2005)

Type	Compensation per vessel (THB)	
	Boat length: < 10 m	>10 m
Retrieval	10,000	25,000
Repair	20,000	70,000
Vessel lost	66,000	200,000

Housing policy gives people with legal land tenure to move to a new alternative house or receive cash compensation to build a house on the existing plot. During the reconstruction and relocation phase, a survey carried out by an NGO network demonstrated that 70% of the people are unwilling to move far from the sea although they are given land tenure and free houses (Thailand Country report, TEC 2006).

As a part of the long-term livelihood recovery, RTG started announcing compensation and rehabilitation funds. The tsunami affected almost 400 fishing communities, and almost 30,000 fishing households along Thailand's Andaman coast. Out of the 5252 M THB tsunami response budget, 1343 M THB was allocated to assist 27,828 fishermen to start fishing and aquafarming activities as soon as possible. This fund helps to repair 3426 small boats and 555 larger fishing boats (DOF 2005) (Table 16.3).

To support small businesses, the Thai government allocated USD 2.5 million by November 2005, and the Ministry of Labour granted USD 5 million to create jobs and upgrade the skills of the workers.

Several of Thailand's most severely impacted areas were important tourist destinations. Therefore, a key priority for the government is to recover and rehabilitate the tourism sector which supports almost 100,000 people in the region directly or indirectly (UNCT). In February 2005 World Tourism Organization convention produced Phuket's action plan to rehabilitate tourism. The Thai government also stimulated a special package of USD 93.7 million to revitalize tourism in the tsunami-affected area which was used to rebuild popular beaches and public infrastructure. The RTG also approved tax relief for the shops and businesses supported by the low price of hotel rooms to promote domestic tourism as well.

16.4 Mitigation Measures for Coastal Protection

Coastal areas, which are potential and currently popular for tourism need attention to protect various coastal resources and small and medium infrastructure associated with the tourism industry. Although various measures can be taken to increase coastal resilience, all measures cannot be implemented easily considering the sentiment of the local community (Wong 2012). The necessity for tourist hotels to be in proximity to the ocean and deal with a hazard like a tsunami with an unknown return period is the usual argument to implement strict measures. On the other hand, it is also difficult to make all coastal infrastructure like hotels, built as tsunami-proof as it incurs a huge cost. Therefore, the tourism industry will always be reliant on local public infrastructure and adaptive measures (Table 16.4).

Apart from the local measures, regional warning systems always remain logical and very important aspects of coastal resilience against tsunamis. The Intergovernmental Oceanographic Commission (IOC) of UNESCO established the intergovernmental coordination group for the Indian Ocean Tsunami Warning and Mitigation System (IOWTM). All of the Indian Ocean countries were participated and divided into six working groups. The Deep-Ocean Assessment and Reporting of Tsunami (DART) also expanded its operation to the Indian Ocean region and connected with the Pacific and Atlantic Ocean systems where real-time data is freely exchanged across the nations, (Bernard 2012). With the help of the National Ocean and Atmospheric Administration Thailand deployed its first tsunami monitoring buoy approximately 965 km west of Phuket Island and later in 2016 a second one was deployed 340 km from Phuket Island (National Disaster Warning Center 2022). Thailand also joined the Intergovernmental Oceanographic Commission and installed more coastal tide gauges in the Indian Ocean region (Koh Miang & Koh Racha) to measure sea level in real time (e.g., www.iocsealevelmonitoring.org/).

Table 16.4 Local measure and implementation barrier/effectiveness

Si no	Local adaptive measures	Implementation barriers/effectiveness
1	Safety zone or setback line: The mean high watermark serves as the minimum distance between the permanent structure and shoreline	Followed by very few large hotels, which is difficult to implement as the beach is the main attraction, it is unnecessary to follow the setback line as the return period and magnitude are unknown for tsunamis (Brikland et al. 2006). Uniform setback line is ineffective as tsunami waves cause considerable damage far inside the inland
2	Tsunami-proof structure. reinforced concrete column, deeper scouring proof structure, Ancher precast floor wall (Dalrymple and Kriebel 2005)	No hotel is truly tsunami-proof, Improvement costs a lot of money to refit with reinforced concrete
3	Natural Barriers: Preservation of coastal Dune and vegetations (Dalrymple and Kriebel 2005) Large scale Modular Planting of Mangroves (Wong 2010)	Not very effective for a disaster like a tsunami
4	Escape route, warning towers, and evacuation shelter	In coastal areas, some routes might be reasonably long before reaching high ground. Although this remains the most implemented solution in Thailand. Eighteen were built on Phuket island, several on the Khao Lak coast

16.5 Post-2004 Tsunami Risk Governance and Policy Implantation in Thailand

Thailand suffers from many natural hazards including floods, drought, tropical storms, earthquakes, and occasionally tsunamis. The comprehensive disaster management system of Thailand consists of Government agencies, Laws, policies, an early warning system, and the standard operating procedures for coordinating and performing inter-agency operations at different steps and stages of the disaster management cycle. Establishment of Disaster Prevention and mitigation (DDPM) under the Ministry of Interior in the mid 1980s, perhaps a significant milestone in coordinating all national and provincial level disasters. During the 1990s, the approaches became much more people-centric, and Thailand started developing community-based early warning systems. Although 2004 Indian Ocean Tsunami came as a shock and pointed out the need for the incorporation of a comprehensive tsunami early warning system. The 2004 Tsunami also helped to shape two legal frameworks, the Disaster Prevention Mitigation Act (2007) and the National Warning and Mitigation System Act (2007), which provide the legal framework for disaster management, outline the roles and responsibilities of various agencies, and establish mechanisms for disaster response and recovery. The DDPM formulated the NDPMP in 2010 to provide an outline for cross-agencies collaboration in all sectors and levels in disaster prevention, mitigation, and management. The plan identifies four strategic mechanisms: (i) impact reduction and prevention, (ii) arrangements for preparedness, (iii) disaster emergency management, and (iv) post-disaster management. Civil society organizations were also identified as supporting agencies for (ii), (iii), and (iv) components.

Although earlier in 2004, there was no preparedness and mitigation plan for tsunami in Thailand, immediately after the event several steps have been taken by the national government and the private sector. The Thai meteorological agency installed sixteen earthquake observation stations throughout Thailand since 2005 (Srivichai et al. 2007). The national government of Thailand improves the capacity of earthquake warning and analysis. At the same time, Thailand's tidal stations in the Indian Ocean (at Kho Miang) were upgraded with the help of external donor agencies (USAID, Asian Disaster Preparedness Center) which assures notification of a tsunami, 15 min before it reaches the mainland.

Reducing the impact of the disaster is the main goal of tsunami mitigation. This foundational component involves understanding the tsunami risk landscape, including the identification of potential tsunami sources (such as undersea earthquakes, landslides, or volcanic eruptions) and the mapping of vulnerable coastal areas. Comprehensive risk assessments incorporate historical data, geological studies, and predictive modeling to evaluate the frequency, magnitude, and potential impact areas of tsunamis. Risk knowledge provides the basis for all subsequent early warning activities, informing the development of warning strategies, evacuation plans, and public education programs.

Risk-informed land use planning is critical for resilient coastal communities. However, it is very challenging to implement a new land use plan for an already developed area. Although the Thai government prepared a new land use plan for several areas such as Patong and Phi Phi to make the community resilient, but measures can't be implemented fully because of strong protests from the local community. Along with that, in terms of community participation and preparedness, hazard maps were developed for highly dense tourist areas like Khao Lak and Kanaka Beach to enhance the awareness of the people. The evacuation plan was also developed by the DDPM and Tambon Administration, and several multipurpose tsunami evacuation shelters were established. The government also constructed several siren towers and planned evacuation drills throughout the year in various locations. Education and awareness are also an important part of community resilience and an important step for capacity building. With this aim, DDPM developed several educational and awareness materials and started distributing them among tourists and the local community (DDPM 2005a).

16.6 Tsunami Risk Management and Challenges

Like many parts of the world, Thailand has a very low probability of occurrence of a very large-scale tsunami. These events have occurred in a distant time and fed away the risk perception. Lack of collective memory of the event makes risk communication for tsunamis both challenging and complex. Unprecedented growth in economic activity and an increase in coastal population make it very difficult to implement any mitigation measures. In the past two decades, several tsunamis such as the Indian Ocean Tsunami in 2004, 2011 Tohoku, and 2018 in Palu, have caught coastal people off guard and shaped how the risk is perceived. The important lesson learned from the past event dictated the need for strengthening coastal community resilience and risk-informed action through policy.

The advancement in tsunami hazard assessment approaches into probabilistic systems has increased risk management and communication barriers many folds and made it difficult to raise awareness because of the complexity of the system. Risk management itself is a comprehensive work divided into two broad categories, (i) Institutional capacity to act on the risk, and (ii) societal capacity to follow the procedure which indeed goes through a continuous cycle of different steps like long-term planning, prevention, capacity building, early warning as well as phases of recovery. Therefore, risk communication becomes an integral part of risk management at each step. Prevention, community preparedness, effective early warning, efficient evacuation and recovery are required to mitigate the socio-economic impact of tsunamis. These steps can be well executed when the risk is well-informed and communicated seamlessly.

Strategies for Coastal Risk Management.

1. Land Use Planning: In response to the 2004 Tsunami, USAID/Asia USAID (DLCD 2015; NTHMP 2001) prepared a guideline for resilient coastal communities where best practices are outlined. Developing coastal forests, leaving buffer zones/no-build zones, moving existing shopping areas to upper floors are some of the best practices to minimize the impact of tsunami. Probabilistic risk assessment studies can also play a huge role in the relocation of important industrial activities outside of the risk zone.
2. Need for Multi-hazard Risk Framework: Strategic planning to enhance resilience must consider possible interaction of the cascading effect to coastal hazards, which should include tsunamis and other rapid and slow onset coastal hazards like cyclones, storm surges, coastal erosion, etc.
3. Physical Mitigation Measures: Physical mitigation measures are important to reduce the impact of tsunamis. This system can be broadly classified into two classes as Nature-Based Solution (NBS) and Engineering Solution. Considering the risk from multiple hazards implementing NBS can be quite effective in enhancing coastal resilience as well as beneficial for preserving the health of the coastal ecosystem (Shuto 1987). On the other hand, the most common engineering solutions are brake water, tsunami walls, gates, etc. Although, alone this structure might not be able to completely mitigate the coastal infrastructure and community from the high waves of tsunami but helps to reduce velocity as well as height as observed in Tohoku in the 2011 Tsunami (Suppasri et al. 2013). Developing a multilayer physical barrier or a combination of grey and green infrastructure might also be an effective solution to minimize the impact. Either way, whatever the solution might be, it is very important that the design has to be risk-informed and supported by a detailed study to withstand multi-hazard scenarios.
4. Capacity building and enhancing collective preparedness: Preparedness in tsunamis requires collective action both from the institute as well as the community. Enhancing knowledge and skill has always been an empirical driver of preparedness and resilience. One of the collective capacities of tsunami is evacuation planning and exercise. It is important to involve the local community not only in regular evacuation exercises but also in the evaluation and planning (MCDEM 2016). It is also important to raise awareness of the popular tourist spots as visitors from outside might not have knowledge of perceived risk and mitigation strategies (Johnston et al. 2007).
5. Tsunami Early Warning System (TEWS): Under the patronage of UNESCO-IOC, TWES covers much of the tsunami-prone coastline. Although this international organization is providing a warning to all national agencies, it is important to build the capacity to monitor and detect the furcating tsunami at the national and community levels. Dissemination of the early warning to the last mile has always been a challenge, therefore, it is important to involve the community at the grassroots level and provide regular training to maintain capacity at the operational level during disaster.
6. Recovery Planning: The concept of Build Back Better was inducted during the early recovery years of the 2004 Tsunami with a predominant emphasis on

building back with better infrastructure and appropriate planning so the impact of the next disaster can be minimized. The recent concept of Building Back Greener adopted by Japan in post-recovery stage also improves overall resilience by adopting NBS as a part of the process (Mabon 2019). Effective use of funds and financial instruments is also crucial in the recovery phase. Particularly introduction of specific catastrophic insurance and other financial protection can help the community at the individual level to reduce financial hardship and stress during recovery.

16.7 Discussion and Conclusion

Despite progress in understanding the tsunami hazard and risk in the last few decades, it is still a grave concern for the safety and security of the coastal population in many regions. In Thailand, the southern provinces are extremely important economically as well as the tourism-based coastal development often outweighs the perception of risk against immediate financial gain, which increases exposure to tsunami hazards. Therefore, a large-scale tsunami (e.g., the 2004 Indian Ocean Tsunami), which has a very low probability of occurrence might give similar consequences even after 20 Years. Risk perception is directly related to how risk is communicated over time and the continuous monitoring of all mitigation measures. The immediate aftermath of the 2004 disaster and devastation drives the community to become vigilant for short-period, but with time, as memory feds away so does the perception of the risk. Therefore, it is imperative to rejuvenate the existing programs and measures that were initiated by the local government institutions and community. Disruptive technologies can significantly help in tsunami risk reduction by integrating real-time IOT sensors, advanced satellite monitoring and the use of advanced data analytics and machine learning technology for tsunami risk modeling. Enhanced communication systems ensure rapid alert, while drones can help to asses primary damage and facilities emergency rescue. Smart Infrastructure, community engagement, and the use of VR (virtual reality) technology in training can significantly bolster preparedness and response efficiency.

In concrete terms, it is important to implement further research on tsunami risk perception and awareness as a part of continuous community resilience building. Continuous development and changes in land use might obsolete the previously designed evacuation plan. Therefore, there is a need to revise the evacuation plan and route and jointly upgrade them with the consensus of the local community. It is also true in the case of evacuation centers in terms of design as well as locations. Based on earlier research, what might be a good and safe location for the centers might not be true with recent changes in demography, risk, and socio-economy. There is also a need to redevelop the standard operating procedure and evacuation plans taking into account newly developed probabilistic Multi-hazard Scenarios.

Returning to normal life after any disaster is often hindered by socio-economic challenges. Therefore, it is important to implement a social safety net as well as a financial instrument to cope with challenges which likely to come after any disaster. In a nutshell, relevant studies on tsunami risk and perception are still needed to build a solid understanding of evacuation behavior during tsunamis. Continuous public education for understanding early warning, standard operating procedures at both institutional and community levels, periodically improved risk communication among locals and tourists as well as improved public information systems is essential to build resilience against tsunamis.

References

ADB (2005) An initial assessment of the impact of the earthquake and tsunami of December 26, 2004 on South and Southeast Asia

Asian Development Bank (2009) ADB Member economies fact sheets. http://www.adb.org/Documents/Fact_Sheets/THA.pdf

Bernard E (2012) Thirteenth annual Roger Revelle commemorative lecture-tsunamis: are we underestimating the risk? Oceanography 25(2):208–218. https://doi.org/10.5670/oceanog.2012.60

Bernard E, Titov V (2015) Evolution of tsunami warning systems and products. Philos Trans R Soc A: Math, Phys Eng Sci 373(2053):20140371. https://doi.org/10.1098/rsta.2014.0371

Brown BE (2005) The fate of coral reefs in the Andaman Sea, eastern Indian Ocean following the Restoration of Coastal Ecosystems and Fisheries in Thailand after Northern Sumatra Earthquake and Tsunami Disasters Sumatran earthquake and tsunami, 26 December 2004. Geogr J 171(4):369–386

Dalrymple RA, Kriebel DL (2005) Lessons in engineering from the tsunami in Thailand. The Bridge 35:4–13

Department of Disaster Prevention and Mitigation (2005a) Report of helping the people from earthquake and tsunami. Office of the secretary of Civil Department. http://61.19.54.131/tsunami/

Department of Land Conservation and Development (DLCD) (2015) Preparing for a Cascadia subduction zone tsunami: A land use guide for Oregon coastal communities. https://www.oregon.gov/lcd/Publications/TsunamiLandUseGuide_2015.pdf

Department of Disaster Prevention and Mitigation (2005b) Save your life from tsunamis, in Thai language (25 pages)

DMCR (2005) Department of Marine and Coastal Resources, Rapid assessment of the tsunami impact on marine resources in the Andaman Sea, Thailand. Phuket Marine Biological Center, Phuket, p 76

DoF (2005) The compensation package for cage farms (Cited 10 June, 2007). http://www.cffp.th.com/tsunami

FAO (2005) Tsunami impact on fisheries and aquaculture in Thailand. Joint report of the NACA, FAO, SEAFDEC and BOBP-IGO

FAO-MOAC (2005) Report of joint FAO/MOAC detailed technical damages and needs assessment mission in fisheries and agriculture sectors in tsunami affected six provinces in Thailand. Food and Agriculture Organization of the United Nations (FAO) in cooperation with Ministry of Agriculture

Fujima K, Tomita T, Honda K, Shigihara Y, Nobuoka H, Hanzawa M, Fuji H, Otani H, Orishimo S, Tatsumi M, Koshimura S (2005) Survey result of Indian Ocean tsunami in The Maldives, Sumatra Tsunami on 26th December 2004. In: Proceeding of the special Asia tsunami session apac 2005 , Seogwipo KAL Hotel, Jeju-do, Korea, pp 37–48

Gubić I (2010) Role and contribution of the UN in recovery of the tsunami affected settlements in Thailand. Facta Univ-Ser: Econ Organ 7(4):455–468

Harada K, Imamura F (2005) Effects of coastal forest on tsunami hazard mitigation—a preliminary investigation. In: Tsunamis, Springer, Dordrecht, pp 279–292

Johnston D, Becker J, Gregg C, Houghton B, Paton D, Leonard G, Garside R (2007) Developing warning and disaster response capacity in the tourism sector in coastal Washington, USA. Disaster Prev Manag, Int J

Kanoglu U, Titov V, Bernard E, Synolakis C (2015) Tsunamis: bridging science, engineering and society. Philos Trans R Soc A: Math, Phys Eng Sci 373(2053):20140369. https://doi.org/10.1098/rsta.2014.0369

Koshimura S, Takashima M (2005) Remote sensing GIS and modeling technologies enhance the synergic capability to comprehend the impact of great tsunami disaster. In: 3rd international workshop on remote sensing for post-disaster response, September, 12–13th, Chiba, Japan, Koshimura, pp 1–6

Mabon L (2019) Enhancing post-disaster resilience by 'building back greener': evaluating the contribution of nature-based solutions to recovery planning in Futaba County, Fukushima Prefecture, Japan. Landsc Urban Plan 187:105–118

Matsutomi H, Hiraishi T, Takahashi T, Matsuyama M, Harada K, Nakusakul S, Supharatid S, Kanbua W, Siwabowon C, Phetdee S, Janchoowong W, Srivichai M (2005) The December 26th , 2004 Sumatra earthquake tsunami, tsunami field survey around phuket, Thailand. http://www.drs.dpri.kyoto-u.ac.jp/sumatra/thailand/phuketsurveye.html

MCDEM (2016) Tsunami evacuation zones. Director's guideline for civil defence emergency management groups, Published by the Ministry of Civil Defence & Emergency Management—New Zealand. www.civildefence.govt.nz/assets/Uploads/publications/dgl-08-16-Tsunami-Evacuation-Zones.pdf

National Disaster Warning Center (2022) History of the installation of tsunami detection buoys in Thailand. https://ndwc.disaster.go.th/ndwc/cms/7088?id=44532

National Warning and Mitigation System Act (2007)

Nidhiprabha B (2007) Adjustment and recovery in Thailand two years after the Tsunami. Asian Development Bank Institute, Tokyo

NOAA (2005) National geophysical data center. http://www.ngdc.noaa.gov/seg/hazard/tsrnsrch_idb.shtml

Noyalhr K (2005) Death toll much higher than government figure. Daily Mirror (Colombo)

(NTHMP) National Tsunami Hazard Mitigation Program (2001). In: NOAA, USGS, FEMA, NSF, Alaska, California, Hawaii, Oregon, and Washington. Designing for Tsunamis: Seven principles for planning and designing for tsunami hazards

NWS TsunamiReady® Program, Working toward a weather-ready nation. https://www.weather.gov/tsunamiready/

ONEP (2006) Office of Natural Resources and Environmental Policy and Planning, Two years after tsunami: restoration of Thailand's natural resources and environment. Ministry of Natural Resources and Environment, Bangkok, Thailand, p 150

Pal I, Ghosh S, Dash I, Mukhopadhyay A (2023) Review of Tsunami early warning system and coastal resilience with a focus on Indian Ocean. Int J Disaster Resil Built Environ 14(5):593–610

Patterson Edward JK, Kulkarni S, Jeyabaskaran R, Lazarus S, Anita M, Venkataraman K, Das SP, Tamelander J, Rajasuriya A, Jayakumar K, Kumaraguru AK (2005) The effects of the 2004 tsunami on mainland India and the Andaman and Nicobar Islands

Rathnayake DK, Kularatne D, Abeysinghe S, Shehara I, Fonseka T, Edirisinghe Mudiyanselage SDJ, Kamalrathne WGCT, Siriwardana C, Alagiyawanna Mohotti Appuhamilage CSB, Dissanayake R (2019) Barriers and enablers of coastal disaster resilience—lessons learned from tsunami in Sri Lanka. Int J Disaster Resil Built Environ 11(2):275–288. https://doi.org/10.1108/IJDRBE-07-2019-0050

Satake K, Okamura Y, Hishikura M, Fujima K (2005) The December 26th, 2004 Sumatra earthquake tsunami, Tsunami field survey around Phuket, Thailand. http://www.drs.dpri.kyotou.ac.jp/sumatra/thailand/phuketsurveye.html
Shuto N (1987) The effectiveness and limit of tsunami control forests. Coast Eng Jpn 30(1):143–153
Srivichai M, Supharatid S, Imamura F (2007) Recovery process in Thailand after the 2004 Indian Ocean tsunami. J Nat Dis Sci 29(1):3–12
Suppasri A, Shuto N, Imamura F, Koshimura S, Mas E, Yalciner AC (2013) Lessons learned from the 2011 Great East Japan tsunami: performance of tsunami countermeasures, coastal buildings, and tsunami evacuation in Japan. Pure Appl Geophys 170(6):993–1018
Suppasri A, Goto K, Muhari A, Ranasinghe P, Riyaz M, Affan M, Mas E, Yasuda M, Imamura F (2015) A decade after the 2004 Indian Ocean tsunami: the progress in disaster preparedness and future challenges in Indonesia, Sri Lanka, Thailand and the Maldives. Pure Appl Geophys 172:3313–3341
Synolakis CE, Kong L (2006) Runup measurements of the December 26 2004. Earthq Spectra 22(3_suppl):S67–S91. https://doi.org/10.1193/1.2218371
Thailand Country report, TEC (2006)
Titov VV, Rabinovich AB, Mofjeld HO, Thomson RE, Gonz_alez FI (2005) The global reach of the 26 December 2004 Sumatra tsunami. Science, 309(5743):2045–2048. https://doi.org/10.1126/science.1114576
Tsuji Y, Kamataki T, Matsutomi H, Murakami Y, Nishimura Y, Sakakiyama T, Tanioka Y (2005) Earthquake, tsunami and damage in Banda Aceh and Northern Sumatra. http://www.tsunami.civil.tohoku.ac.jp/sumatra2004/report.html
Tsuji Y, Namegaya Y, Matsumoto H, Iwasaki S, Kanbua W, Srivichai M, Meesuk V (2006) The 2004 Indian tsunami in Thailand: Surveyed run-up heights and tide gauge records. Earth Planets Space 58:223–232
United Nations Development Program (2005) The human toll. http://www.tsunamispecialenvoy.org/country/humantoll.asp
WCDR (World Conference on Disaster Reduction).
Wong PP (2010) Coastal communities in Asia Pacific: Vulnerability and adaptation (abstract). In: CoastalZone Asia Pacific conference—CZAP 2010, 18–22 October 2010, Bangkok, pp 13–14
Wong PP (2012) Impacts, recovery and resilience of Thai tourist coasts to the 2004 Indian Ocean tsunami. Geol Soc, Lond, Spec Publ 361(1):127–138
World Bank Report (2005) Indonesia: preliminary damage and loss assessment: the December 26, 2004 natural disaster. www.siteresources.worldbank.org/INTINDONESIA/Resources/Publication/280016110613030 5439/damage_assessment.pdf
Yasuda T, Arikawa T, Imamura F (2005) Field survey Report on 2004 Indian Ocean tsunami along the Southwestern Coast of Sri Lanka. Comprehensive analysis of the damage and its impact on coastal zones by the 2004 Indian Ocean Tsunami, pp 32–48

Chapter 17
Development and Progress of Community-Based Disaster Risk Reduction in Malaysia

Takako Izumi and Eriko Motoyama

Abstract The historic 2004 Indian Ocean Tsunami left a profound impact on Malaysia. Its consequences were far-reaching, including a large loss of life, physical damage, and a range of other issues, such as environmental degradation, socioeconomic setbacks, and health problems. In 2018, the project "Strengthening the DRR Capacity to Improve the Safety and Security of Communities by Understanding Disaster Risk (SeDAR)" was initiated by Tohoku University in Japan in collaboration with the SDMU, the Malaysia Civil Defense Force (APM), and Universiti Teknologi Malaysia (UTM) to equip local governments and community leaders with the skills and know-how to build a grassroots DRR program from the bottom-up. This project was funded by Japan International Cooperation Agency and has had a tremendous impact on communities, significantly changing the mindsets of the community members who eventually led the DRR projects. This chapter shares the processes and achievements of the SeDAR project as a case study and as best practice for implementing community-based DRR (CBDRR), which any local or national government can duplicate.

Keywords Disaster risk reduction · Community-based disaster risk reduction · Early warning · Evacuation drill

17.1 Introduction

The historic 2004 Indian Ocean Tsunami left a profound impact on Malaysia. Its consequences were far-reaching, including a large loss of life, physical damage, and a range of other issues, such as environmental degradation, socioeconomic setbacks, and health problems (Ahmadun et al. 2020). This disaster, which claimed 80 lives in

---

T. Izumi (✉)
International Research Institute of Disaster Science (IRIDeS), Tohoku University, Sendai, Japan
e-mail: izumi@irides.tohoku.ac.jp

E. Motoyama
Servis EDA Pvt. Ltd., Kuala Lumpur, Malaysia

R. Shaw et al. (eds.), *Two Decades from the Indian Ocean Tsunami*, Disaster Risk Reduction, https://doi.org/10.1007/978-981-96-2669-4_17

Malaysia, catalyzed the advancement of disaster risk reduction (DRR) measures in the country. This led to the identification of tsunami hazard zones and the proactive establishment of the National Disaster Management Agency (NADMA) in October 2015, which took the reins on crisis and disaster management from the National Security Council (NSC) (CFE-DM 2020). In Indonesia, a similar response was seen, with the establishment of the National Agency for Disaster Countermeasures (BNPB) in January 2008 to oversee disaster management mitigation, response, and recovery.

From 2000 to 2024, Malaysia experienced 78 natural hazards, and there were 393 in Indonesia in the same period (EM-DAT). Malaysia's hazard risks are lower than those of other disaster-prone Asian countries, but they are not to be underestimated. Flooding in Malaysia in 2021 caused 56 casualties. Historically, disaster management in Malaysia has been a top-down government responsibility, with Malaysians being heavily reliant on government assistance in cases of emergency (Rosmadi et al. 2023). The inefficiencies accruing to a rigid and hierarchical, top-down disaster management system are among the most important for Malaysia (ADPC 2020). However, a significant shift has been made toward a community-based approach (Ahmadun et al. 2020). This evolution can clearly be seen in the establishment of the Selangor Disaster Management Unit (SDMU) in the state of Selangor in 2015, marking Malaysia's first state-level disaster management unit. Two other disaster management units have been established in the states of Penang and Sarawak.

In 2018, the project "Strengthening the DRR Capacity to Improve the Safety and Security of Communities by Understanding Disaster Risk (SeDAR)" was initiated by Tohoku University in Japan in collaboration with the SDMU, the Malaysia Civil Defense Force (APM), and Universiti Teknologi Malaysia (UTM) to equip local governments and community leaders with the skills and know-how to build a grassroots DRR program from the bottom-up. This project was funded by the Japan International Cooperation Agency and has had a tremendous impact on communities, significantly changing the mindsets of the community members who eventually led the DRR projects. This chapter shares the processes and achievements of the SeDAR project as a case study and as best practice for implementing community-based DRR (CBDRR), which any local or national government can duplicate.

17.2 Why is CBDRR So Important?

The field of DRR is constantly evolving, and CBDRR is at the forefront of this innovation. CBDRR has been recognized globally as a crucial approach for effectively reducing the impacts of natural hazards (Wu et al. 2023). CBDRR has a long history, evolving in the 1960s (Imperiale and Vanclay 2023) and flourishing since the late 1980s (Lee et al. 2022). It is still considered an effective, essential tool with regard to all DRR measures.

A survey of the most effective DRR innovations was conducted among DRR specialists. Respondents were asked to select the three measures, including products and approaches, choosing from 30 options, such as geographic information system

(GIS), drones, SNS, earthquake early warning, a national platform for DRR, indigenous DRR technology, etc. CBDRR received the largest number of votes in this survey (Izumi et al. 2019).

CBDRR tends to emphasize a bottom-up approach (Khudri et al. 2024); however, a synthesis of top-down and bottom-up approaches is needed if an initiative is to be entirely successful. The approach cannot be pursued by governments and communities separately (McLennan 2018). Top-down development includes resource allocation, collection, and adaptation of scientific knowledge and techniques (Wu et al. 2023). Bottom-up approaches emphasize the strength and wisdom of local stakeholders and encourage their participation in all processes, from risk assessment to designing and implementing activities, so that issues and challenges can be confronted in a way that incorporates the perspective of those who experience them regularly (Imperiale and Vanclay 2023; Partidario et al. 2022).

In addition, various challenges have been identified to the successful implementation of the approach, such as limited resources, language and communication barriers, power dynamics, and weak social capital (Khudri et al. 2024). While the importance of incorporating local knowledge and the active role of local communities are widely discussed, participatory action continues to be neglected (Partidario et al. 2022). Community voices and inputs continue not to be reflected in many planning documents produced by actors funded by the government and international donors (Lassa et al. 2018).

Taking these challenges into account, it is essential for CBDRR activities to do the following:

1. Secure a budget to initiate the CBDRR program and design a course of activities lasting at least 2–3 years.
2. Assign a team or designated office to coordinate, implement, and manage the CBDRR process and program.
3. Ensure the involvement of community leaders and local government along with relevant agencies involved in the sciences and in emergency response (e.g., civil defense, police, or fire department) and build relationships among the key players. Enhance their knowledge of DRR and risk management.
4. Train leaders at the local level who can effectively educate their communities on DRR.
5. Conduct science-based risk assessment/identification and assign leaders to communicate the risks to the public.
6. Conduct town-watching (becoming familiar with the risks surrounding a community); create community maps to identify hotspots, safe zones, and other areas of interest; and identify solutions to address risks.
7. Educate communities on early warning systems and communicate how they work.
8. Create maps to show evacuation centers and what to do when warning sirens activate. Conduct evacuation drills (a regular event, yearly or every six months).
9. Plan and implement DRR activities with the local government and communities, supported by experts, and ensure the sustainability of the activities through local community ownership.

17.3 The SeDAR Project

The SeDAR project takes a unique approach to communities, instilling a science-based understanding of disaster risk in community leaders, members, and local authorities, having them work together to develop the DRR activities and programs that are best suited to their knowledge and needs. In addition, two universities from Japan and Malaysia are joining this project. They played a key role, emphasizing the importance of data, science, and technology in DRR, identifying future risks as well, based on scientific evidence. In DRR, science and technology are crucial for the early detection of disaster scenarios, prevention, recovery, and management in minimizing losses (Khan et al. 2020). Certain important big data technologies can easily be used in the different phases of disaster management to enhance resilience, such as remote sensing imagery, social media data, crowdsourced data, GIS, and mobile metadata (Sarker et al. 2020). The Sendai Framework for DRR adopted in 2015 at the United Nations World Conference on DRR included four priority actions to be urgently considered in DRR strategies in each country. Among them is "Understanding disaster risks" (UN 2015). For this purpose, it is essential to understand future risks, in particular, using numerous data and modeling/simulations. This was a major activity in the SeDAR project.

17.3.1 SeDAR Program Flow

This project included five steps: 1. Understanding risk using scientific evidence/analysis; 2. Training local leaders; 3. Increasing communities' knowledge of disaster risks and preparedness; 4. Planning and implementing DRR activities with the participation of communities; 5. Localizing and sustaining programs (Fig. 17.1).

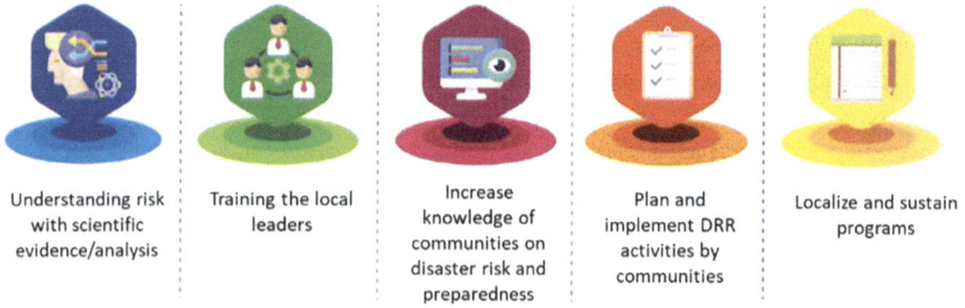

Fig. 17.1 Project steps

17.3.1.1 Understanding Risk Using Scientific Evidence/Analysis

The first activity in this project was collecting disaster data in the state of Selangor to analyze future risks of landslides and floods, both of which are major natural hazards in Malaysia. For landslides, hazard maps were created under the Slope Hazard and Risk Mapping (PBRC) project under the Department of Mineral and Geoscience (JMG), Malaysia. The SeDAR project used PBRC outputs, including slope maps and landslide hazard maps, to better understand the root causes of future mass movements and slope failures. Using LiDAR, it is possible to observe bare terrain underneath dense forest cover. For instance, by stripping away vegetative cover, shaded relief maps derived from LiDAR digital elevation models expose the terrain cover to show past or current human activities and geological or hydrological processes. In this project area, shaded relief maps or hillshade views show that what appears to the naked eye to be a natural slope is riddled with potential failures.

For flood risks, data on past flood events, data on failure/breaching of the bund, visual evidence obtained during flooding, long-term tidal and rainfall data, and the geographic features of the area were used to identify possible food scenarios and to visualize flood events. Long-term rainfall data was analyzed to determine the months with higher rainfall. The result suggests that locals should be alert from November to December. It was also found that a flood situation could worsen in case of a bund breach, which could happen at any time of the year. In addition, a special database was established using GIS to clarify the geography of the area, which is a flat plain consisting mainly of oil palm plantations and villages, easily accessible through a network of roads. From the photos, it was noted that in the event of high tides, several scenarios would result in flooding in the area: 1. Tidal control gate failure, 2. Bund breach/breaking, 3. Bund overtopping, or 4. Heavy rainfall and pump failure.

These risk studies highlighted how heavily scientific study relies on the availability of data and information, producing three observations:

a. Data availability

   - The data needed for disaster risk analysis are limited;
   - Data are being collected but are not effectively used;
   - Capacity development is required for understanding data and its importance;
   - Investment in data acquisition is also necessary;
   - A platform for data sharing is needed.

b. Need to create hazard maps

   - Hazard and risk maps help governments prioritize high-risk slops for mitigation or rehabilitation;
   - Federal and state governments have the responsibility to identify risks using science;
   - Hazard maps must be updated due to changing land use and mitigation measures;
   - Standardization of methodology is needed for landslide and flood hazard and risk assessment.

c. Need for risk communication

- Hazard experts must instruct communities on how to read and use hazard maps;
- Hazard maps must be used to dispel community misconceptions and assumptions regarding slopes;
- Early warning systems should be deployed in high-risk areas;
- The government should share maps for DRR purposes.

These results are included in the Disaster Risk Report: Understanding Landslide and Flood Risks for Science-based DRR in the state of Selangor (Fig. 17.2), which was submitted to the Governor of Selangor (Fig. 17.3).

The results given in the report were shared at risk communication workshops at the local level, which district government officials and community leaders attended (Fig. 17.4).

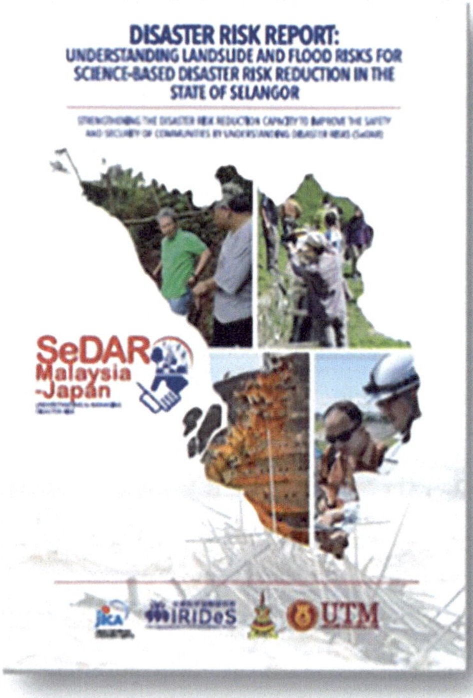

Fig. 17.2 Disaster risk report

Fig. 17.3 Governor of Selangor state holding the report

Fig. 17.4 Community workshop at Ulu Klang and Hulu Langat

17.3.1.2 Training of Trainers for Local Leaders

This project included two layers of training programs. First, DRR experts trained state government officials, who became master trainers. Second, state government officials trained district government officials and community leaders. Before conducting the trainers' training, a trainers' guidebook for planning and implementing CBDRR programs was developed (Fig. 17.5). This guidebook was designed to raise awareness of local disaster history, introduce DRR concepts, understand hazards and risks, and demonstrate how to plan, implement, and evaluate community-led programs.

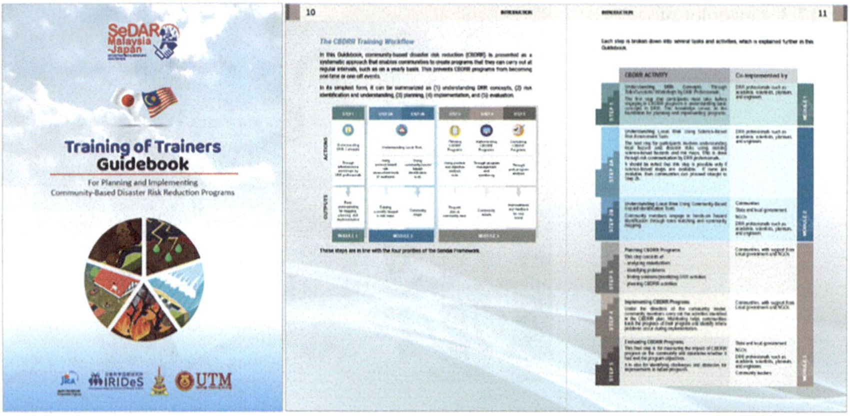

Fig. 17.5 Training of trainers guidebook

17.3.1.3 Increase Communities' Knowledge of Disaster Risks and Preparedness

The trainers who were trained in Step 2 conducted community workshops to share their knowledge and information on CBDRR with the other members of their communities (Fig. 17.6). The agenda included sessions on Why we need CBDRR, Understanding hazards and risks in our community, and What is next: Planning and taking action.

Following this, town-watching and community mapping exercises were conducted, to which community members were invited (Fig. 17.7). Each group created a map based on the findings (positive/negative) and observations (dangerous, safe, useful, and unique) during town-watching.

At the end of the exercise, the participants identified issues and challenges in their communities and proposed possible solutions. They addressed the lack of community awareness and preparedness for disasters and the lack of information, dialogue, and discussion between local governments and their communities. To address these

Fig. 17.6 Community workshops

Fig. 17.7 Town-watching

issues, the participants proposed setting up a resource center in the community to share and disseminate hazard-related matters.

17.3.1.4 Plan and Implement DRR Activities by Communities

Through this project, the communities will plan and conduct CBDRR activities themselves. To obtain ideas and suggestions concerning potential CBDRR activities in each community, DRR experts from Japan were invited to discuss and exchange views on the potential and best-suited activities in relation to the issues and capacities of the different communities (Fig. 17.8). For flood-prone communities, drawing on the presentation and inputs from the participants, the DRR experts noted the issues they raised into potential actions such as 1. Canal/drainage system management, 2. Bund/dyke management, 3. Public warning systems, 4. Evacuation including signage and an evacuation center, 5. Awareness/monitoring/capacity, and 6. Digital platforms for showcasing activities/reporting to authorities. For landslide-prone communities, the issues noted by the experts were 1. Lack of awareness and education concerning slopes and landslide risks in the area, 2. Need for dialogue with stakeholders doing development in the area, 3. Preparedness drills and inclusivity, and 4. Risk communication requires science communication to grasp real risks.

Fig. 17.8 Experts meetings

From the discussion and inputs from the expert meetings, different communities decided on the CBDRR activities to be conducted by themselves (Table 17.1).

Community A: Landslide

(a) Civil defense emergency response training

This took place as a half-day workshop on emergency response training for community members (Fig. 17.9). The program included basic first aid, an overview of evacuation center management, an overview of emergency rescue, and a basic overview of firefighting. The session as a whole combined theory, tips, demonstrations, and hands-on practice. The activity profoundly affected the mindset of the community members. They learned some essential skills for responding to emergencies and how community members can help each other during a disaster.

Table 17.1 Lists of community DRR activities

	Community A	Community B	Community C	Community D
Target hazard	Landslide	Landslide	Flood	Flood
Activity 1	Civil defense emergency response training	Community slope monitoring workshop	Community evacuation drill	Plogging: Trash cleanup to prevent flash flood
2	Setting up of the community DRR resource center	DRR for school children: DRR workshop and Development of DRR mural poster	Development and distribution of information sharing materials: (1) Evacuation center location and (2) Public warning system procedures	Development and distribution of information sharing materials: (1) Evacuation center location and (2) Public warning system procedures
3				Community evacuation drill

Fig. 17.9 Civil defense emergency response training

Fig. 17.10 Resource center

(b) Setting up of the community DRR resource center

A community DRR resource can provide access to information and resources related to DRR, landslides, and the environment (Fig. 17.10). The center set up in this project was housed in a former government community hall donated to this project and the community for this purpose. Technical government agencies donated resource materials such as posters and maps on scientific information. This was truly a collaboration between the local government, communities, and the technical government agencies. The mission of the center includes:

- Providing a communal space for learning about the physical environment in the community;
- Building local capacity through training;
- Providing access to science-based material and resources;
- Serving as a meeting place for discussions and dialogues between technical agencies and communities.

The center forms a physical legacy and showcases the knowledge imparted through the SeDAR program. It is run and managed entirely by community volunteers and a team of tour guides who show visitors around the center gallery.

Community B: Landslide

(a) Community slope monitoring workshop

This one-day workshop instructed community members on how to monitor slopes by using visual inspection and certain early warning signs for landslides (Fig. 17.11). As there are no early warning systems for landslides in effect in this project area, community members were educated regarding a visual inspection system that could be valuable for detecting potential landslides. Speakers from the Public Works Department (JKR) and the Department of Mineral and Geoscience (JMG) provided basic knowledge of geology, an engineering/geotechnical explanation of the causes of landslides, and basic knowledge of slopes. Following these lectures, the participants visited the landslide-affected sites for a visual inspection. Upon returning from the site visits, the participants shared their findings and engaged in discussion with the two agencies. Routine visual inspection and monitoring of slopes as a form of early warning against potential landslides can be engaged in by communities and does not require local government action.

(b) DRR for Schoolchildren–1. DRR workshop and 2. Development of DRR mural poster

(b)–1 DRR Workshop

This workshop targeted students aged 15–17 years (Fig. 17.12). Its format was a combination of a presentation or lecture and a hands-on activity/game. It was essential to keep the lecture short and to include sufficient hands-on activities to keep the student audience focused and engaged. The presentation topics included the following: Understanding landslides and floods, Concepts in DRR, and preparation and response. The students were equipped with knowledge of DRR and what they could do to prepare for future disasters. The District Department of Education showed a strong interest in conducting similar programs at other schools in the district.

(b)–2 Development of DRR mural poster

The DRR Mural Poster for Schoolchildren is a wall poster that measures 7 × 1.8 m and covers the length of a corridor within the premises of the community's high school (Fig. 17.13). The poster is divided into three sections: (1) Hazards becoming disasters due to high vulnerability and low capacity, (2) how to reduce vulnerability and increase capacity, and (3) DRR actions that students can take. The mural poster has become a showcase piece in a hallway converted into an Environment Corridor featuring materials and resources with an environment theme.

Community C: Flood

(a) Community evacuation drill

This drill was a two-day event: a pre-drill briefing and orientation and an evacuation drill (Fig. 17.14). While the area is prone to coastal, river, and flash flooding, no drills or DRR activities have previously been conducted. Hence, a community evacuation

Fig. 17.11 Community slope monitoring workshop

Fig. 17.12 School workshop

Fig. 17.13 DRR mural poster

drill was required in the project area. The pre-drill briefing shared information on upcoming activities for the drill, a description of the sirens to be sounded, the actions to take, and an explanation of the logistics and the day of the drill. Families were encouraged to join the drill as a family, and the drill included vulnerable people, such as older people, persons with disabilities, and others, to understand their support needs. After the drill, a feedback session was organized. The feedback from the participants included: the flood siren can be heard only for 500 m; many identified open drains that they could fall into and get swept away; many potential obstacles or changes in an actual flood situation, such as impassible rods due to rushing waters, floating debris, and live electricity cables in the floodwaters, and strong currents to wade through. Before this drill, 70% said they knew very little or nothing about the evacuation process. The communities and state government agencies have established a stronger working relationship based on this drill, and community leaders realized they could conduct drills on their own initiative without waiting for the government

Fig. 17.14 Flood evacuation drill

to initiate it, and the community members recognized the needs of vulnerable groups during the evacuation drill process.

(b) Development of Information-Sharing Materials

(b)–1 Posters and refrigerator magnets for locations of evacuation center locations and public warning system procedures

Many community residents either did not know or only vaguely knew the locations of the evacuation centers and the course of action they should in response to a flood siren or alert was activated. Therefore, it was decided to develop an instruction poster (Fig. 17.15) and refrigerator magnets to show the names and locations of the evacuation centers in the villages and the public warning system procedures. The government had formally established no public warning procedures for communities. This work identified public warning steps in collaboration with chief village leaders, other community leaders, district offices, and the state/district Department of Irrigation and Drainage. The evacuation locations were also confirmed and verified with the assistance of agencies. These posters were put up in public places, such as community halls, schools, and mosques. In addition, the contents of the posters were made into A5-sized fridge magnets to be placed in handy, prominent places in the residents' homes, such as on a refrigerator door. This was done to ensure that this information would not be lost or misplaced.

(b)–2 Signage for EVACUATION CENTERS

Many community members in the sub-district were unaware of the locations of the evacuation centers within their areas. Signs measuring 2 × 3 feet were designed, printed, and distributed for installation at evacuation centers (Fig. 17.16). These signs were co-designed with considerable input from village leaders. The signs were reviewed and verified by the agencies that own and manage the evacuation center, premises such as the Klang District and Land Office, Klang Municipal Council, Selangor Department of Education, and Selangor Islamic Religious Department. The signs were also endorsed by the National Disaster Management Agency (NADMA).

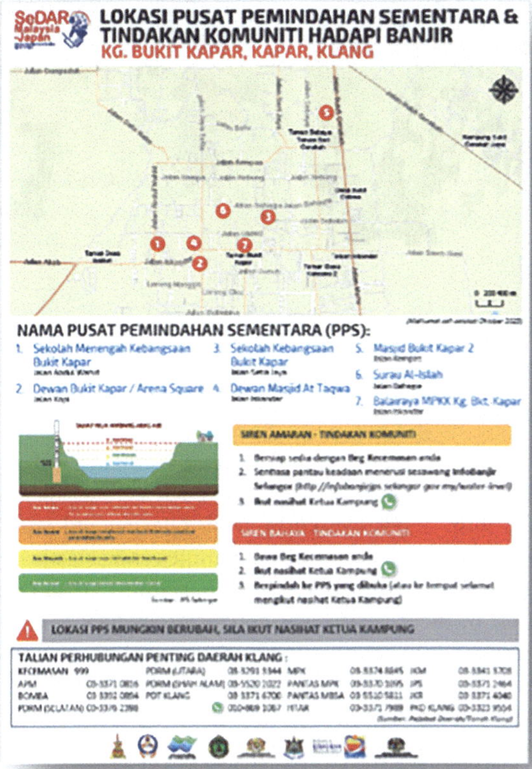

Fig. 17.15 Evacuation center/early warning information poster

Fig. 17.16 Sign for evacuation center

Fig. 17.17 Plogging

Community D: Flood

a. Plogging: Trash cleanup to prevent flash floods

A trash cleanup was held as a 1-day family event, including quizzes, interactive games, and light physical activities to teach how DRR actions such as "plogging" can help mitigate the flash floods caused by trash littering and drainage clogging (Fig. 17.17). The event was family-themed, and families enthusiastically participated in the plogging activity. They were also interested in learning about DRR, as this was a new concept to most of them, and it was beneficial, as they lived in flood-prone areas. The DRR briefing was interactive and well-received by the participants. Both students and parents enthusiastically participated in the Q&A sessions that tested the participants' knowledge from the briefing. Overall, the participants understood the DRR briefing and answered the questions well.

b. Community evacuation drill

This drill followed the same procedure as that conducted in community C. Community leaders began to engage in proactive actions in feedback sessions following the drill. One example is in the preparation of a list of people who require assistance during an evacuation, such as the physically disabled and elderly, which they identified as a need after conducting a community evacuation drill. It was decided to create a form for residents who require assistance in community evacuation to keep a record of the kind of support needed in an emergency; in this way, community members can provide the necessary assistance to the most vulnerable in time for their safety evacuation.

c. Development and Distribution of Information-Sharing Materials: 1. Evacuation Center Location and 2. Public Warning System Procedures

(c)–1 Posters and refrigerator magnets for evacuation center locations and public warning system procedures

The same types of posters and refrigerator magnets were developed to identify the locations of evacuation centers and to clarify the procedures followed for the public warning system. As with community C, the residents did not know or only vaguely knew the locations of the evacuation centers in their communities. In addition, they did not know what to do after the flood siren or alert was activated.

(c)–2 Signage for evacuation centers

Signage for evacuation centers was also developed, and signs were put up at the designated evacuation centers. Knowing the location of the evacuation centers is crucial in case immediate evacuation becomes necessary.

17.3.1.5 Localize and Sustain Programs

The SeDAR project produced three significant impacts on the localization and sustaining of programs.

(a) A Mechanism Was Established to Enable Local Governments and Communities to Plan and Implement CBDRR

This project has established a mechanism to plan and implement CBDRR by local stakeholders. Toward the end of the project, the state government announced its commitment to using an annual budget to replicate CBDRR activities in other districts in Selangor, using the same process and methodology developed in the SeDAR project. The SDMU plans, designs, and implements activities in collaboration with other government agencies, such as the Malaysia Civil Defense Force. District government officials and community leaders who trained as DRR trainers through the SeDAR project play an essential role as trainers in future trainings and workshops in other areas in Selangor.

The institutionalization of DRR at the state level, and the involvement of local governments, agencies, and community leaders, ensured the sustainability and continuity of DRR activities, one of the goals of the SeDAR program (Fig. 17.18). SeDAR created a pathway for the Selangor State Government to institutionalize and formalize DRR infrastructure at the state level.

In response to the program's achievements, the Selangor State Government has designated it as a model for future DRR initiatives in the state. To maintain the continuity and sustainability of the SeDAR program, the Chief Minister of Selangor has committed to allocating funds for training of trainers programs to be carried out in nine districts within the state at the 2024 Budget State Assembly Conference.

Fig. 17.18 Institutionalized mechanism

(b) Change in Mindset/DRR Becoming a Part of Everyday Life

In the SeDAR project, community members changed their mindset and attitudes, resulting in proactive behavior and practices incorporating DRR into their regular community activities. Community leaders shared their experience in this project at the 2023 experience-sharing symposium. The changes that were made included (1) increased knowledge of DRR, (2) understanding the causes of disasters in their communities, (3) recognizing the importance of inclusive, communal effort, (4) strengthening the collaboration between residents and community leaders and between the authorities and the communities, (5) increasing awareness of environmental changes and issues, and (6) understanding the effectiveness and benefits of the use of the scientific data in explaining disaster risks. In discussions and surveys conducted by community members, it was found that before SeDAR, the participants had not heard of DRR as a concept and certainly had never done anything in relation to DRR. Time and time again, community leaders lamented that community members were apathetic and indifferent to proactive actions to prevent or mitigate the impacts of disasters in their areas.

(c) Created a Network of DRR Collaborators at the Government and Community Level

The SeDAR program created a DRR network of many players with diverse profiles and skills to contribute to a synergy of DRR planning and implementation. While the SeDAR project initially involved only a few players—IRIDeS, SDMU, UTM, and APM—many other agencies have joined in, such as the Department of Mineral and Geoscience, the Department of Irrigation and Drainage, the Public Works Department, the Department of Education, Selangor Islamic Religious Department, and others. Likewise, community leaders have recognized the importance of communal effort, such that the entire community, not just its leaders, must cooperate and commit to the project. Inclusivity was a key point that they recognized.

In SeDAR, government agencies and communities have worked together in forging a deepened community–government working relationship. Communities are anticipated to maintain close relationships with government agencies through maintaining open lines of communication and working together. Additionally, communities must continue receiving technical and funding support from the government to sustain DRR at the community level.

Government agencies have expressed similar aspirations in the project. Thanks to the continued collaboration between government and communities, DRR activities can be sustained long after the SeDAR program has ended.

17.4 Conclusion

While this project had several achievements, some challenges were faced during the implementation of the SeDAR program, as follows:

- Collecting and accessing the data necessary for risk identification and modeling;
- Changing the mindset that DRR and management are solely the government's responsibility alone;
- Access to early warning and clear risk communication;
- Understanding the relationship between environmental issues and disasters;
- Low interest in pre-disaster work and high interest in post-disaster outcomes.

DRR is a long process that can best be achieved in a systematic series of actions and steps. It is highly recommended that a DRR program be designed in 2–3 years to fully incorporate the concepts of DRR at the local level. While carrying out only one of the activities is effective for raising DRR awareness, its impact and effectiveness will be limited. This project prioritized building and maintaining a strong partnership with stakeholders to ensure their involvement and collaboration throughout the project. The program's leadership and coordination at the government and local levels are critical. It is especially vital to have community involvement and allow community members to take ownership of the program, as it plays a crucial role in maintaining knowledge and managing DRR activities over the long term. It is also occasionally necessary to obtain guidance and input from DRR experts to improve the quality of activities.

The recommendations from the projects for the success of future CBDRR are as follows:

1. Set up a data hub to ensure the collection and storage of data, share them for risk analysis and identification, and strengthen the application of science and technology.
2. Increase disaster education at schools and raise public awareness on DRR.
3. Strengthen early warning systems and enhance risk communication.
4. Provide government support in communities to initiate CBDRR and combine DRR into their everyday lives.
5. Increase communication between local governments and communities.
6. Provide budgetary allocations for investing in DRR activities in both structure and nonstructural measures.
7. Design a DRR program having a timeframe of at least 2–3 years.
8. Consider inclusivity in developing future CBDRR.

References

Ahmadun F-R, Wong MMR, Said AM (2020) Consequences of the 2004 Indian Ocean Tsunami in Malaysia. Saf Sci 121:619–631

Asian Disaster Preparedness Center (2020) Disaster risk reduction in Malaysia: Status Report 2020

Center for Excellence in Disaster Management and Humanitarian Assistance (2022) Malaysia: Disaster management reference handbook

Imperiale AJ, Vanclay F (2023) Re-designing social impact assessment to enhance community resilience for disaster risk reduction. Clim Action Sustain Dev, Sustain Develop 32:1571–1587

Izumi T, Shaw R, Djalante R, Ishiwatari M, Komino T (2019) Disaster risk reduction and innovations, progress in disaster science vol 2, p 100033

Khan A, Gupta S, Gupta SK (2020) Multi-hazard disaster studies: Monitoring, detection, recovery, and management, based on emerging technologies and optional techniques. Int J Disaster Risk Reduct 47:101642

Khudri HBM, Akma SNS, Ruhaida A, Hisyam KK (2024) Factors contributing to participation in community-led DRR programs in Malaysia. Disaster Adv 17(8)

Lassa J, Boli Y, Nakmofa Y, Fanggidae S, Ofong A, Leonis H (2018) Twenty years of community-based disaster risk reduction experience from dryland village in Indonesia. J Disaster Risk Stud 10(1)

Lee C-C, Huang K-C, Kuo S-Y, Lin Y-J, Ke K-Y, Pan T-Y, Tai L-L, Cheng C-K, Shih Y-L, Lai H-T, Ke B-H (2022) Genter matters: The role of women in community-based disaster risk management in Taiwan. Int J Disaster Risk Reduct 77:103046

McLennan BJ (2018) Conditions for effective coproduction in community-led disaster risk management. Int J Volunt Nonprofit Organ, 31(2)

Partidario M, Saad G, Monteiro MB, Dias J, Martins R, Ramos IL, Ribeiro H, Teixeira M, de Freitas MBC, Antunes C (2022) Using participatory mapping to foster community-based disaster risk reduction in forest fire-prone areas: the case of Monchique in Portugal. Fire 5:146

Paschalia YPM, Doondori AK, Sekunda MS, Budiana I (2022) Revitalization of disaster mitigation management and community empowerment through Community-Based Disaster Riks Reduction (CBDRR). J Penelit Pendidik IPA 8(5)

Rosmadi HS, Ahmed MF, Mokhtar MB, Lim CK (2023) Reviewing challenges of flood risk management in Malaysia. Water 15:2390

Sarker MdNI, Peng Y, Yiran C, Shouse RC (2020) Disaster resilience through big data: Way to environmental sustainability. Int J Disaster Risk Reduct 51:101769

United Nations (2015) Sendai framework for disaster risk reduction 2015–2030

Chapter 18
India's Progress on Disaster Risk Reduction Over the Past 20 Years

Sujata Saunik and Rajib Shaw

Abstract Over the past 20 years, India has made significant progress in mitigating the impacts of natural disasters, induced by events like the Indian Ocean Tsunami. This chapter focuses on the progress made since the tsunami and other natural disasters. It discusses the development of policy frameworks, community engagement and technological innovations. It aimed at reducing disaster impacts including improved early warning systems, resilient infrastructure, and enhanced community preparedness. Particularly in regions prone to cyclones, floods, and earthquakes. The chapter also highlights collaborative efforts among government bodies, NGOs and international partners, emphasizing knowledge exchange and proactive risk management. Reflecting on India's journey since the tsunami, this document illustrates resilience, innovation, and collective action, offering insights into global disaster mitigation strategies.

Keywords Disaster management act 2005 · National disaster management authority · Sendai framework for disaster risk reduction · Community-based disaster risk management · Geographic information systems · Integrated coastal zone management

Abbreviations

BPR	Bottom Pressure Recorders
CBDRM	Community-Based Disaster Risk Management
CRZ	Coastal Regulation Zone
DRR	Disaster Risk Reduction
GIS	Geographic Information Systems
ICZM	Integrated Coastal Zone Management
IMD	Indian Meteorological Department

S. Saunik · R. Shaw (✉)
Graduate School of Media and Governance, Keio University, Fujisawa, Japan
e-mail: shaw@sfc.keio.ac.jp

R. Shaw et al. (eds.), *Two Decades from the Indian Ocean Tsunami*, Disaster Risk Reduction, https://doi.org/10.1007/978-981-96-2669-4_18

NDMA	National Disaster Management Authority
UNDRR	United Nations Office for Disaster Risk Reduction
UNISDR	United Nations International Strategy for Disaster Reduction

18.1 Introduction

Picture this: In October 1999, Odisha suffered from Cyclone Paradip, leaving a trail of destruction in its wake. The devastation was profound, with widespread loss of life and livelihoods. Despite facing challenges, Odisha did not give in to despair. Fast forward to 2013, and history seemed poised to repeat itself as Cyclone Phailin advanced towards the coastline of Odisha with force. Yet, this time, the narrative took a different turn. Armed with lessons learned from past disasters and strengthened by proactive measures, Odisha's disaster management apparatus swung into action. Community-based early warning systems were activated, empowering local residents to evacuate to safer ground well ahead of the cyclone's landfall. Coastal shelters, built with disaster resilience in mind, provided sanctuary to thousands, shielding them from the fury of Phailin's wrath. Although Phailin was one of the most powerful cyclones to strike India in recent years, its impact was surprisingly minimal. Lives were saved, livelihoods preserved, and communities emerged stronger in the aftermath.

This trajectory of improvement continued into 2021 when Cyclone Yaas threatened the region. By this time, technological advancements and improved coordination frameworks had further enhanced Odisha's preparedness. The state implemented rigorous evacuation protocols, including COVID-19 safety measures, and utilized digital tools for real-time tracking and response. The strategic deployment of resources, combined with an increased focus on sustainable infrastructure and community training, ensured that Cyclone Yaas, while severe, resulted in minimized human and economic losses, showcasing a significant evolution in disaster management over the decades.

The 2004 Indian Ocean Tsunami and Earthquake further underscored the need for a robust disaster management framework in India. The tsunami, which struck with little warning, resulted in an unprecedented scale of devastation, claiming thousands of lives across coastal regions. This catastrophic event was a wake-up call that led to a paradigm shift in India's approach to disaster management. It highlighted the urgency of establishing comprehensive early warning systems, enhancing coastal zone management, and building resilient infrastructure. The lessons learned from the tsunami significantly influenced the formulation of subsequent policies and strategies.

India's journey in disaster management has been transformative, marked by significant legislative and technological advancements that have fundamentally altered its approach to handling natural calamities. The enactment of the Disaster Management Act in 2005 was a turning point, leading to the formation of the National Disaster

Management Authority (NDMA) and a transition from reactive to proactive disaster management. This legislative framework was further bolstered by the National Policy on Disaster Management in 2009, which introduced the "Build Back Better" principle. This policy emphasizes community resilience, early warning systems, and integrating risk reduction into urban and rural development planning. The first National Disaster Management Plan (NDMP) in 2016, revised in 2019, expanded the scope to include new hazards like the glacial lake outburst floods and biological emergencies. It outlined specific, time-bound responsibilities for various stakeholders, ensuring alignment with the Sendai Framework for Disaster Risk Reduction.

18.2 Policy and Institutional Frameworks

The enactment of the Disaster Management Act in 2005 represented a fundamental change in India's approach to disaster management, establishing the National Disaster Management Authority (NDMA). This legislation introduced a proactive disaster management system that emphasizes prevention, mitigation, and preparedness over mere response to disasters. The strategic shift aims to reduce disaster risks significantly, focusing on comprehensive planning and community involvement at all levels of governance.

National Policy on Disaster Management of 2009

The National Policy on Disaster Management of 2009 expands upon these principles by emphasizing a preventive strategy for disaster risk management (DRM). The policy advocates for the "Build Back Better" approach during recovery, rehabilitation, and reconstruction efforts. It highlights the importance of enhancing community resilience, establishing robust early warning systems, and integrating risk reduction measures into the planning and development of both urban and rural areas. This approach ensures that disaster risk reduction is a core consideration in all development activities.

National Disaster Management Plan (NDMP)

The initial National Disaster Management Plan (NDMP), developed by the NDMA in 2016 and updated in 2019, now addresses additional hazards including thunderstorms, lightning, squalls, dust storms, hailstorms, Glacial Lake Outburst Floods (GLOF), heatwaves, and biological and public health emergencies. The updated NDMP outlines specific, time-bound responsibilities for all relevant ministries, departments, agencies, and stakeholders, ensuring alignment with the timelines of the Sendai Framework for Disaster Risk Reduction.

Guidelines and Implementation

The NDMA employs a mission-driven approach, implementing various initiatives with the support of institutions at the national, state, and local levels. It issues guidelines for specific disasters and themes, including capacity development and public awareness, which serve as the foundation for plan development. These guidelines are crafted through a participatory and consultative process involving stakeholders, ensuring the creation of comprehensive and actionable strategies.

Institutional Mechanisms and Functions of NDMA

As the apex authority on disaster management in India, the NDMA is tasked with establishing policies, plans, and guidelines to ensure a prompt and effective response to disasters. Its statutory duties include approving the national disaster plan and the plans created by various ministries and government departments in line with this national framework. The NDMA also sets guidelines for state authorities in developing their own state-specific plans and advocates for funding provisions for disaster mitigation. Furthermore, the NDMA coordinates the enforcement and implementation of disaster management policies and plans, while also supporting community-based disaster preparedness, medical readiness, psychosocial care, and trauma management.

Enhanced Institutional Frameworks and Administrative Systems

Recent developments have significantly improved disaster risk governance in India. For instance, the effective response to Cyclone Phailin in 2013 highlighted the improvements in institutional frameworks and administrative systems. The establishment of State Disaster Management Authorities (SDMAs) and District Disaster Management Authorities (DDMAs) has decentralized disaster management, ensuring better coordination and implementation at the state and district levels. Additionally, the involvement of local self-governments and community-based organizations has strengthened disaster preparedness and response at the grassroots level.

18.3 Technological Advancements in Disaster Response

Advancements in technology have been pivotal in enhancing India's disaster preparedness and response. Geographic Information Systems (GIS) and remote sensing have revolutionized risk assessment and management practices, offering capabilities for real-time monitoring and efficient data management crucial for disaster response. These technologies have enabled authorities to map disaster-prone areas, assess vulnerabilities, and plan mitigation measures effectively. For example, GIS has been used to create detailed hazard maps that identify areas at risk of earthquakes, floods, and landslides, facilitating targeted interventions and resource allocation.

The Indian Ocean Tsunami Warning System, established in response to the 2004 tsunami, has become a critical component of the regional disaster response strategy, enabling timely evacuations and reducing potential casualties. This system integrates seismic stations, Bottom Pressure Recorders (BPR), and tide gauges to provide an effective early warning system (Samarajiva et al. 2005). The success of this system in recent years has been evident, as timely alerts and accurate predictions have significantly reduced the loss of life and property during tsunamis.

The integration of IT and GIS has supported the development of decision-support systems and real-time management platforms that facilitate efficient coordination and quick decision-making during disasters. Such systems use spatial data sets from various sources, enhancing the capabilities of tsunami warning centers by providing detailed vulnerability mappings and decision-making tools (Schiermeier and Witze 2009). For instance, during the 2015 Nepal earthquake, GIS-based tools were used to map the extent of the damage and prioritize areas for rescue and relief operations, demonstrating the critical role of technology in disaster management.

In addition to the advancements mentioned above, the utilization of artificial intelligence (AI) and machine learning (ML) algorithms presents a significant value addition in enhancing India's disaster preparedness and response capabilities. These technologies can analyze vast amounts of data from various sources, including satellite imagery, social media feeds, and weather forecasts, to provide predictive analytics and early warning systems for potential disasters. AI and ML algorithms can identify patterns and trends that might be missed by human analysts, improving the accuracy of predictions and enabling proactive measures to mitigate risks.

By incorporating AI and ML into existing GIS and remote sensing frameworks, India can further improve its ability to forecast, mitigate, and respond to natural calamities with greater accuracy and efficiency. AI-powered systems can also support real-time decision-making during disasters, helping authorities to allocate resources more effectively and coordinate response efforts. For example, AI algorithms can optimize evacuation routes based on real-time traffic data and predict the spread of wildfires using weather and vegetation data, ensuring timely and effective interventions.

Moreover, AI-powered decision-support systems can assist authorities in making informed decisions during crisis situations, optimizing resource allocation and response strategies for maximum impact. These systems can analyze historical data on past disasters to identify best practices and lessons learned, guiding the development of more effective disaster management plans. Additionally, AI and ML can enhance community resilience by providing personalized risk assessments and preparedness recommendations to individuals and households, empowering them to take proactive measures to protect themselves and their property.

The integration of advanced technologies such as GIS, remote sensing, AI, and ML into India's disaster management framework represents a significant step forward in building a more resilient and responsive system. As these technologies continue to evolve, they will play an increasingly important role in reducing the impact of disasters and safeguarding communities across the country (Agrawal 2019).

18.4 Community and Local-Level Initiatives

Community involvement has significantly bolstered local capacities in disaster preparedness and response. Nationwide programs like **Community-Based Disaster Risk Management (CBDRM)** focus on empowering communities through education, training, and capacity-building activities. These initiatives have led to the development of local disaster management plans, the conducting of mock drills, and the establishment of community shelters. Research emphasizes that understanding local knowledge and social networks enhances disaster preparedness and response capabilities, which are pivotal for effective community resilience. The success of these community-driven approaches is evident in the increased resilience observed during recent cyclones in Odisha and Andhra Pradesh, where community readiness and early evacuations significantly reduced death tolls (Shaw 2014; Niekerk et al. 2018).

Also in addition to this, incorporating indigenous knowledge systems (IKS) can further enhance the effectiveness of local capacities in disaster preparedness and response. Indigenous communities often possess unique insights, practices, and traditional coping mechanisms that have been developed over generations to deal with environmental hazards. By integrating IKS into CBDRM programs, authorities can tap into this valuable reservoir of knowledge to enhance the resilience of vulnerable communities. Furthermore, fostering partnerships between local communities, academic institutions, and governmental agencies can facilitate the documentation, preservation, and dissemination of indigenous knowledge, ensuring its incorporation into mainstream disaster management strategies. Embracing IKS not only strengthens community resilience but also promotes cultural diversity and inclusivity in disaster risk reduction efforts.

18.5 Coastal Zone Management and Climate Adaptation

Coastal areas, which house a large portion of India's population, face significant vulnerabilities to cyclones, sea-level rise, and tsunamis. The Coastal Regulation Zone (CRZ) notifications have been essential in safeguarding these areas by regulating coastal development activities. Efforts in **Integrated Coastal Zone Management (ICZM)** focus on conserving and sustainably managing coastal resources. Additionally, mangrove restoration and coral reef conservation are crucial in protecting shorelines and supporting marine biodiversity. Engaging community stakeholders in making coastlines more resilient to coastal storms has proven necessary, with an emphasis on incorporating community assets into disaster resilience planning (Bostick et al. 2017).

In addition to the strategies mentioned, leveraging ecosystem-based approaches can offer a significant value addition in enhancing the resilience of coastal areas to natural hazards. Ecosystem-based adaptation (EbA) involves harnessing the natural functions of ecosystems such as mangroves, wetlands, and dunes to reduce the impacts of coastal hazards. For example, mangrove forests act as natural barriers, absorbing the energy of waves and reducing coastal erosion during cyclones and tsunamis. Wetlands serve as natural sponges, absorbing excess water during storms and helping to mitigate flooding. By integrating EbA into ICZM efforts, India can enhance coastal resilience while simultaneously conserving biodiversity and promoting sustainable development. Moreover, fostering partnerships between government agencies, local communities, and environmental organizations can facilitate the implementation of EbA initiatives, ensuring that coastal ecosystems continue to provide vital services for both people and nature.

18.6 Urban Risk Reduction

Urban centers, characterized by high population densities and complex infrastructures, are focal points for disaster risk reduction efforts. The enforcement of stringent building codes, urban planning regulations, and the development of disaster-resilient infrastructure have been central to mitigating urban risks. The Smart Cities Mission incorporates disaster risk management into the broader framework of urban development, focusing on building resilient cities equipped to handle various urban challenges, including disasters. Research highlights that integrating urban resilience into disaster risk management and risk reduction can bridge gaps between sustainable development goals and disaster preparedness, emphasizing the need for coherent frameworks that incorporate risk assessments and resilience indicators into urban planning. The absence of such indicators can hinder the monitoring of evolving disasters and underlying risk drivers, suggesting that a preparedness framework that includes these aspects is crucial for building sustainable and resilient urban environments (Etinay et al. 2018).

18.7 Enhancements in Early Warning Systems

The modernization of the Indian Meteorological Department (IMD) and the expansion of the network of Doppler radars have greatly improved the accuracy and timeliness of weather forecasts and early warnings. These systems are crucial for effective disaster preparedness, enabling authorities to issue timely alerts and conduct evacuations, thus minimizing human casualties and economic losses. Further improvements in early warning systems could incorporate transdisciplinary risk mapping and building codes that consider multi-hazard scenarios to strengthen urban resilience, ensuring a comprehensive approach to disaster risk management that addresses not only immediate but also long-term urban planning and infrastructure needs (Bendito 2020).

18.8 Financial Strategies for Risk Reduction

Financial strategies for risk reduction are critical components of effective disaster management frameworks. These strategies involve a range of mechanisms aimed at ensuring adequate financial resources are available before, during, and after a disaster. By addressing the financial aspects of disaster risk reduction, governments and organizations can enhance resilience and support quicker recovery. Over the last two decades, India has made significant progress in strengthening its financial preparedness for disasters. This section explores how funding is channeled for DRR initiatives at the state level, specifically focusing on the role of the State Disaster Management Authority (SDMA) and its Technical Advisory Committee (TAC) in ensuring funds reach the last mile for on-ground implementation.

Disaster Response Funds

The Government of India has established the National Disaster Response Fund (NDRF) and the State Disaster Response Fund (SDRF) to provide immediate financial assistance to states affected by disasters. These funds are designed to ensure that resources are readily available to manage disaster response and recovery efforts without delay. The funds can be utilized for relief measures such as providing food, shelter, and medical assistance to affected populations, as well as for rebuilding critical infrastructure.

State Disaster Management Fund (SDMF)

The 15th Finance Commission (XFC) recommended a substantial allocation of 20% of the total State Disaster Risk Management Fund (SDRMF) for State Disaster Management Funds (SDMF) during the period 2021–22 to 2025–26. These funds are crucial for financing DRR activities at the state level.

Fund Allocation for Mitigation and Capacity Building: The SDMF allocations are divided into two primary components.

Mitigation Component (20%)

This component addresses structural and non-structural mitigation works as outlined by the National Disaster Management Authority (NDMA) guidelines. Examples include investments in cyclone shelters, land procurement for the National Disaster Response Force (NDRF), and other initiatives that reduce the risk of disasters.

Capacity Building and Training Component (10%)

This component supports DRR-related training programs, capacity-building exercises, and institutional strengthening activities. It also covers DRR initiatives that fall outside the purview of the SDRMF.

Funding Allocation Process

A well-defined procedure ensures transparency and accountability in allocating funds for mitigation activities:

1. District Proposals: Districts submit proposals endorsed by the District Disaster Management Authority (DDMA) to the SDMA in a format prescribed by the NDMA.
2. Project Appraisal Committee (PAC) Review: A PAC, chaired by the Honourable Minister of Disaster Management, Relief and Rehabilitation, conducts a preliminary assessment of the proposals.
3. State Executive Committee (SEC) and SDMA Approval: Proposals with the PAC's recommendations are presented before the SEC and SDMA for approval.
4. Detailed Project Report (DPR) Preparation: Following preliminary approval from the SDMA, the proposing agency prepares a DPR with technical sanction from the designated technical authority.
5. Technical Appraisal Committee (TAC) Scrutiny: The TAC, chaired by a subject matter expert, meticulously scrutinizes the DPRs.
6. Final Approval by SEC and SDMA: Based on the TAC's recommendations, the SEC and SDMA provide final approval for the DPRs.
7. Administrative Approval and Implementation: Upon receiving approval, the SDMA issues administrative sanctions for initiating the mitigation work.

Insurance Schemes

Insurance schemes are another crucial financial tool for disaster risk reduction. Agricultural insurance, for example, protects farmers from losses due to crop failure caused by natural disasters. The Indian government's Pradhan Mantri Fasal Bima Yojana (PMFBY) is a notable example, offering coverage to farmers at subsidized premium rates. This scheme not only provides financial protection to farmers but also encourages them to adopt better farming practices and invest in high-yield crops, thereby enhancing overall agricultural productivity and resilience.

Community-Based Financial Mechanisms

Disaster risk reduction strategies emphasize the importance of community-based financial mechanisms. Community savings groups, microfinance institutions, and local disaster funds can provide immediate financial support to affected individuals and families. These mechanisms ensure that financial assistance reaches those most in need quickly and efficiently. For instance, grain banks established in disaster-prone areas help communities store food grains during non-disaster periods, ensuring food security when disasters strike.

International Cooperation and Funding

International cooperation and funding also play a vital role in enhancing financial strategies for disaster risk reduction. Countries often collaborate with international organizations such as the United Nations, World Bank, and International Monetary Fund to secure funding and technical assistance for disaster preparedness and response initiatives. These collaborations can provide access to advanced technologies, expertise, and financial resources that might not be available domestically.

Integrating DRR into Development Planning

Incorporating disaster risk reduction into development planning and budgeting is crucial for effective and sustainable risk management. This requires dedicating part of national and local budgets to disaster risk reduction initiatives, such as constructing resilient infrastructure, performing risk assessments, and setting up early warning systems. By integrating disaster risk reduction into financial planning, governments can enhance long-term resilience and mitigate the overall economic impact of disasters.

The SDMA, in collaboration with the TAC, plays a critical role in ensuring effective financial management for DRR initiatives in India. The allocation process is designed to prioritize evidence-based mitigation strategies while ensuring transparency and accountability. This multi-tiered approach helps ensure that funds reach the last mile for on-ground implementation, ultimately strengthening India's preparedness for disasters.

18.9 International Cooperation and Global Best Practices

India's DRR efforts are closely aligned with international standards and practices, particularly through the adoption and integration of the **Sendai Framework for Disaster Risk Reduction 2015–2030**. The Framework emphasizes a historic shift from disaster management to disaster risk management, focusing on the underlying drivers of disaster risk, such as poorly planned urban growth and the impacts of climate change. In the context of the Indian Railways, for instance, the Sendai Framework has been crucial in shaping their DRR policy, although there are noted gaps, particularly in areas like railway construction (Ghosh and Sar 2021).

India's active engagement in international forums and collaborations with global organizations, such as the **United Nations Office for Disaster Risk Reduction (UNDRR)**, has facilitated substantial knowledge exchange and the enhancement of national capabilities. The nation's commitment to the Sendai Framework was highlighted in its leading role at regional meetings, such as the International Strategy for Disaster Reduction Asia Partnership meeting in Bangkok, where regional implementation strategies were discussed (Dickinson et al. 2016). Additionally, the integration of health and DRR has seen significant advances under the Framework, enhancing both national and regional resilience and preparedness (Wright et al. 2020).

India's approach exemplifies a model of international cooperation, emphasizing not only the adoption of global frameworks but also active participation in shaping and leading DRR initiatives on the international stage. This cooperative stance has been pivotal in strengthening the nation's overall disaster resilience, aligning with global best practices and contributing to a safer, more prepared society.

18.10 Conclusion

India's disaster management framework is characterized by proactive policies, technological advancements, community involvement, and international cooperation, representing a robust model for reducing disaster risks and enhancing resilience. Institutional frameworks have been significantly strengthened with the establishment of State and District Disaster Management Authorities, enhancing coordination and implementation at local levels. Technological advancements have played a pivotal role, with Geographic Information Systems (GIS) and remote sensing revolutionizing risk assessment and management practices. The Indian Ocean Tsunami Warning System, along with the integration of artificial intelligence (AI) and machine learning (ML) into disaster management frameworks, has enhanced India's ability to predict, mitigate, and respond to disasters.

Community-based initiatives like Community-Based Disaster Risk Management (CBDRM) and the integration of Indigenous Knowledge Systems (IKS) have empowered local communities, leading to more effective disaster preparedness and response. Coastal zone management efforts, including Integrated Coastal Zone Management (ICZM) and ecosystem-based adaptation strategies, have improved resilience against coastal hazards. Urban risk reduction initiatives under the Smart Cities Mission and the modernization of early warning systems by the Indian Meteorological Department (IMD) have significantly improved disaster preparedness in urban areas.

Financial strategies, including disaster response funds and insurance schemes, ensure that resources are readily available to manage disaster response and recovery efforts. International cooperation, particularly through the adoption of the Sendai Framework for Disaster Risk Reduction, has aligned India's disaster risk reduction efforts with global best practices. By integrating advanced technologies, fostering community involvement, and adhering to international standards, India continues to build a more resilient and responsive disaster management system. This comprehensive approach safeguards lives and livelihoods, ensuring that the nation is better prepared to face the increasing frequency and severity of natural disasters.

References

Agrawal N (2019) Innovative use of technology (GIS) for disaster risk reduction: A case study. In: Zutshi B, Ahmad A, Srungarapati AB (eds) Disaster risk reduction. Palgrave Macmillan, Singapore. https://doi.org/10.1007/978-981-10-8845-2_9

Bendito A (2020) Grounding urban resilience through transdisciplinary risk mapping. Urban Transform 2. https://doi.org/10.1186/s42854-019-0005-3

Bostick T, Holzer T, Sarkani S (2017) Enabling stakeholder involvement in coastal disaster resilience planning. Risk Anal 37. https://doi.org/10.1111/risa.12737

Dickinson C, Aitsi-Selmi A, Basabe P, Wannous C, Murray V (2016) Global community of disaster risk reduction scientists and decision makers endorse a science and technology partnership to support the implementation of the Sendai framework for disaster risk reduction 2015–2030. Int J Disaster Risk Sci 7:108–109. https://doi.org/10.1007/s13753-016-0080-y

Financial assistance during disaster—A complete handbook (2024) Assam state disaster management authority. https://asdma.assam.gov.in/documents-detail/financial-assistance-during-disaster-a-complete-handbook

Etinay N, Egbu C, Murray V (2018) Building urban resilience for disaster risk management and disaster risk reduction. Procedia Eng 212:575–582. https://doi.org/10.1016/J.PROENG.2018.01.074

Ghosh S, Sar A (2021) The Sendai framework for disaster risk reduction and disaster management plan of Indian railways. Disaster Adv. https://doi.org/10.25303/148da7921

Niekerk D, Nemakonde L, Kruger L, Forbes-Genade K (2018) Community-based disaster risk management, pp 411–429. https://doi.org/10.1007/978-3-319-63254-4_20

NDMA Annual Report 2022–23. https://ndma.gov.in/sites/default/files/PDF/Reports/Annual_Report_NDMA_2022-2023.pdf

Pal I, Ghosh T, Ghosh C (2017) Institutional framework and administrative systems for effective disaster risk governance—Perspectives of 2013 Cyclone Phailin in India. Int J Disaster Risk Reduct 21:350–359. https://doi.org/10.1016/J.IJDRR.2017.01.002

Samarajiva R, Knight M, Anderson P, Zainudeen A (2005). National early warning system: Sri Lanka—A participatory concept paper for the design of an effective all-hazard public warning system. https://doi.org/10.2139/SSRN.1554782

Schiermeier Q, Witze A (2009) Tsunami watch. Nature 462:968–969. https://doi.org/10.1038/462968a

Shaw R (2014) Disaster risk reduction and community approaches, pp 3–20. https://doi.org/10.1007/978-4-431-54246-9_1

Wright N, Fagan L, Lapitan J, Kayano R, Abrahams J, Huda Q, Murray V (2020) Health emergency and disaster risk management: Five years into implementation of the Sendai framework. Int J Disaster Risk Sci 11:206–217. https://doi.org/10.1007/s13753-020-00274-x

Chapter 19
Bangladesh's Progress on Disaster Risk Reduction Over the Past 20 Years

Md. Shamsuzzoha, Md. Fuad Hasan, Md. Rasheduzzaman, Md. Foisal Talukder, Tofael Ahamed, and Rajib Shaw

Abstract Over the past two decades, Bangladesh has significantly advanced disaster risk reduction (DRR), underpinned by comprehensive policies and legislative frameworks. Initiated by the Disaster Management Act of 2012, the country has established a robust institutional framework, including the Disaster Management Bureau (DMB), to coordinate DRR efforts at national and local levels. Strategic initiatives like the Standing Orders on Disaster (SOD) of 1999 revised in 2010 and 2019, respectively, the National Plan for Disaster Management (NPDM) 2021–2025, and the updated Bangladesh Climate Change Strategy and Action Plan (BCCSAP) of 2022 have guided initiative-taking measures throughout the emergency management cycle's mitigation and adaptation including total eleven thematic areas. Leveraging innovative technologies, projects like the Mobile Vulnerability Assessment and Mapping (mVAM) of the World Food Programme (WFP) have identified high-risk areas for targeted interventions. Community-based and managed programs, notably the Cyclone Preparedness Programme (CPP) since 1972, have empowered local populations, focusing on women's involvement, resulting in noteworthy progress on DRR over the past 20 years. Investments in infrastructure, shelters, and early warning systems, implementing Bangladesh Delta Plan (BDP) 2100 have notably reduced disaster fatalities, positioning Bangladesh as a global role model for comprehensive DRR approaches through appropriate science and technology-based disaster

Md. Shamsuzzoha (✉) · Md. F. Hasan · Md. Rasheduzzaman · Md. F. Talukder
Department of Emergency Management, Faculty of Environmental Science and Disaster Management, Patuakhali Science and Technology University, Dumki, Patuakhali, Bangladesh
e-mail: zohageo@pstu.ac.bd

Md. Shamsuzzoha
Graduate School of Life and Environmental Sciences, University of Tsukuba, Tsukuba, Japan

Md. F. Hasan
Department of Environment Protection, German University Bangladesh, Gazipur, Bangladesh

T. Ahamed
Institute of Life and Environmental Sciences, University of Tsukuba, Tsukuba, Japan

R. Shaw
Graduate School of Media and Governance, Keio University, Fujisawa, Japan

R. Shaw et al. (eds.), *Two Decades from the Indian Ocean Tsunami*, Disaster Risk Reduction, https://doi.org/10.1007/978-981-96-2669-4_19

science and emergency management (DSEM) education for achieving sustainable development goals (SDGs).

Keywords Disaster management · Disaster risk reduction · Disaster science · Emergency management · Sustainable development goals

19.1 Introduction

Climate vulnerabilities and their impacts in Bangladesh vary significantly across regions, timescales, and socio-economic conditions. For example, during the monsoon season, the country often faces excessive water causing floods, while in the dry season, water scarcity leads to droughts (Akter and Sultana 2022; Wright et al. 2019). The coastal areas are particularly vulnerable to salinity intrusion and tropical cyclones (Huq et al. 2015). Central floodplains are prone to flooding, the northwestern region frequently experiences droughts, and the northeastern part is susceptible to flash floods (Islam et al. 2022). These diverse climate vulnerabilities necessitate specific adaptation measures tailored to each type of disaster. Bangladesh has historically been regarded as one of the most disaster-prone countries in the world, with serious effects resulting from both natural and human-caused disasters (Shamsuzzoha et al. 2024a; Shamsuzzoha and Al-Maruf 2012). Environmental disasters, whether natural or human-induced, have massive impacts on society and the environment, leading to human and material losses. The country frequently experiences severe cyclones, such as those in 1970, 1991, 2007, and 2009, leading to significant loss of life and property (Rahman et al. 2017). Regular and devastating floods, notably in 1988 and 1998, impact vast areas annually. Tornadoes, with major events in 1969, 1974, 1977, and 1989, and droughts, occurring significantly between 1949 and 1991, further exacerbate the situation (Braun and Aßheuer 2011; Karim and Mimura 2008). Earthquakes, historically significant in 1548 and 1897, and landslides, particularly post-2007, pose serious threats (Alam and Ray-Bennett 2021; Choudhury 2002). Riverbank erosion remains a recurrent issue, while fire and building collapses are rising, highlighting human-induced risks (Rahman and Gain 2020; Zhang and Nakagawa 2018). These disasters collectively contribute to extensive damage, affecting millions of people, particularly in coastal and flood-prone areas, necessitating robust mitigation and management strategies to enhance resilience and reduce vulnerability (Shamsuzzoha and Ahamed 2024a; Shamsuzzoha et al. 2021). In the last two decades, Bangladesh has made significant strides in disaster risk reduction to address its diverse climate vulnerabilities and the impacts of various natural and human-induced hazards (Habiba et al. 2013; Kulatunga et al. 2013; Mannan et al. 2021; Nasreen et al. 2023; Perwaiz et al. 2020). Recognizing the need for disaster-specific adaptation measures, the country has developed and implemented robust mitigation and management strategies (Khan 2024). These efforts include strengthening policy frameworks, enhancing community-based disaster preparedness, improving infrastructure, and leveraging technological advancements for early warning systems (Akhand 2003; Habib et al. 2012; Syed et al. 2021). In collaboration with non-governmental organizations and international agencies, the government has focused on building resilience in vulnerable regions, such as coastal areas prone to cyclones and salinity intrusion, central floodplains, the drought-affected northwest, and the flash flood-prone northeast (Matin and Taher 2001; Rahman and Tasnim 2023). Through these concerted efforts, Bangladesh aims to mitigate the adverse effects of disasters, safeguard lives and property, and promote sustainable development in the face of ongoing and emerging challenges (GED 2020; Salequzzaman and Gorana 2016).

19.1.1 Administrative Information of Bangladesh

Bangladesh's administrative structure is divided into eight first-order divisions: Barishal, Chattogram, Dhaka, Khulna, Rajshahi, Rangpur, Mymensingh, and Sylhet. Each division is named after its main city, which acts as the administrative capital (BBS 2024). These divisions are divided into sixty-four districts, or the '*zila*' boundary (BND) at the second-order level. The districts are subsequently divided into around five hundred sub-districts called '*thana*' or '*upazila*', which are ultimately combined into unions (Fig. 19.1). The 8 divisions and 64 districts exist, but they play a small role in public policy (GoB 1972; MoDMR 2015). Local administration is overseen by upazilas, municipalities or town councils, city corporations (metropolitan municipal corporations), and union councils. Bangladesh's administrative structure consists of five layers of government (PCGN 2023).

19.1.2 Geographical Settings and LULC of Bangladesh

Bangladesh, located in South Asia, is known for its diverse geographical settings and complicated land use and land cover (LULC) patterns, making it vulnerable to various natural disasters (Fig. 19.2) (Shamsuzzoha and Ahamed 2024b). The country covers around 147,570 square kilometers and has a low-lying topography, with over two-thirds of its land less than 5 m above sea level (Hasan et al. 2021; Siddeqa et al. 2023). The intricate network of rivers, including the Ganges, Brahmaputra, and Meghna, produces one of the world's largest deltas, which is productive yet prone to flooding and riverbank erosion (Mahamud and Takewaka 2018; Shamsuzzoha and Ahamed 2023). With its substantial seasonal rainfall, the monsoon climate exacerbates the flood hazards. Coastal locations are especially vulnerable to tropical cyclones and storm surges, while inland areas are threatened by riverine flooding and landslides, particularly in the Chittagong Hill Tracts (Alam and Ray-Bennett 2021). Sea level rise (SLR) is a great warning for the country among all exterior threats and drivers (Ali 1996; Bhuiyan and Dutta 2012; Karim and Mimura 2008; Woodruff et al. 2013). Agriculture dominates the LULC in Bangladesh, with paddy

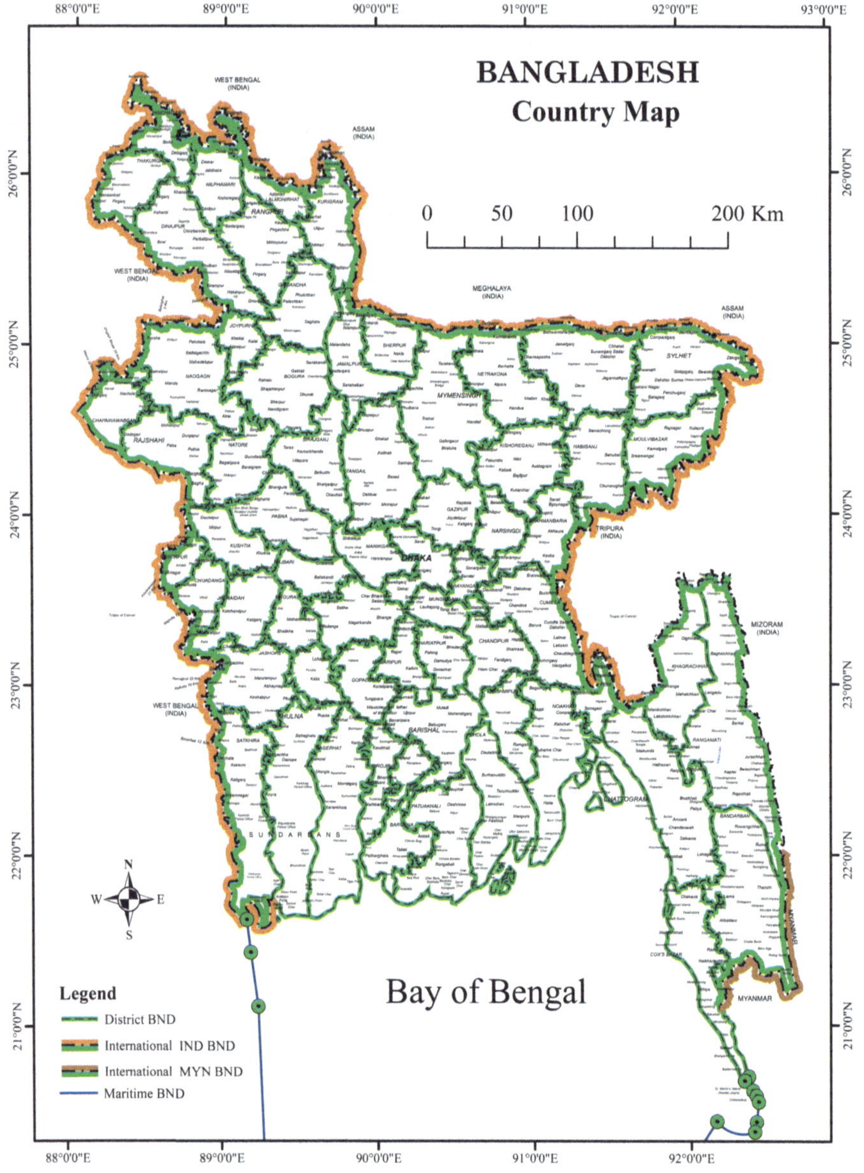

Fig. 19.1 Bangladesh administrative map

fields, croplands, and rural villages taking up substantial amounts of land (Shamsuzzoha et al. 2022). Urban areas are quickly expanding, putting additional strain on land resources, and increasing vulnerability to disasters due to poor infrastructure and unplanned development (Shamsuzzoha et al. 2018a, b). Additionally, deforestation and land degradation in steep areas increase the risk of landslides (Arab et al. 2022). The combination of these geographical and LULC characteristics makes Bangladesh a catastrophe-prone country, demanding strong disaster risk reduction and management plans (Alam and Affiliate 2010; Alam et al. 2024).

19.1.3 State of Disaster Preparedness in the Early 2000s

In the early 2000s, Bangladesh's disaster preparedness, particularly for atmospheric hazards, was significantly shaped by its geographical and socio-economic landscape (Haque et al. 2022). Located in tropical Asia with a concave coastline and shallow continental shelf, the country was highly vulnerable to tropical cyclones, storm surges, and floods, with historical data from 1775 to 1997 showing a high mortality rate from such disasters (Matin and Hasan 2021; Shamsuzzoha and Ahamed 2023). The April 1991 cyclone, which caused substantial loss of life and property, underscored this vulnerability (Choudhury 2002). The cyclone warning system had evolved to incorporate satellite imagery and computer methods for better forecasting, with warnings disseminated through various media channels (Akhand 2003). The Cyclone Preparedness Program (CPP), established after the 1970 cyclone disaster, played a crucial role in community-level warning dissemination and preparedness coordination (Habib et al. 2012). However, challenges remained, particularly in integrating human responses to warnings and ensuring public awareness. Around 15 million people in coastal and offshore areas were at particular risk, with existing cyclone shelters unable to accommodate the entire population, highlighting the need for more shelters and public education on disaster preparedness (Haque 1997). The country's socio-economic conditions, including traditional values and a lack of coordinated disaster management, exacerbated the impacts of these hazards. Bangladesh's disaster management framework involves a multi-tiered organizational structure coordinating efforts from national to local levels, with various committees responsible for distinct aspects of disaster response, including prevention, mitigation, preparedness, and response (Rahman et al. 2017; Shaw et al. 2013a, b). Efforts to improve disaster preparedness included constructing more cyclone shelters, implementing public awareness programs, and enhancing infrastructure to mitigate disaster impacts. Recent technological advancements played a significant role in enhancing early warning systems, making them more accurate and dependable (Hossain 2022; Syed et al. 2021). Despite the progress made, challenges persisted, such as the inadequate number of shelters and the continuous need for public education to improve disaster preparedness. Long-term recovery and rehabilitation efforts require better integration with national development plans to ensure sustainability and resilience (Ahmed et al. 2016; GoB 2017; Graveline and Germain 2022; Mukherjee

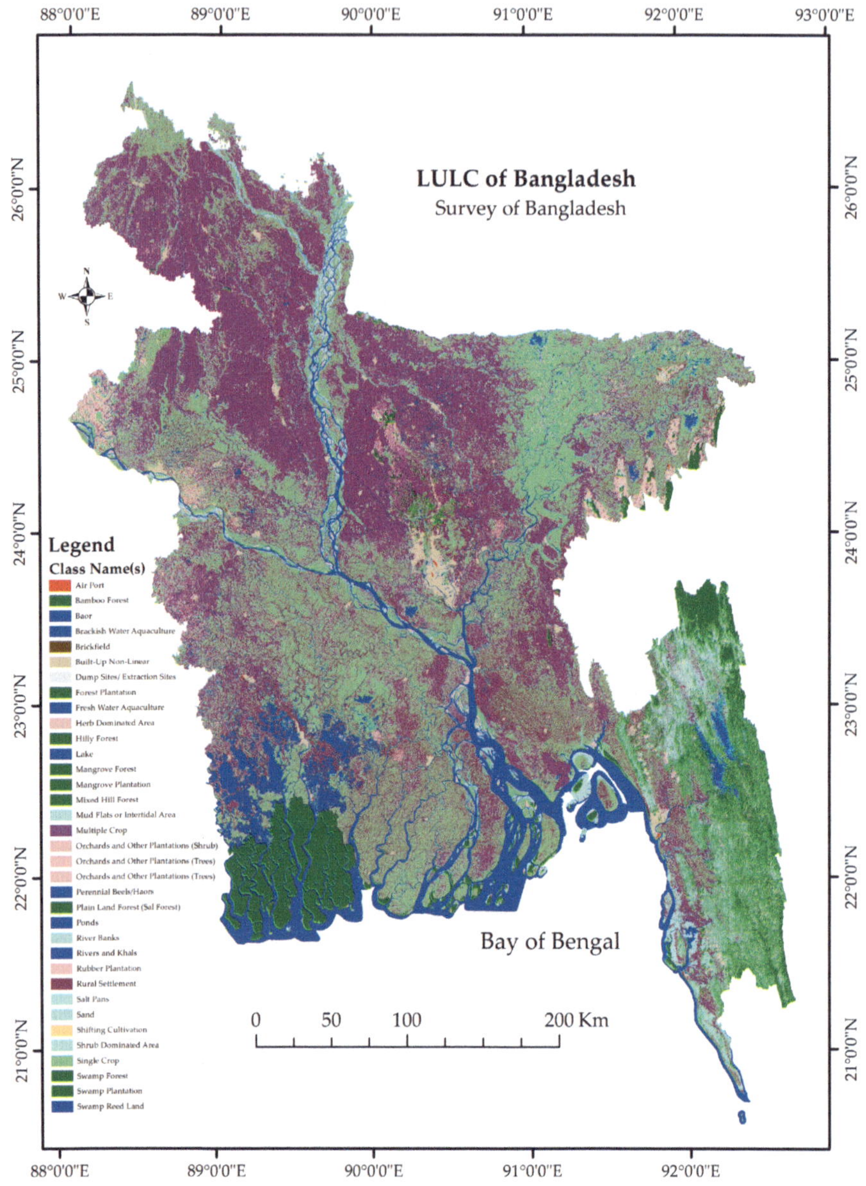

Fig. 19.2 Land use and land cover (LULC) of Bangladesh

et al. 2022). While Bangladesh has made significant strides in disaster preparedness and management, continuous efforts were necessary to address existing vulnerabilities and improve resilience, highlighting the importance of integrating technological advancements, public education, and coordinated management strategies into a comprehensive disaster management framework (Alam and Affiliate 2010; Chakma 2021; GoB 2017; MoF 2020).

19.2 Policy and Institutional Framework

19.2.1 Creation and Evolution of the MoDMR

In Bangladesh, the Ministry of Food and Disaster Management began implementing the Hyogo Framework for Action 2005–2015 to build resilience against disasters. Consequently, disaster risk reduction and climate change risk issues were integrated into all development sectors, leading to the update of the Standing Orders on Disaster (SOD) in 2010 (GoB 2010). The ministry was then transformed into the Ministry of Disaster Management and Relief (MoDMR) in 2012 (GoB 2020b). The vision of MoDMR since 2012 has been to reduce the danger to people, particularly the poor and disadvantaged, from natural, environmental, and human-caused hazards to a manageable and acceptable humanitarian level, while also developing an efficient emergency response system. The ministry is also responsible for implementing national risk reduction changes, transitioning from traditional response and relief to a complete risk reduction culture, and prioritizing food security as critical to community resilience (GoB 1972, 2008, 2012, 2017, 2020a). Here the transformation of the ministry related to disaster management and risk reduction is given (Fig. 19.3).

19.2.2 Lead Government Agencies in Disaster Management

The disaster management system in Bangladesh is structured around three primary bodies. These are the National Disaster Management Council (NDMC), the Inter-Ministerial Disaster Management Coordination Committee IMDMCC, and the MoDMR (Alam and Ray-Bennett 2021; GoB 2010, 2020b; Mahin Uddin et al. 2022). The NDMC led by the Prime Minister makes strategic disaster management decisions; the IMDMCC facilitates coordination among various ministries; and the National Disaster Management Advisory Committee focuses on policy development and provides advice (MoDMR 2015; Shammin et al. 2022). The NDMC serves as the top decision-making authority for disaster management in Bangladesh. The MoDMR acts as the Secretariat to the NDMC and oversees the coordination of disaster response efforts across all agencies at the national level (Alam and Ray-Bennett 2021; CFE-DMHA 2020). Figure 19.4 illustrates the disaster management

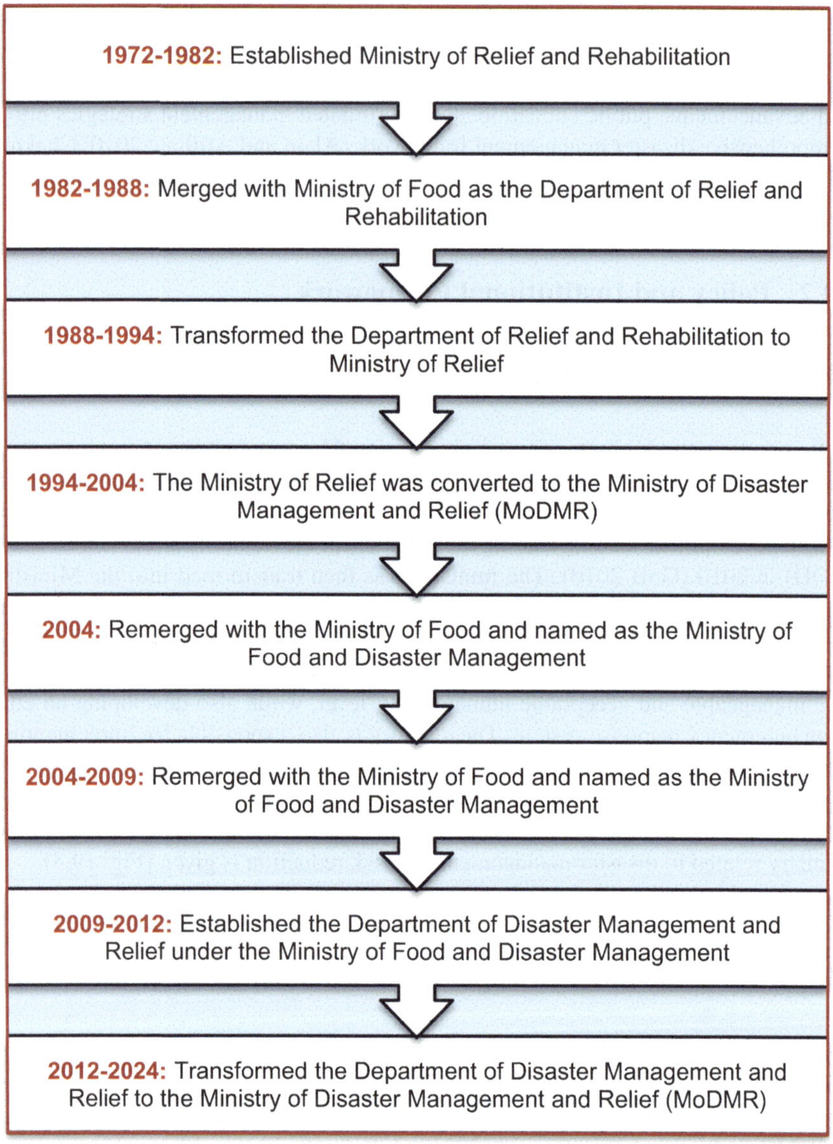

Fig. 19.3 Transformation of MoDMR from 1972 to 2024

coordination structure in Bangladesh. At the top of the organizational chart is the NDMC, with various departments, committees, and boards reporting directly to it. This includes a top-down institutional framework from the National Disaster Management Council headed by the honorable Prime Minister to the Union Disaster Management Committee headed by an elected Chairman (GoB 2010, 2020b).

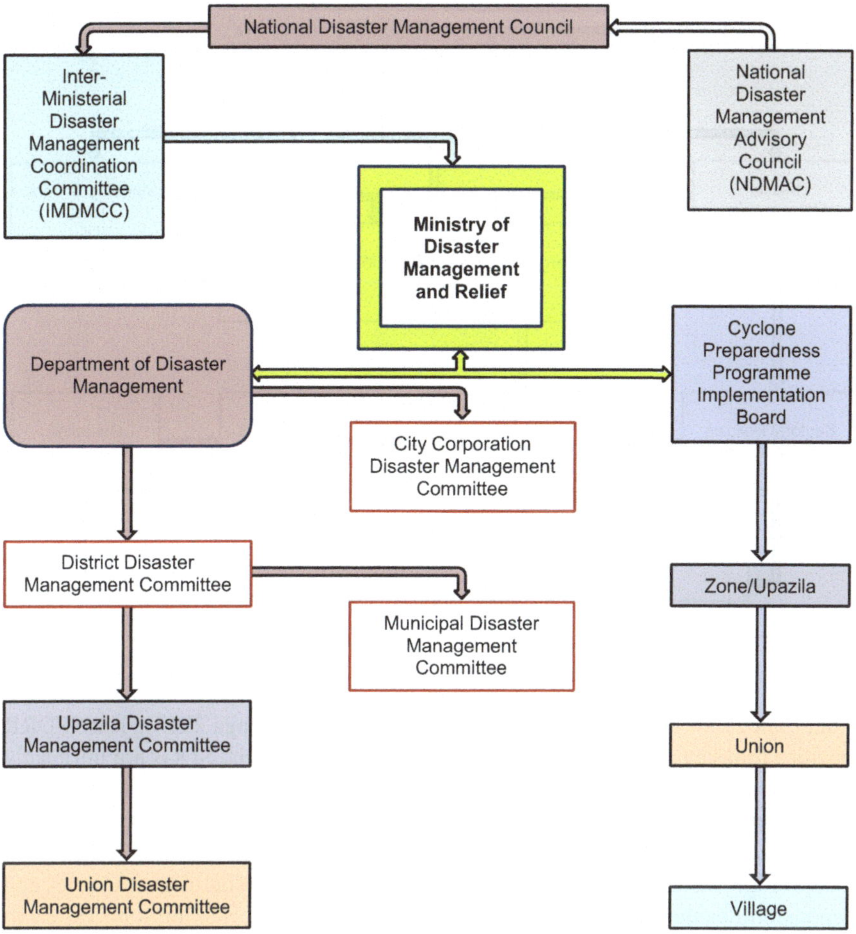

Fig. 19.4 Disaster management institutions in Bangladesh

19.2.3 Disaster Management Policy Structure and Plans

Bangladesh's disaster risk management legal framework has transitioned from a reactive to an initiative-taking stance, now serving as a cornerstone for the country's disaster risk management programs (Ahmed 2019; GoB 2010, 2017, 2020b, 2020a) (Fig. 19.5). This framework, which includes laws, policies, rules, orders, and guidelines, underpins all initiatives related to disaster risk reduction (DRR) and Emergency Response Management (ERM) (Alam and Ray-Bennett 2021; CFE-DMHA 2020; Shammin et al. 2022). For effective disaster management, it is essential to align this framework with the Sendai Framework for Disaster Risk Reduction (SFDRR), Sustainable Development Goals (GED 2020; UN 2024) (SDGs), the Paris Agreement on Climate Change, and national strategies such as the Five-Year Plan and

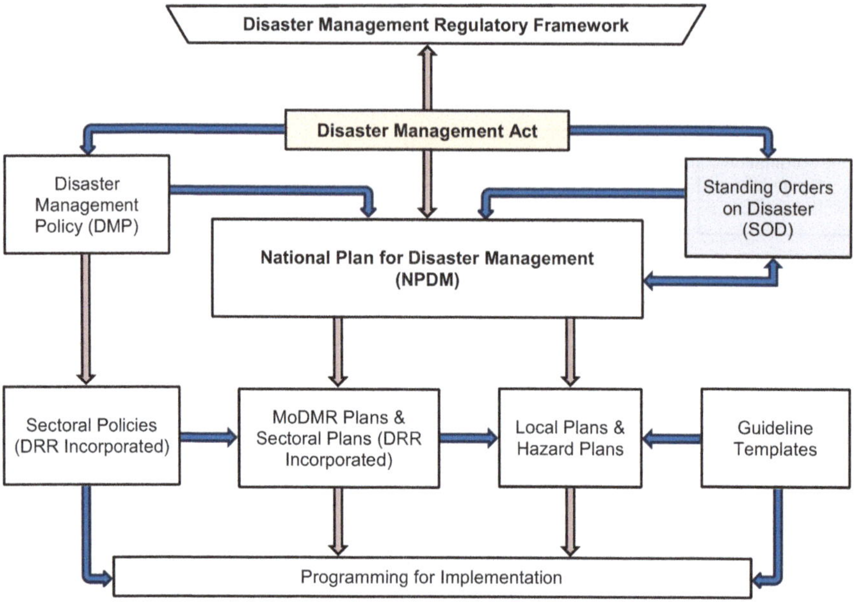

Fig. 19.5 Disaster management regulatory framework in Bangladesh

Delta Plan 2100 (Alam et al. 2018; Terwisscha Van Scheltinga 2013). The SFDRR specifies four key action priorities and seven targets, while the SDGs are intricately connected to disaster risk reduction (GED 2020; UN 2023, 2024). The Paris Agreement underscores the role of climate change in increasing disaster severity, especially in deltaic areas like Bangladesh. Thus, coordinating these international frameworks is vital (Fig. 19.5). The Standing Orders on Disasters (SOD) delineate the roles and responsibilities that reflect these frameworks' main themes, making it necessary to create medium and long-term disaster risk management and risk-informed development plans that address the six critical areas identified in the Delta Plan 2100 (Alam et al. 2018; Choudhury et al. 2012; GED 2020; GoB 2010, 2020b; Terwisscha Van Scheltinga 2013; UN 2024).

19.2.4 The Disaster Management Act (2012)

The Disaster Management Act of 2012 provides a legal framework for disaster risk reduction and emergency response management in Bangladesh. This Act establishes an organizational structure for disaster management at both the national and municipal levels, as well as the responsibilities of all government departments and committees involved in the disaster management system (GoB 2012).

19.2.5 National Disaster Management Policy (2015)

The national catastrophe management policy of 2015 describes Bangladesh's national approach to catastrophe risk reduction and emergency management. It provides a strategic framework and sets forth the essential principles for disaster management, detailing the overarching national objectives and strategies for controlling disasters (MoDMR 2015).

19.2.6 Standing Orders on Disaster (2019)

The SOD is an important part of Bangladesh's disaster management regulatory system. The SOD was first issued in 1997 and amended in 1999, 2010, and 2019 (GoB 2010, 2020b). It serves as the legal underpinning for disaster management actions. It describes the duties and responsibilities of committees, ministries, and other organizations involved in disaster risk reduction and emergency management, as well as the steps required to implement Bangladesh's Disaster Management Model (GoB 2010, 2020b). The SOD informs all relevant parties of their responsibilities at various levels of disaster management. To enable efficient execution, the SOD requires that all ministries, divisions, and agencies create their Action Plans suited to their specific duties. The NDMC and the IMDMCC manage disaster-related operations on a nationwide level (Fig. 19.4). At the district and sub-district levels, coordination is managed by the respective District, *Upazila/Thana*, Union, and Ward Disaster Management Committees, with the Department of Disaster Management providing support to facilitate these processes (GoB 2020b).

19.2.7 The National Plan for Disaster Management (NPDM) 2021–2025

The Bangladesh National Plan for Disaster Management (NPDM) is a strategic umbrella plan that serves as a general guideline for relevant sectors and disaster management committees at all levels of disaster management (Fig. 19.4) to develop and implement their plans (GoB 2017, 2020a, b). NPDM 2021–2025 is Bangladesh's strategic framework to enhance disaster resilience. It builds on the achievements and lessons of previous NPDMs (2010–2015, 2016–2020), transitioning from reactive to initiative-taking DRR (GoB 2017, 2020a). The NPDM 2021–2025 aligns with national, regional, and international frameworks, including the SFDRR, Delta Plan 2100, and the 8th Five-Year Plan (Alam et al. 2018; GoB 2020a; Mannan et al. 2021; Terwisscha Van Scheltinga 2013). The plan focuses on saving lives and reducing economic losses through comprehensive disaster cycle management, incorporating DRR, humanitarian response, and emergency recovery. It includes 34 core targets for implementation in collaboration with various stakeholders. Key achievements of NPDM 2016–2020 included early warning systems, gender-inclusive DRR, and civil-military coordination. However, gaps remain in the whole disaster cycle management and tackling urban disasters (Ahmed 2019; GoB 2017, 2020a; Khan 2024; Perwaiz et al. 2020; Shaw et al. 2013a, b, c; UN 2023).

NPDM 2021–2025 adopts a 'whole-of-society' approach, emphasizing inclusivity and a gender-responsive strategy. It aims to guide the implementation of the Disaster Management Act 2012 and the SOD 2019, promoting risk-informed planning and public–private investments (GoB 2012, 2020a, b). The plan addresses emerging risks like climate change and rapid urbanization, crucial for sustainable development. The NPDM 2021–2025 outlines eight strategic directions, including upgrading DM programs, improving governance, social protection, inclusive development, and resilient post-disaster recovery. It follows SFDRR's guidelines for DRR (UNDRR 2015). The plan's implementation is structured into three phases: a preparatory year (2021), initiation of new actions (2021–2022), and a peak activity period (2023–2025). This adaptive document includes periodic reviews and stakeholder feedback to ensure its relevance and effectiveness in building a resilient Bangladesh (Mannan et al. 2021).

19.2.8 Sendai Framework for Disaster Risk Reduction (SFDRR) 2015–2030

The SFDRR 2015–2030, adopted in Sendai, Japan, provides a global approach to enhancing disaster resilience, which aligns closely with Bangladesh's disaster management efforts. The SFDRR emphasizes an initiative-taking approach to disaster risk reduction, focusing on reducing risk and damage at personal, community, and national levels (UNDRR 2015). This is particularly relevant for Bangladesh, given its vulnerability to natural disasters and climate change (Mannan et al. 2021). By aligning its strategies with the SFDRR, Bangladesh aims to mitigate disaster impacts, enhance resilience, and achieve the Sustainable Development Goals (SDGs) (Fig. 19.6).

Bangladesh's disaster management strategy incorporates the SFDRR's four priorities (UNDRR 2015):

i. *Understanding Disaster Risk*: Bangladesh has developed comprehensive disaster risk assessments and maps to better understand and address risks at various levels.
ii. *Strengthening Disaster Risk Governance*: The country has established a robust institutional framework, including the NDMC and various disaster management committees, to ensure effective governance and coordination.
iii. *Increasing Investment in Disaster Risk Reduction*: Bangladesh integrates disaster risk reduction measures into national and local development plans, such as the Delta Plan 2100, and allocates resources to enhance resilience.

iv. *Enhancing Preparedness and Response*: The country focuses on improving disaster preparedness and response capabilities, including community-based initiatives, and updating policies and plans to ensure effective recovery and reconstruction.

19.2.9 NGOs as Disaster Management Partners

Non-Governmental Organizations (NGOs) in Bangladesh play a pivotal role in disaster preparedness by conducting risk assessments, developing early warning systems, and facilitating community education (Matin and Taher 2001). They collaborate closely with local communities to create awareness about disaster risks and preparedness measures. Through training programs, workshops, and simulation exercises, NGOs help communities build resilience and develop evacuation plans. In the aftermath of a disaster, NGOs are often among the first responders (Rahman and Tasnim 2023). They provide emergency relief such as food, water, medical care, and temporary shelter. NGOs like BRAC, the Red Crescent Society, and CARE Bangladesh have well-established protocols for rapid response and coordination, ensuring that aid reaches affected areas swiftly (Alam and Ray-Bennett 2021). Their extensive networks and local knowledge enable them to deliver assistance efficiently even in the most remote or severely affected regions (Dhungana et al. 2023; Sattar et al. 2020).

NGOs play a critical role in long-term recovery and rehabilitation efforts, in addition to providing immediate assistance. They aim to reconstruct infrastructure, restore livelihoods, and offer psychological help to those affected (Hossain et al. 2020; Tambe 2022). Organizations such as Save the Children and Oxfam focus on

Fig. 19.6 Sustainable development goals (SDGs) (UNDRR 2015)

rehabilitating educational facilities, supporting displaced families, and promoting sustainable livelihoods. Their work helps communities to recover and rebuild better, incorporating lessons learned from past disasters. NGOs are also involved in disaster risk reduction and mitigation efforts. They engage in projects that aim to reduce the vulnerability of communities to future disasters (Haque et al. 2022; Neeraj 2022; Shuvo et al. 2022). This includes initiatives like constructing flood-resistant homes, improving drainage systems, and implementing reforestation projects to stabilize riverbanks. By promoting sustainable practices and resilience-building measures, NGOs contribute to reducing the overall impact of disasters. Effective disaster management necessitates collaboration among multiple stakeholders, including government agencies, international organizations, and local communities (Matin and Taher 2001; Rahman and Tasnim 2023; Shamsuzzoha et al. 2018a, b). NGOs frequently serve as mediators, promoting communication and collaboration among various entities. They participate in coordination forums, communicate information, and collaborate with government departments to ensure a united and efficient response (Rahman and Tasnim 2023).

19.2.10 "Build Back Better" Concept-Based Priority No. Four of SFDRR in Bangladesh

Disasters Throughout The Last Few Decades Have Taught us important lessons about resilience. Otherwise, reconstructing communities to pre-disaster norms will reintroduce previous vulnerabilities and expose them to further destruction from future calamities. The world is now attempting to develop disaster-resilient communities, and numerous frameworks have been established. The global SFDRR 2015–2030 was initiated in response to the Hyogo Framework for Action (HFA), which was completed earlier this year (Mannan et al. 2021; UNDRR 2015). The SFDRR builds on the HFA's tremendous success in galvanizing worldwide efforts and commitment to addressing the mounting losses caused by catastrophic events. Furthermore, 10 years of DRR operations at the global, regional, national, and local levels have yielded several lessons and best practices. As a result, the SFDRR can create a more nuanced, diverse, and targeted policy (MoDMR 2015). The SFDRR broadens the scope of disaster risk reduction, aiming for a more diversified and inclusive kind of resilience (Table 19.1). The SFDRR offers a new concept for making disaster-affected communities more resilient to disasters. Build Back Better in recovery, rehabilitation, and reconstruction is regarded as a requirement for sustainable development (Fig. 19.6). Reducing catastrophe risk requires shared responsibility between governments and essential stakeholders, as well as their dedication, goodwill, expertise, experience, and resources for better rebuilding in Bangladesh (Tables 19.1 and 19.2).

Table 19.1 Build back better concept in DRR in Bangladesh

Improvement of structural designs	(a) Promote multi-hazard assessments structural designs code and the availability of resources (b) Using traditional technologies and traditional materials (c) Enforce and educate the community about DRR (d) Providing training and preparing a database of skilled builders
Land-use planning	(a) Land divided into risk zones based on multi-hazard (b) Integrated hazard maps containing up-to-date multi-hazard information (c) Revised land-use plans before reconstruction (d) Improved safety for the relocated people
Improving the social recovery	(a) Trained and educated the policymaker about the local community and their background (b) Provide psychological support (c) Organize group activities to maintain the focus on permanent reconstruction (d) The Government to be clear with the community constraints
Improving the economic recovery	(a) Collected local data through the surveys of the community (b) Identify the beneficiary and parallel economic activities (c) Building the self-reliance and a sustainable livelihood (d) Establishing sustainable forms of assistance (e) Providing cash grants
Legislation and regulation	(a) The agreement calls for legislation to enforce rehabilitation programs that align with BBB principles (b) Assist recovery activities to speed up the recovery process (c) Legislative provisions to assist livelihood recovery (d) Simplified permits and subsidized.

19.3 Technological Advancements and Infrastructure Development

19.3.1 Introduction of Science and Technology in DRR of Bangladesh

Science and technology have transformed many aspects of society in the twenty-first century, including health, agriculture, energy, development, communications, education, social interactions, and security (Basher 2013). Disaster risk reduction relies heavily on scientific knowledge and evidence-based techniques, and the application of science and technology can coexist (Rathore 2016). For disaster risk reduction, scientific capacities must be broadened to include all relevant scientific and technical matters, where science is defined broadly to include the natural, environmental, social, economic, health, and engineering sciences, and the technological means of

Table 19.2 DRR role of different stakeholders in Bangladesh

Civil society, volunteers, and community-based organizations	(a) Engage in the implementation of the build-back better activities in DRR (b) Contribute to and support the improvement of the structural designs (c) Advocate for resilient communities and strengthen synergies across groups, as appropriate
Academia, scientific, and research entities	(a) Focus on the disaster risk factors and scenarios (b) Increase research for regional, national, and local applications of building back better concept (c) Support action by Local Government Body (LGD) and community people
Business, professional groups, and private financial institutions	(a) Engage in awareness-raising and training for building self-reliance and a sustainable livelihood (b) Engage in and support research and innovation as well as technological development (c) Share and disseminate knowledge, practices, and non-sensitive data about DRR
Media (Bangladesh radio, Bangladesh television)	(a) Rising public awareness and understanding of the build-back better concept (b) Disseminate accurate and non-sensitive disaster risk, hazard, and disaster information and cooperation with a different organization

confronting them have expanded greatly (Chaiya et al. 2022; Inan et al. 2023; Islam et al. 2013a). People's approach to disaster preparedness, including their use of science and technology, is determined by how they perceive disasters and risk. The social sciences have been essential in shaping new perspectives on risk, vulnerability, and poverty, as well as the risk process and human involvement in risk accumulation (Fakhruddin et al. 2022; Ogra et al. 2021). The social sciences also provide a fundamental understanding of risk perception, individual risk-taking behavior, and decision-making processes. They are actively involved in the integrated modeling of global development, industrial transformation, urbanization, and other elements of global change. Social science principles and approaches are critical for risk management practice, particularly in community contexts (Nasreen et al. 2023; Shaw et al. 2013a, b, c). Bangladesh has had a successful experience with the application of science and technology in DRR from the HFA 2005–2015 and the use of science and technology in DRR is also continued by following the SFDRR 2015–2030 (Mannan et al. 2021; UNDRR 2015). HFA 2005–2015 for building the resilience of nations and communities to disasters became the global outline of the disaster risk reduction approach in Bangladesh. The Government of Bangladesh (GoB) gives equal importance to both structural as well as non-structural mitigation measures (GoB 2020b; MoDMR 2015; MoF 2020; UNDRR 2015).

19.3.2 *Examples of Application of Science and Technology in DRR in Bangladesh*

The progression of science and technology to natural hazards and the coping mechanisms to the natural hazards over the past years helps us to understand the mechanisms of natural hazards of atmospheric, geologic, hydrological, and biological origins and to analyze the conversion of these hazards into disasters (Shaw et al. 2018). The scientific and technological disciplines which are involved include basic and engineering sciences, natural, social, and human sciences. The critical information currently provided on global climate change and its implication on the global environment is the fruit of this progress. The application of science and technology can reduce losses of lives and property, as illustrated by the following examples (Shaw et al. 2016).

Bangladesh government has coordinated various relevant stakeholders for interrelated social, economic, and environmental risks and vulnerabilities by using science and technology through the National Platform for Disaster Risk Reduction (NPDRR) to support the identification of priority needs around DRR, advice for allocating resources, presenting timetable for actions and monitoring and reviewing the implementation of DRR activities in line with the SFDRR 2015–2030 (Table 19.2). Now this country provides technical support and advice for effective integration into all national, regional, and international policies and programs and catalyzes national consultations and consensus building for DRR to facilitate the allocation of resources from donors, development banks, financial institutions, and United Nations (UN) agencies that are not represented in their respective countries (Basher 2013; Shaw et al. 2016, 2018). Some examples of the application of science and technology are aiming at ensuring every preparedness measure and reducing disaster risks in Bangladesh, under the rules and regulations of the MoDMR of the Government of the People's Republic of Bangladesh (Table 19.3).

19.3.3 *Examples of Comprehensive Disaster Management in Bangladesh*

To enhance resilience against natural disasters and climate change, the Comprehensive Disaster Management Program (CDMP) was launched in 2004 with support from the Department for International Development (DFID) and the United Nations Development Programme (UNDP) (Islam et al. 2021; NSET 2011). CDMP emphasized

Table 19.3 Application of science and technology in DRR in Bangladesh

Organization	Example of responsibilities
Ministry of Disaster Management and Relief (MoDMR)	(a) Hazard prediction, forecasting, and warning services (b) Proper management of emergency services (c) Improvement of search and rescue strategies (d) Capability enhancement for economic disaster risk reduction (e) Building multipurpose cyclone shelter
Flood Forecasting and Warning Center (FFWC)	(a) Conduct flood risk assessment (b) Develop a long-term risk reduction action plan (c) Identify the flood-prone areas in Bangladesh (d) Conducting research in extending the lead time flood forecasting (e) Link with regional flood forecasting sources
Cyclone Preparedness Program (CPP)	(a) Control room set up and prepare volunteer team (b) Maintain contact with BMD and other offices (c) Collect information and disseminate warning messages (WM) (d) Alert the responsible person and advise the CPP volunteer (e) Conduct rescue work in coordination with the liable authority (f) Provide first aid and assist liable authority with relief distribution
Bangladesh Meteorological Department (BMD)	(a) Development of sectorial risk management and policy strategies (b) Budgetary provision guarantee and continuous improvement of cyclone forecasting and warning system (c) Continuous monitoring of weather conditions (d) Dissemination of all information on a regional basis (e) For the quickest dissemination improving the equipment
Bangladesh Space Research and Remote Sensing Organization (SPARRSO)	(a) Develop required satellite images and mapping (b) Support sectorial risk M & P Strategies and ensure the budgetary provision for risk reduction activities (c) Improvement of image support to Disaster Forecasting and Warning System (DFWS) (d) Human resource development on Remote Sensing (RS) and Geographic Information System (GIS) technology and technological activities

(continued)

Table 19.3 (continued)

Organization	Example of responsibilities
Bangladesh Betar (Radio Bangladesh) (BdB) & Bangladesh Television (BTV)	(a) Disseminate the WMs to evacuate the vulnerable people (b) Prepare the rescue team during alert, warning, and disaster stage (c) Broadcast weather forecast from all broadcasting stations and announce warning signals (d) Continues the broadcast without any interval
Fire Service and Civil Defense (FSCD)	(a) Design and implement an awareness program (b) Improve the understanding and train up volunteers (c) Implementation of a coordinated action plan regarding the safety of assets and lives (d) Evacuation and rescue of people and livestock (e) Relief and rehabilitation-related work

pre-disaster risk reduction, significantly reducing deaths and damage. CDMP Phase I (2004–2009) introduced community-based disaster risk management (CBDRM), which was expanded in Phase II (2010–2014) to cover more unions in vulnerable districts (MoDMR 2010; NSET 2011). Local-level implementation included strengthening Disaster Management Committees (DMCs) and creating Community Risk Assessments (CRA) and Risk Reduction Action Plans (RRAP). The Local Disaster Risk Reduction Fund (LDRRF) supported these initiatives financially. CDMP II introduced four CBDRM intervention types: comprehensive interventions addressing multiple hazards, disaster-resilient habitats for high-risk villages, small-scale DMC-led projects funded by LDRRF, and specific interventions for unique challenges like post-Cyclone Aila saline intrusion (MoDMR 2010). Additionally, various NGOs like Oxfam, CARE, and BRAC have been active in promoting disaster preparedness and resilience at the community level, with initiatives like food-for-work programs and infrastructure improvements to withstand future disasters (Matin and Taher 2001; Rahman and Tasnim 2023). Bangladesh has made significant strides in disaster risk reduction, with a pronounced focus on community-based disaster management (CBDM). This approach has evolved in response to the increasing frequency and severity of natural disasters impacting the region, particularly in the coastal areas. CBDM emphasizes the active participation of local communities in identifying, analyzing, and managing disaster risks. This shift from a top-down approach to a more inclusive, bottom-up strategy aims to empower communities to take charge of their safety and resilience (Ahmed et al. 2016; Habiba et al. 2013; Islam 2013; Van Niekerk et al. 2018).

The HFA, adopted in 2005, played a crucial role in shaping Bangladesh's disaster management policies. It stressed the importance of developing and strengthening

CBDM systems. In practice, this has involved engaging communities at every stage of disaster management, from prevention and mitigation to preparedness, response, and recovery. Participatory risk assessments, where community members map hazard-prone and vulnerable areas, have become a cornerstone of this approach. This method not only leverages local knowledge but also fosters a sense of ownership and responsibility among community members. One key aspect of CBDM in Bangladesh has been the focus on sustainable livelihood frameworks, which integrate risk reduction into broader development goals (Kumar et al. 2020). This comprehensive approach recognizes that reducing vulnerability to disasters is intertwined with addressing poverty, environmental degradation, and social inequality (Islam et al. 2021). For example, programs supported by the United Nations Centre for Regional Development (UNCRD) have aimed to improve the safety of critical community infrastructure, such as schools, and to disseminate best practices in disaster risk management at the grassroots level (Shi et al. 2020). Despite these advancements, challenges remain. The effectiveness of CBDM relies heavily on the continuous engagement and capacity-building of community members (Islam et al. 2021; MoDMR 2010; NSET 2011). It also requires adequate support from governmental and non-governmental organizations to ensure that communities have the necessary resources and knowledge to implement disaster risk reduction measures effectively. The integration of scientific risk assessments with community-driven initiatives is essential to create robust and resilient systems capable of withstanding future disasters (Shaw et al. 2016).

19.3.4 Infrastructure Development and Economic Growth of the Past 20 Years

As part of structural mitigation measures, the government has so far built 3,976 cyclone shelters in coastal regions (Miyaji et al. 2017) and 272 flood shelters (Islam et al. 2020a, b). So far, a coastal embankment of approximately 5,700 km in length has been built to protect coastal territory from flooding by tidal waves and storm surges along with different water management structures (BWDB 2020). As part of disaster training and public awareness, almost 45,000 people have been trained through 721 courses (Imran et al. 2016), workshops, and seminars (Choudhury et al. 2006). Bangladesh has had significant economic expansion over the last two decades (Macrotrends 2024; WB 2024) (Table 19.4), which is likely to have influenced its progress in disaster risk reduction. Between 2004 and 2024, the country's Gross Domestic Product (GDP) increased from US dollars 65.11 billion to US dollars 455 billion, indicating a large rise in financial resources available for disaster risk reduction measures. This economic expansion was accompanied by an increase in per capita income from US dollars 469 to US dollars 2,646, showing higher living standards that can boost community resilience to catastrophes through better housing, education, and healthcare. The generally rapid growth rates, particularly between 2006 and 2019, which frequently exceeded 6%, undoubtedly supported ongoing investments in disaster risk reduction infrastructure such as flood defenses, cyclone shelters, and early warning systems. With a better economy, Bangladesh may invest in vital disaster mitigation infrastructure such as upgraded roads, bridges, and water management systems, lowering communities' vulnerability to natural disasters. Furthermore, increasing GDP may be associated with the improvement of disaster management institutions, enhanced governmental capability, and improved collaboration with NGOs and international partners to ensure successful disaster response and recovery. Economic growth also allows for more community-based disaster risk reduction programs, which educate and train local communities in disaster preparedness and response, developing a resilient culture (Rasheduzzaman et al. 2020). Improved technology for early warning systems, weather forecasting, and disaster monitoring, funded by increased GDP, is critical for prompt and successful disaster response. Furthermore, increased per capita income reflects education and public awareness advances, allowing communities to understand and implement catastrophe risk reduction strategies. Despite the economic downturn between 2020 and 2023 caused by global issues such as the COVID-19 pandemic, Bangladesh's concentration on disaster risk reduction programs is likely to have helped offset the effects of such disruptions. Table 19.4 shows the economic growth scenario of Bangladesh over the past 20 years (Islam et al. 2020a, b; Tabassum et al. 2023).

19.3.5 Mortality Rate from Major Cyclones Since 1991

Cyclones have historically posed a severe threat to Bangladesh, frequently resulting in catastrophic loss of life. Table 19.5 shows the fatality rates from major cyclones in the region since 1991, demonstrating both success and challenges in disaster risk reduction over time. The disastrous cyclone of 1991, known as Marian or 1991's Cyclone left a frightening death toll of 138,882, highlighting Bangladesh's vulnerability to natural catastrophes. In contrast, more recent cyclones have resulted in significantly reduced fatality tolls, demonstrating advances in disaster preparedness and response (Rahman et al. 2017; Wahiduzzaman and Paul 2023). For instance, the 2020 Amphan cyclone, one of the most powerful in recent years (Rashida et al. 2022), claimed 26 lives and the most recent cyclone Remal claimed 10 lives, a considerable decrease from previous decades (Dasmahapatra 2024). Cyclones such as Sidr in 2007 (3,363 deaths) and Aila in 2009 (190 deaths) highlight the increasing effectiveness of disaster management measures (Rahman et al. 2017). The ongoing decline in mortality rates, such as those in Yaas (2021) with only three deaths and Sitrang (2022) with 35 deaths, demonstrates Bangladesh's improved capabilities in early warning systems, evacuation planning, and resilient infrastructure (BDRCS 2022; IFRC 2021) (Table 19.5). This achievement illustrates the country's effective disaster risk reduction initiatives, which are critical for limiting the effects of cyclones and saving lives (BDRCS 2022; Dasmahapatra 2024; IFRC 2021, 2024; Rahman et al. 2017; Rashida et al. 2022; Wahiduzzaman and Paul 2023).

Table 19.4 Economic growth of Bangladesh in the past 20 years

Year	GDP (Billion US dollars)	Per capita (US dollars)	Growth rate (%)
2004	65.11	469	5.24
2005	69.44	493	6.54
2006	71.82	504	6.67
2007	79.61	552	7.06
2008	91.63	630	6.01
2009	102.48	699	5.05
2010	115.28	777	5.57
2011	128.64	856	6.46
2012	133.36	877	6.52
2013	149.99	974	6.01
2014	172.89	1,109	6.06
2015	195.08	1,236	6.55
2016	265.24	1,660	7.11
2017	293.75	1,816	6.59
2018	321.38	1,963	7.32
2019	351.24	2,122	7.88
2020	373.90	2,233	3.45
2021	416.26	2,458	6.94
2022	460.20	2,688	7.10
2023	446.35	2,528	5.82
2024	455.01	2,646	5.70

Source (Macrotrends 2024; WB 2024)

Table 19.5 Death toll from the major cyclones since 1991

Year	Cyclone's name	Death toll
1991	Marian or 1991's cyclone	138,882
1997	May 1997 Bangladesh cyclone	550
2007	Sidr	3363
2009	Aila	190
2019	Fani	17
2019	Bulbul	179
2020	Amphan	266
2021	Yaas	03
2022	Sitrang	35
2024	Remal	16

19.3.6 National Climate Finance Mechanism and National Adaptation Plan of Bangladesh

The administration of Bangladesh founded the Bangladesh Climate Change Trust Fund (BCCTF) in the fiscal year 2009–2010 from self-funds, to become an ancestor nation among other developing states. From this fund, about 851 projects have been undertaken with speculation of about 490 million US dollars to set strategic strategies and action plans of the Bangladesh Climate Change Strategy and Action Plan (BCCSAP) which principally focus on adaptation, mitigation, damage, loss assessment, technology transfer, capacity development based on different climate change-related research (MoEFCC 2022a, b). As supplementary climate research financing from the national budget, the Government of Bangladesh also additionally implemented a Climate Fiscal Framework (CFF) in 2014 and restructured it to make climate-related issues all-encompassing in the Public Financial Management (PFM) system of the country in 2020. The national budget (55.82% of the total budget) is cumulatively allocated to the accounts of important twenty-five Ministries, Divisions, and Departments for the fiscal year of 2022–2023. Additionally, for 2022–2023, the allotment related to climate change is 8.07% of the national budget. In the past 8 years, the allocation to climate-relevant research and initiatives has increased by 2.5 times in the year 2015–2016 from 12,163 crore Bangladesh Taka (BDT) (about US dollars 1.44 billion) to in the year 2022–2023, respectively, 30,531.98 crore BDT (about US dollars 3.08 billion) (MoEFCC 2022a).

Bangladesh has just developed the Draft National Adaption Plan (NAP), which will increase the country's aspirations for adaptation. A total of eight main sectors are covered by the Draft NAP, and they are agriculture; fisheries, aquaculture, and livestock; water resources; urban areas; ecosystem, wetlands, and biodiversity; capacity development, research, and innovation; policy and institutions; disaster, social safety, and security; and further eleven climatic stress areas. For the years 2023–2050, an estimated US dollars 230 billion would be spent on implementing 113 interventions, of which a total of 90 are high-priority and 23 are moderate-priority under the total of eight thematic areas. In the prospect of the plan in developing countries, the NAP will anticipate serving as the principal strategic proclamation used by the United Nations Framework Convention on Climate Change (UNFCCC) practices to conduct adaptation measures (GoB 2020a; MoDMR 2015; MoEFCC 2022a).

19.3.7 Information and Communication Technologies (ICT) Enhances DRR Strategies

Application of Information and Communication Technologies (ICT) in Bangladesh refers to the Bangladeshi government's ambition to transform the country into a digitally advanced society by 2021. The major purpose is to use ICT to promote economic development, transparency, accountability, and enhanced citizen services.

Critical components include human resource development, citizen connectivity, digital government, and business-related ICT applications. This digital improvement of the nation aims to improve education, reduce poverty, improve health care, and expand economic prospects using e-governance and digital platforms (Mazumdar and Alharahsheh 2020; Shuvo et al. 2022). Over the past two decades, Bangladesh has made considerable progress in disaster risk reduction, due to comprehensive, and foreign assistance through the enhancement of ICT applications (Shuvo et al. 2022). Therefore, ICT applications have played a key role in this progress by incorporating digital disaster management plans and DRR strategies. Here are some important ways in which information and communication technologies have contributed to disaster risk reduction purposes (Hossain 2022; Syed et al. 2021):

i. *Enhanced communication and early warning systems*: The incorporation of ICT has enhanced communication networks, allowing for faster broadcast of early warning messages to at-risk areas. Mobile phones, internet services, and social media platforms have become critical instruments for warning residents of approaching calamities, enabling prompt evacuation and preparation.
ii. *Data collection and analysis*: Digital platforms have made acquiring and analyzing data about weather patterns, flood levels, and other environmental aspects easier. This data-driven strategy has increased the accuracy of disaster predictions and response plans, thereby mitigating the effects of natural disasters.
iii. *E-governance and coordination*: The e-governance platforms have improved cooperation among various government entities involved in disaster management. Digital platforms have allowed for improved resource allocation, faster response times, and more effective management of relief efforts.
iv. *Capacity building and training*: The human resource development component of ‘Digital Bangladesh’ concentrated on training people in ICT skills. This includes disaster management personnel training, which provides them with the skills and expertise they need to use digital technologies in disaster response and recovery activities.
v. *Remote sensing and GIS technology*: The use of RS and GIS has improved the capacity to monitor and assess disaster-prone areas. These technologies give real-time data and visualizations, contributing to the planning and implementation of catastrophe risk reduction strategies.

19.4 Future DRR Directions

19.4.1 Developed and Sustainable Country Perspective

The Bangladesh government is working on a forward-looking program aimed at propelling the country to a developed and sustainable country, building on the foundation of digital applications based on ICT enhancement (GED 2020; Hossain 2022; Pal and Sarker 2023; UN 2015, 2024). This ambitious concept is built on highlighting

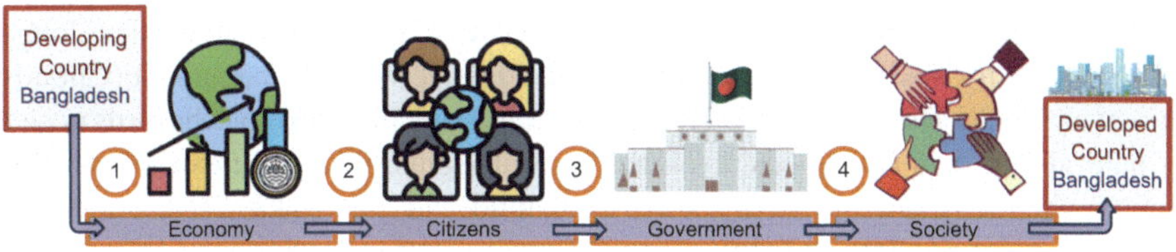

Fig. 19.7 Perspective of transformation as a developed and sustainable country

the country's economy, citizens, government, and society (Fig. 19.7) (Pal and Sarker 2023).

It outlines a set of projects and policies with different timelines. From 2023 to 2025, efforts will be focused on improving ICT skills among Bangladeshis, implementing a comprehensive ICT strategy, developing a national e-market for procurement, establishing digital job platforms, and expanding public services alongside a paperless government (Hossain 2022). Shortly, particularly between 2023 and 2031, significant projects will include the adoption of universal digital identities (IDs), the introduction of digital curriculum, advances in healthcare, and the promotion of blended learning approaches. Looking ahead to be a developed country, the vision seeks to revolutionize land management, postal services, the judiciary, and social safety nets (Pal and Sarker 2023). It also includes the modernizing of police services and the establishment of a fintech accelerator. It encompasses functional areas such as boosting internet connectivity, employing AI for traffic management, and enhancing healthcare through body sensors. Furthermore, the plan calls for the integration of technology in agriculture and education via blended learning. Image detection technology will improve waste management, and vegetation index assessment to increase civil service efficiency (Shamsuzzoha et al. 2024b). It also includes the creation of village transportation systems that utilize environmentally friendly electric vehicles and charging infrastructure. Communication for behavior change and service delivery will be prioritized through social co-creation, mobile solutions, and enhanced cybersecurity safeguards (Hossain 2022; Syed et al. 2021). This comprehensive strategy reflects an integrated approach to using technology and innovation to address various aspects of life in Bangladesh, with a clear focus on disaster risk reduction and resilience, in line with the goals outlined in the smart transition from developing Bangladesh to developed Bangladesh as a developed and sustainable country (Fig. 19.7) (Mazumdar and Alharahsheh 2020; Pal and Sarker 2023).

19.4.2 Implementing Bangladesh Delta Plan 2100

As Bangladesh moves closer to its objective of becoming a middle-income country, it faces several issues that demand a strong disaster risk reduction plan. The Bangladesh Delta Plan 2100 is crucial to this plan, as it provides a visionary framework meant to manage the delta region's multiple challenges (Alam et al. 2018; Choudhury et al.

2012; Terwisscha Van Scheltinga, 2013). The Delta Plan is a comprehensive, long-term strategy for sustainable development via adaptive water management (Choudhury et al. 2012). This approach is critical for addressing the enormous hazards posed by natural disasters like floods, cyclones, and droughts, which are exacerbated by climate change (Crawford et al. 2021; Mahamud and Takewaka 2018; Shamsuzzoha and Ahamed 2023, 2024b). The plan aims to improve the delta's resilience and living standards by combining several policy sectors and establishing effective institutional frameworks (MoDMR 2015). The Delta Plan, launched in 2011 after extensive consultations with key stakeholders and international organizations such as the Embassy of the Kingdom of the Netherlands and financial institutions such as the Asian Development Bank and the World Bank, represents a commitment to a comprehensive vision for the future. It lays out a clear path for the delta's growth up to the year 2100, emphasizing the necessity of "no-regret" solutions that oversee existing problems while planning for future change (Alam et al. 2018; Choudhury et al. 2012; Terwisscha Van Scheltinga 2013).

a. The government has approved the Bangladesh Delta Plan 2100, a comprehensive 100-year strategic plan with the goal of progressive, sustainable growth through an adaptive delta management procedure.
b. The strategy seeks to establish wealthy, safe, and climate-resilient delta management by reducing susceptibility to natural disasters and boosting resilience to climate change. Its goals are to confirm economic growth, environmental sustainability, and long-term food and water security.

The Delta Plan's funding approach is crucial, as it demands the inclusion of an investment plan inside a medium-term budgeting framework (MTBF). This methodology will help to prioritize projects, identify funding sources, and establish execution deadlines. The plan intends to develop a durable foundation for sustainable land and water resource management by combining government funds, corporate investments, and community contributions (Alam et al. 2018). In terms of disaster risk reduction, the Delta Plan serves as a strategic framework for future efforts by integrating existing plans and guaranteeing a coordinated approach to controlling disaster risks mainly dominated by agricultural land use and riverine land cover is vulnerable to floods, tidal surges, and other extreme hydro-climatic events (Figs. 19.8 and 19.9). Its execution will be critical in determining Bangladesh's ability to negotiate the delta region's complex shoreline-changing issues while maintaining long-term resilience and sustainability (Fig. 19.10) (Alam et al. 2018; Terwisscha Van Scheltinga 2013).

19.4.3 Future Directions to Disaster Science and Emergency Management

Bangladesh aims to further integrate DRR strategies into national development plans, including the Delta Plan 2100 and the SDGs. Future priorities include enhancing

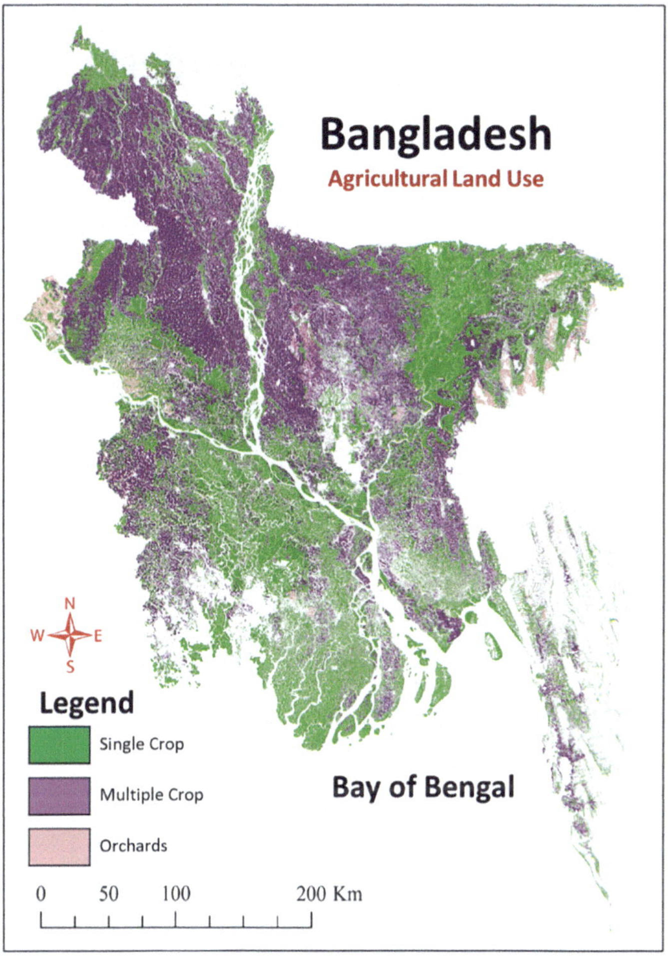

Fig. 19.8 Agricultural land use represents the key factor of the country's economy

urban disaster resilience, expanding gender-inclusive DRR strategies, and strengthening civil-military coordination (Ikeda 1995). Continuous investment in science and technology, coupled with a focus on sustainable infrastructure development and community empowerment, will be key to maintaining and enhancing Bangladesh's progress in disaster risk reduction. Additionally, there will be an emphasis on public–private partnerships to foster innovative solutions and ensure comprehensive risk-informed planning (Alam et al. 2018; GoB 2017, 2020a, 2020b; Islam et al. 2013; MoEFCC 2022b).

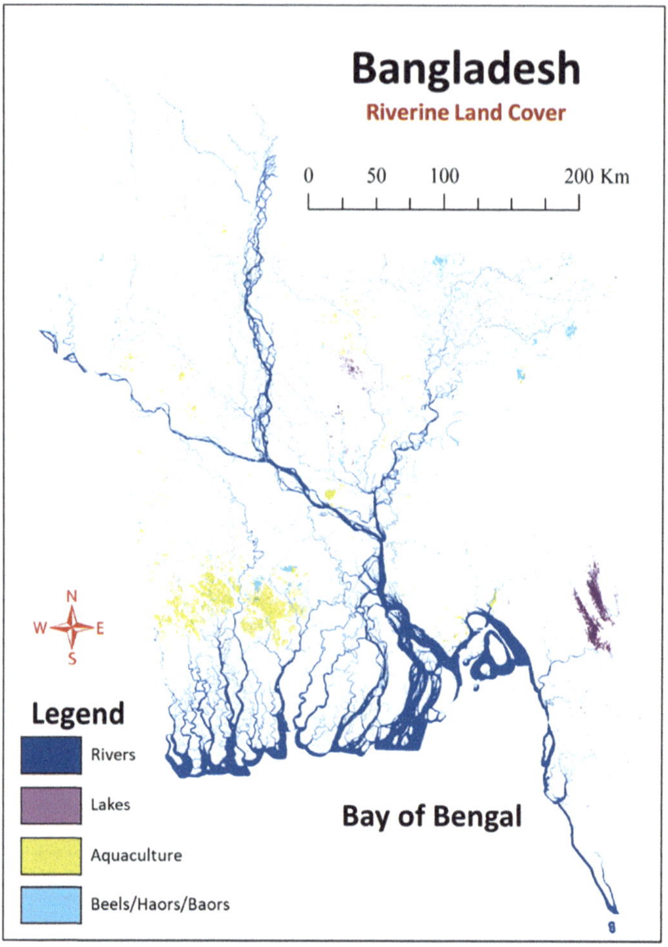

Fig. 19.9 Riverine land cover represents a country that is vulnerable to floods, tidal surges, and other extreme hydro-climatic events

Figure 19.11 illustrates an integrated emergency management cycle that emphasizes the application of disaster science in DRR. This cycle prioritizes DRR alongside disaster resilience and sustainable development, ensuring timely and effective measures for the population. The cycle uses color coding to represent distinct stages: red indicates an emergency or catastrophic situation that can be mitigated by yellow, indicates reduced risk. Green symbolizes a normal stage characterized by adequate preparedness (Fig. 19.11). In the preparedness phase, comprehensive planning, policy, and legal frameworks are established through community risk assessments and social impact evaluations (Bardsley 2018; Rahman et al. 2017). This stage also includes emergency communication systems with advanced prediction and forecasting capabilities, as well as effective evacuation procedures and timely alerts and

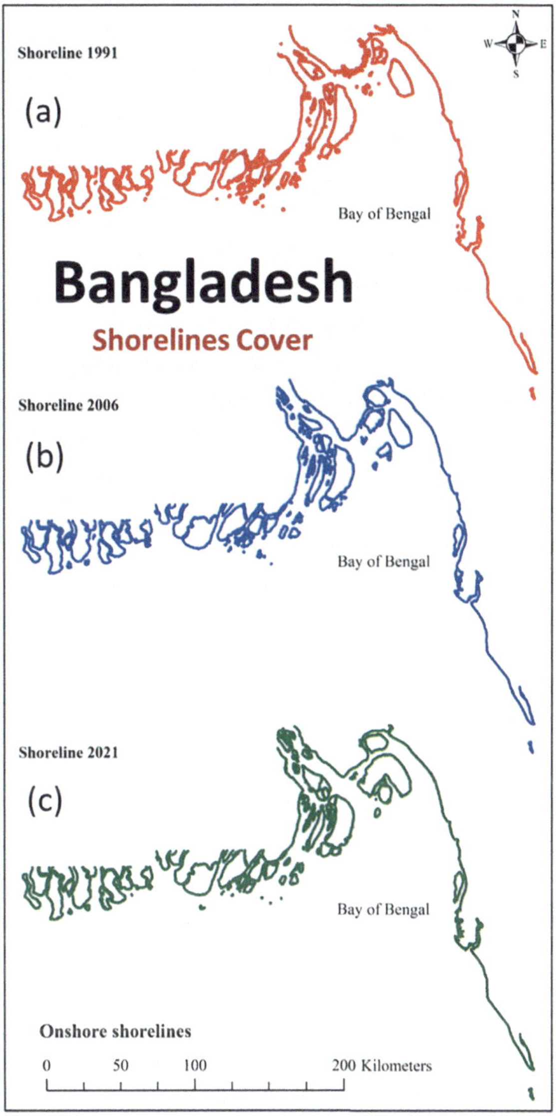

Fig. 19.10 Shoreline alternations represent a disaster-prone coastal setting of the Bangladesh delta

warnings (Akhand 2003; Haque 1997; Raj et al. 2010; Syed et al. 2021). During an emergency, the immediate response is crucial and must adhere to humanitarian standards, ensuring human security and rights from the alert phase through to rehabilitation (Talukder Shamsuzzoha and Hasan 2018). The recovery stage involves relief, aid, and assistance, as well as search and rescue operations and volunteering efforts until normalcy is restored (Fernandez and Ahmed 2019; Haque 1997; Sadik et al. 2018). Post-recovery, the impacts of the disaster should be leveraged as opportunities

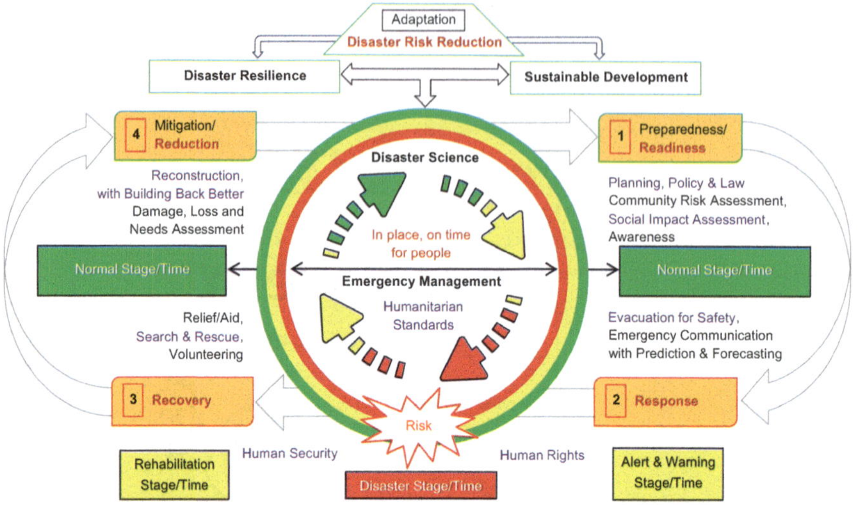

Fig. 19.11 Introducing an integrated emergency management cycle based on disaster science application in DRR

for sustainable development (Tambe 2022; UN 2024). In the mitigation or reduction phase, disaster science plays a pivotal role in reconstruction efforts, focusing on building back better towards adaptation (Chowdhury et al. 2022; Rahman et al. 2017; Sattar et al. 2020). This involves assessing damage, loss and needs to enhance preparedness and readiness for future events (Feng and Cui 2021; Shamsuzzoha et al. 2021; Talukder et al. 2018).

19.5 Discussion

Over the past 20 years from 2003 to 2023, Bangladesh has made great progress in disaster risk reduction (MoEFCC 2022a). The restructuring of the MoDMR has improved risk mitigation and emergency response activities (GoB 2017, 2020a). Early warning systems have been strengthened by technological advancements such as increased satellite imaging and forecasting capabilities (Hossain 2022; Syed et al. 2021). Economic growth and the appropriate technology for DRR programs have integrated modern expertise into disaster risk reduction initiatives, allowing for improved data access and infrastructure resilience (GoB 2020a). Significant advancements, such as cyclone and flood shelters, have improved disaster preparedness. Community-based programs, such as the CDMP, have strengthened local communities. Alignment with international frameworks, such as the Sendai Framework, as well as an emphasis on socioeconomic aspects and gender-responsive initiatives, all contribute to increased resilience (Habib et al. 2012; Haque et al. 2022; Shaw et al. 2018; UNDRR 2015). Bangladesh has created the necessary policies and regulatory frameworks and made admirable efforts to streamline institutional and regulatory settings to realize the goal of climate-resilient sustainable development (Alam and Hoque 2022; MoDMR 2015; Wright et al. 2019). To combat climate change and disaster management, the government has developed the following policies, strategies, and programs over time (Table 19.6). Despite progress, Bangladesh's disaster risk reduction is limited by inadequate infrastructure, uneven community preparedness, and technological obstacles (MoDMR 2010, 2015). Socioeconomic inequities and climate change consequences exacerbate efforts, while incorporating disaster risk reduction into larger development goals remains difficult, resulting in fragmented implementation and inconsistent success.

As a neighboring country, India has also made progress, focusing on policy frameworks and technological innovations since the Indian Ocean Tsunami, though implementation at regional levels remains incremental (Chakrabarty 2022; Ogra et al. 2021; Parikh and Hota 2022). Both nations emphasize community engagement and

Table 19.6 Major policies, plans, and programmes to address, environmental protection, climate change, and disaster management in Bangladesh at the national level over the past 20 years

Policy response	i. Bangladesh Environment Conservation Act, 1995, (Updated in 2010) ii. Standing orders on disaster, 1997 (updated in 2010 and 2019) iii. Renewable energy policy of Bangladesh, 2008 iv. Bangladesh Climate Change Trust Act, 2010 v. Disaster Management Act, 2012 vi. National disaster management policy, 2015 vii. Nationally Determined Contributions (NDC), 2015 (Updated in 2021) viii. Bangladesh Biodiversity Conservation Act, 2015 ix. Ecologically critical area management rules, 2016 x. Bangladesh Climate Fiscal Framework (BCFF), 2020 xi. National Strategy on Internal Displacement Management, 2021 xii. Solid waste management rules, 2021 xiii. Hazardous Waste (E-waste) Management Rules, 2021 xiv. Nationally Determined Contributions (NDC), 2021 xv. Air Pollution (Control) Rules, 2022
Plan and programme	i. Bangladesh Climate Change Strategy and Action Plan (BCCSAP), 2009 (updated in 2022) ii. Bangladesh National Action Plan for Reducing SLCPs, 2012 (Updated in 2018) iii. National Adaptation Plan (NAP), 2022 iv. Mujib Climate Prosperity Plan (MCPP), 2022–2041 v. Plan of action to implement Sendai framework for disaster risk reduction, 2015–2030 vi. National plan for disaster management, 2021–2025 vii. Bangladesh energy efficiency and conservation master plan up to 2030 viii. National action plan for clean cooking, 2020–2030 ix. Bangladesh delta plan, 2100 x. National solar energy roadmap, 2021–2041

resilient infrastructure, but challenges persist in translating national policies into effective local actions and fully integrating DRR into development plans (Chaiya et al. 2022; Ogra et al. 2021).

On the other hand, Thailand's approach, coordinated by the National Disaster Prevention and Mitigation Plan, focuses on risk assessment, early warning systems, and a structured national disaster management framework overseen by the National Disaster Prevention and Mitigation Committee (Chaiya et al. 2022; Kitagawa 2020). In contrast, Bangladesh has revamped its Ministry of Disaster Management and Relief to improve risk reduction and emergency response by using innovative technologies through sustainable and ICT-based programs (Hossain 2022). Both countries prioritize community-based programs, with Thailand establishing partnerships and Bangladesh adhering to international standards such as the Sendai Framework, which reflects a holistic, multi-sectoral approach to disaster risk reduction.

Meanwhile, China and Bangladesh have both created comprehensive disaster risk reduction frameworks, although their techniques differ greatly (MoDMR 2010; NSET 2011; Su et al. 2023). China's National Master Plan for Responding to Public Emergencies describes a centralized and well-coordinated response process that involves many levels of government and integrates civilian and army forces for disaster relief (Chaiya et al. 2022). This plan divides public emergencies into four categories: natural disasters, accidents, public health events, and social security events. In contrast, Bangladesh's approach has been revolutionary, with an emphasis on improving the Ministry of Disaster Management and Relief, incorporating sophisticated science and technology through DRR initiatives and sustainable development program, and aligning with international frameworks like the SFDRR. Both countries emphasize the necessity of coordination and community-based programs, but China's approach is more centralized, whilst Bangladesh's is marked by technology integration and international alignment.

Also, Japan has a Basic Disaster Management Plan based on the Disaster Countermeasures Basic Act, which gives responsibilities to national and municipal governments, public enterprises, and other entities (Shaw et al. 2016, 2018; Shaw et al. 2009). It comprises specific preparations for several catastrophe stages that are coordinated by the Central Catastrophe Management Council, which consists of the prime minister, ministries, and experts (Chaiya et al. 2022). In contrast, Bangladesh's disaster risk reduction framework has evolved under the MoDMR, with a focus on technical improvements and international frameworks such as the Sendai Framework (GoB 2020a, 2020b; MoDMR 2015; MoEFCC 2022b; UNDRR 2015). Bangladesh favors technology integration and global alignment, whereas Japan focuses on thorough planning and cooperation at all levels of government.

Meanwhile, Nepal's disaster risk reduction has typically focused on response and relief, but the recently enacted Disaster Risk Reduction and Management Act (2017) marks a shift toward a more holistic catastrophe management cycle (Nepal et al. 2018). However, regulatory gaps, institutional inadequacies, and contradictions among several statutes impede practical implementation. In contrast, Bangladesh has expanded its disaster risk reduction through technological integration and alignment with international frameworks such as the Sendai Framework, which focuses on preparedness, response, and rehabilitation. While Nepal struggles to implement its policies, Bangladesh benefits from a better-coordinated strategy led by the MoDMR, as well as technological efforts such as Digital Bangladesh.

19.6 Conclusions

Bangladesh has had enormous success in DRR practices by combining technology improvements, institutional reforms, community-based approaches, and infrastructure development. The shift from a reactive to an initiative-taking disaster management paradigm shows a comprehensive and forward-thinking plan for mitigating the effects of natural disasters and strengthening national resilience. In the early 2000s, Bangladesh's disaster preparedness was heavily influenced by its geographical weaknesses and socioeconomic status. Because of the country's substantial risk of tropical cyclones, storm surges, and floods, strong early warning systems and community preparedness activities were developed. The CPP and the incorporation of satellite images and computer modeling into forecasting systems made important advances in disaster response. Despite these advances, difficulties such as insufficient shelter capacity and the need for improved public education remained. The reorganization of the Ministry of Food and Disaster Management into the Ministry of Disaster Management and Relief (MoDMR) in 2012, as well as later modifications to the SOD, signaled a move toward a more organized and systematic approach to disaster management. The formation of different councils and committees, such as the National Disaster Management Council (NDMC) and the Inter-Ministerial Disaster Management Coordination Committee (IMDMC), allowed for a more coordinated response to disaster risk reduction and climate change adaptation. Bangladesh's approach was improved by its integration with international frameworks like the Sendai Framework for Disaster Risk Reduction (SFDRR), which focused on identifying disaster risk, strengthening governance, investing in resilience, and enhancing preparedness.

Community-based disaster risk reduction (CBDRR) has been beneficial in fostering grassroots resilience. Initiatives such as the comprehensive disaster management program have empowered local communities by providing risk assessments, action plans, and financial assistance, boosting disaster preparedness and mitigation efforts. Non-governmental organizations (NGOs) have supplemented these efforts by providing immediate aid, long-term recovery, and risk control projects. Infrastructure development, such as the construction of cyclone and flood shelters, coastal embankments, and drainage systems, has improved Bangladesh's natural disaster preparedness. These efforts, together with technological breakthroughs and public awareness initiatives, have helped to make society more resilient. Bangladesh's future DRR strategy would prioritize integrating disaster risk reduction into national development objectives, improving urban resilience, and supporting gender-inclusive methods. The emphasis on science and technology, sustainable infrastructure, and public–private partnerships will be critical in preserving and accelerating development.

Acknowledgements The first authors would like to thank the Ministry of Science and Technology (MoST), Government of Bangladesh (GoB), and the Bangabandhu Science and Technology Fellowship (BSTF) program for sponsoring a doctoral program scholarship at the University of Tsukuba, Japan. The Environmental Systems Research Institute (ESRI) is also acknowledged by the authors for providing the ArcGIS® 10.8.2 software used for map preparation of this study. For providing necessary open-source data, reports, and works of literature used in this study, the authors cordially thank respective ministries and organizations.

Author Contributions Literature Review, Maps and Figures Preparation, Writing of Original Draft and Corresponding Author: M. S.; Literature Collection, Tables preparation and Writing of Original Draft: M. F. H.; Report Collection and Analytical Review: M. F. T. and M. R.; Editing, and Supervision: T. A. and R. S. After reading the final draft of the document, all authors gave their approval for publication of the manuscript.

Funding This research did not receive any external funding.

Data Availability Statement Upon a reasonable request, the corresponding author will make the dataset created and analyzed for this study public; however, there are restrictions on the data's repeatability for reuse and commercially reliable details.

Conflicts of Interest Regarding the research, the authors say they have no conflicts of interest.

References

Ahmed I (2019) The national plan for disaster management of Bangladesh: Gap between production and promulgation. Int J Disaster Risk Reduct 37:101179. https://doi.org/10.1016/j.ijdrr.2019.101179

Ahmed B, Kelman I, Fehr HK, Saha M (2016) Community resilience to cyclone disasters in coastal Bangladesh. Sustain (Switz), 8(8). https://doi.org/10.3390/su8080805

Akhand MH (2003) Disaster management and cyclone warning system in Bangladesh. In: Early warning systems for natural disaster reduction. Springer, Berlin, Heidelberg, pp 49–64. https://doi.org/10.1007/978-3-642-55903-7_8

Akter N, Sultana Z (2022) Climate change and disaster management in Bangladesh. In: Singh A (ed) International handbook of disaster research. Springer Nature Singapore, Singapore, pp 1–23. https://doi.org/10.1007/978-981-16-8800-3_136-2

Alam E, Affiliate DC (2010). Cyclone disaster vulnerability and response experiences in Coastal Bangladesh. https://doi.org/10.1111/j.03613666.2010.01176.x

Alam A, Hoque MM (2022) Local level collaborative governance for pandemic responses: Unpacking a case in Bangladesh. Policy & GovAnce Rev 6(3):207–228. https://doi.org/10.30589/pgr

Alam E, Ray-Bennett NS (2021) Disaster risk governance for district-level landslide risk management in Bangladesh. Int J Disaster Risk Reduct, 59. https://doi.org/10.1016/j.ijdrr.2021.102220

Alam S, de Heer J, Choudhury G (2018) Bangladesh delta plan 2100. Dhaka. Retrieved from http://www.plancomm.gov.bd/site/files/0adcee77-2db8-41bf-b36b-657b5ee1efb9/Bangladesh-Delta-Plan-2100

Alam KF, Shamsuzzoha M, Arab ST, Pangaribuan IJP, Ahamed T (2024) Application of remote sensing in the analysis of climate extremes due to global climate change. In: Ahamed T (ed)

Remote sensing application II. New frontiers in regional science: Asian perspectives, vol 77. Springer, Singapore, pp 1–21. https://doi.org/10.1007/978-981-97-1188-8_1
Ali A (1996) Vulnerability of Bangladesh to climate change and sea level rise through tropical cyclones and storm surges
Arab ST, Islam MM, Shamsuzzoha M, Alam KF, Muhsin N, Noguchi R, Ahamed T (2022) A review of remote sensing applications in agriculture and forestry to establish big data analytics. In: Ahamed T (ed) Remote sensing application. New frontiers in regional science: Asian perspectives, vol 59. Springer, Singapore, pp 1–24. https://doi.org/10.1007/978-981-19-0213-0_1
Bardsley DK (2018) Indigenous knowledge and practice for climate change adaptation. In: Encyclopedia of the anthropocene. Elsevier, pp 359–367. https://doi.org/10.1016/B978-0-12-809665-9.09797-4
Basher R (2013) Science and technology for disaster risk reduction: A review of application and coordination needs. Geneva. Retrieved from https://www.caneus.org/undrr/resources/3.Science-and-Technology-for-Disaster-Risk-Reduction.pdf
BBS (2024) Statistical Yearbook Bangladesh 2023. Dhaka. Retrieved from http://nsds.bbs.gov.bd/storage/files/1/Stat_Yearbook-23.pdf
BDRCS (2022) Cyclone Sitrang, Bangladesh. Dhaka. Retrieved from https://bdrcs.org/wp-content/uploads/2022/10/BDRCS-Sitrep2-Cyclone-Sitrang_1600_261022.pdf
Bhuiyan MdJAN, Dutta D (2012) Analysis of flood vulnerability and assessment of the impacts in coastal zones of Bangladesh due to potential sea-level rise. Nat Hazards 61(2):729–743. https://doi.org/10.1007/s11069-011-0059-3
Braun B, Aßheuer T (2011) Floods in megacity environments: Vulnerability and coping strategies of slum dwellers in Dhaka/Bangladesh. Nat Hazards 58(2):771–787. https://doi.org/10.1007/s11069-011-9752-5
BWDB (2020) Bi-annual environment monitoring report for January–June 2020. Dhaka. Retrieved from http://ceip-bwdb.gov.bd/Tech_Report/Env/9BAEAR.pdf
CFE-DMHA (2020) Bangladesh: Disaster management reference handbook. Joint Base Pearl Harbor—Hickam, Honolulu. Retrieved from https://reliefweb.int/report/bangladesh/disaster-management-reference-handbook-2020-bangladesh
Chaiya C, Pal I, Pinthong P (2022) Paradigm of disaster risk reduction: A comparative study of five Asian countries. Soc Sci Asia, Off J Natl Res Counc Thail Conjunction J Polit Gov 8(3):1–33. https://doi.org/10.14456/ssa.2022.17
Chakma S (2021) An empirical study on the community based disaster management framework with the cyclone preparedness program in Bangladesh. Kobe University, Kobe. Retrieved from https://hdl.handle.net/20.500.14094/D1008181
Chakrabarty S (2022) Factors affecting governance aspect of disaster management: Comparative study of the Sundarbans in India and Bangladesh. In: Chatterjee U, Akanwa AO, Kumar S, Singh SK, Dutta Roy A (eds) Ecological footprints of climate change: Adaptive approaches and sustainability. Springer International Publishing, Cham, pp 219–245. https://doi.org/10.1007/978-3-031-15501-7_9
Choudhury WA, Quraishi FA, Haque Z (2006) Mental health and psychosocial aspects of disaster preparedness in Bangladesh. Int Rev Psychiatry 18(6):529–535. https://doi.org/10.1080/09540260601037896
Choudhury GA, van Scheltinga CT, van den Bergh D, Chowdhury F, de Heer, Hossain M, Karim Z (2012) Preparations for the Bangladesh delta plan: Alterra Report 2300, pp 1–90. Retrieved from https://research.vu.nl/en/publications/preparations-for-the-bangladesh-delta-plan
Choudhury AM (2002) Managing natural disasters in Bangladesh. In: Dhaka meet on sustainable development in Bangladesh: Achievements, opportunities and challenges at Rio+10. Bangladesh Unnayan Parishad, Dhaka, pp 1–19. Retrieved from https://www.dramchoudhury.info/files/publications/ManagingNaturalDisasterInBD.pdf

Chowdhury MA, Hasan MK, Islam SLU (2022) Climate change adaptation in Bangladesh: Current practices, challenges and the way forward. J Clim Chang Health. Elsevier Masson s.r.l. https://doi.org/10.1016/j.joclim.2021.100108

Crawford TW, Rahman MK, Miah MG, Islam MR, Paul BK, Curtis S, Islam MS (2021) Coupled adaptive cycles of shoreline change and households in deltaic Bangladesh: Analysis of a 30-year shoreline change record and recent population impacts. Ann Am Assoc Geogr 111(4):1002–1024. https://doi.org/10.1080/24694452.2020.1799746

Dasmahapatra D (2024) Impact of the tropical cyclone "REMAL" along the Coast of Bangladesh: Perspective analysis. Int J Innov Res Multidiscip Field 10(6):112–122. https://doi.org/10.2015/IJIRMF/202406012

Dhungana G, Pal I, Ghimire R, Dhungana RK, Tuladhar N (2023) Foundation of indigenous knowledge theory for disaster risk reduction. In: Multi-hazard vulnerability and resilience building: Cross cutting issues. Elsevier, pp 347–361. https://doi.org/10.1016/B978-0-323-95682-6.00016-4

Fakhruddin B, Kintada K, Hassan Q (2022) Understanding hazards: Probabilistic cyclone modelling for disaster risk to the Eastern Coast in Bangladesh. Prog Disaster Sci, 13. https://doi.org/10.1016/j.pdisas.2022.100216

Feng Y, Cui S (2021) A review of emergency response in disasters: Present and future perspectives. Natural Hazards. Springer Science and Business Media B.V. https://doi.org/10.1007/s11069-020-04297-x

Fernandez G, Ahmed I (2019) "Build back better" Approach to disaster recovery: Research trends since 2006. Progress in Disaster Science. Elsevier Ltd. https://doi.org/10.1016/j.pdisas.2019.100003

GED (2020) Sustainable development goals: Bangladesh Progress Report 2020. Dhaka. Retrieved from https://www.undp.org/bangladesh/publications/bangladesh-sustainable-development-goals-sdgs-progress-report-2020

GoB (2008) Cyclone Sidr in Bangladesh: Damage, loss and needs assessment for disaster recovery and reconstruction. Retrieved from https://reliefweb.int/report/bangladesh/cyclone-sidr-bangladesh-damage-loss-and-needs-assessment-disaster-recovery-and

GoB. Disaster Management Act, Pub. L. No. 34, 173441 (2012) Dhaka, Bangladesh: Bangladesh gazette, Government of the people's Republic of Bangladesh (GoB). Retrieved from https://legislativediv.portal.gov.bd/sites/default/files/files/legislativediv.portal.gov.bd/page/5a6bca14_6a2e_44e4_b155_c8147d1edd65/23.%20Disaster%20Management%20Act%2C%202012.pdf

GoB (1972) The constitution of the people's Republic of Bangladesh (Act No. of 1972). Government of the People's Republic of Bangladesh (GoB), Bangladesh. Retrieved from https://nnsop.gov.bd/storage/files/file-2023-02-03-63dce64f56e0c.pdf

GoB (2010) Standing orders on disaster 2010. Dhaka. Retrieved from https://ddm.portal.gov.bd/sites/default/files/files/ddm.portal.gov.bd/policies/11a33a04_f11d_44fa_a235_48d2f47521b7/2020-01-13-12-23-a4a05f9a58a819021a6d591667fa4ad6.pdf

GoB (2017) National plan for disaster management (2016–2020): Building resilience for sustainable human development. Dhaka. Retrieved from https://modmr.portal.gov.bd/sites/default/files/files/modmr.portal.gov.bd/policies/0a654dce_9456_46ad_b5c4_15ddfd8c4c0d/NPDM(2016-2020)%20-Final.pdf

GoB (2020a) National plan for disaster management (2021–2025): Action for disaster risk management towards resilient nation. Dhaka. Retrieved from https://modmr.portal.gov.bd/sites/default/files/files/modmr.portal.gov.bd/page/a7c2b9e1_6c9d_4ecf_bb53_ec74653e6d05/NPDM2021-25%20DraftVer5_23032020.pdf

GoB (2020b) Standing orders on disaster 2019. Dhaka. Retrieved from https://reliefweb.int/report/bangladesh/government-people-s-republic-bangladesh-standing-orders-disaster-2019

Graveline MH, Germain D (2022) Disaster risk resilience: Conceptual evolution, key issues, and opportunities. Int J Disaster Risk Sci 13(3):330–341. https://doi.org/10.1007/s13753-022-00419-0

Habib A, Shahidullah M, Ahmed D (2012) The Bangladesh cyclone preparedness program. A vital component of the nation's multi-hazard early warning system. In: Institutional partnerships in multi-hazard early warning systems. Springer, Berlin Heidelberg, pp 29–62. https://doi.org/10.1007/978-3-642-25373-7_3

Habiba U, Shaw R, Abedin MA (2013) Community-based disaster risk reduction approaches in Bangladesh. In: Shaw R, Mallick F, Islam A (eds) Disaster risk reduction approaches in Bangladesh. Springer, Tokyo, pp 259–279. https://doi.org/10.1007/978-4-431-54252-0_12

Haque CE (1997) Atmospheric hazards preparedness in Bangladesh: a study of warning, adjustments and recovery from the April 1991 cyclone. Nat Hazards 16(2/3):181–202. https://doi.org/10.1023/A:1007942712838

Haque A, Haider D, Rahman MS, Kabir L, Lejano RP (2022) Building resilience from the grassroots: The cyclone preparedness programme at 50. Int J Environ Res Public Health 19(21). https://doi.org/10.3390/ijerph192114503

Hasan MM, Islam R, Rahman MdS, Ibrahim M, Shamsuzzoha M, Khanam R, Zaman AKMM (2021) Analysis of land use and land cover changing patterns of Bangladesh using remote sensing technology. Am J Environ Sci 17(3):64–74. https://doi.org/10.3844/ajessp.2021.64.74

Hossain B, Sohel MS, Ryakitimbo CM (2020) Climate change induced extreme flood disaster in Bangladesh: Implications on people's livelihoods in the char village and their coping mechanisms. Prog Disaster Sci, 6. https://doi.org/10.1016/j.pdisas.2020.100079

Hossain M (2022) ICTs for sustainable development in Bangladesh. In: Hossain M (ed) Digital transformation and economic development in Bangladesh: Rethinking digitalization strategies for leapfrogging. Springer Nature Singapore, Singapore, pp 275–283. https://doi.org/10.1007/978-981-19-2753-9_13

Huq N, Hugé J, Boon E, Gain AK (2015) Climate change impacts in agricultural communities in rural areas of Coastal Bangladesh: A tale of many stories. Sustain (Switz) 7(7):8438–8460. https://doi.org/10.3390/su7078437

IFRC (2021) Bangladesh: Cyclone YAAS. Geneva. Retrieved from https://reliefweb.int/report/bangladesh/bangladesh-cyclone-yaas-final-report-n-mdrbd027

IFRC (2024) Bangladesh: Cyclone Remal. Geneva. Retrieved from https://reliefweb.int/report/bangladesh/bangladesh-cyclone-remal-operation-update-emergency-appeal-no-mdrbd035

Ikeda K (1995) Gender differences in human loss and vulnerability in natural disasters: A case study from Bangladesh. Indian J Gend Stud 2(2):171–193. https://doi.org/10.1177/097152159500200202

Imran M, Hossain MI, Sakib NH (2016) An evaluation of disaster management methodologies in Bangladesh. In: 1st AISD international multidisciplinary conference on sustainable development. Australian Institute for Sustainable Development, Dhaka, pp 61–75. https://doi.org/10.13140/RG.2.1.3938.6648

Inan DI, Beydoun G, Othman SH (2023) Risk assessment and sustainable disaster management. Sustainability (Switzerland). Multidisciplinary Digital Publishing Institute (MDPI). https://doi.org/10.3390/su15065254

Islam K (2013) Community Based Disaster Management in Bangladesh. Khulna University, Khulna. Retrieved from https://www.researchgate.net/publication/333972153_Community_Based_Disaster_Management_in_Bangladesh

Islam MM, Jannat A, Al Rafi DA, Aruga K (2020a) Potential economic impacts of the COVID-19 pandemic on South Asian economies: A review. World 1(3):283–299. https://doi.org/10.3390/world1030020

Islam S, Chu C, Liew L, Smart JCR (2020b) Distributing flood shelters for disaster risk reduction: Exploring the practices in Bangladesh from a political economy perspective. Disaster Prev Manag: Int J 29(3):322–339. https://doi.org/10.1108/DPM-02-2019-0060

Islam MM, Matsushita S, Noguchi R, Ahamed T (2022) A damage-based crop insurance system for flash flooding: A satellite remote sensing and econometric approach. Asia-Pac J Reg Sci 6(1):47–89. https://doi.org/10.1007/s41685-021-00220-9

Islam A, Shaw R, Mallick F (2013) Bangladesh climate change strategy and action plans. In: Shaw R, Mallick F, Islam A (eds) Climate change adaptation actions in Bangladesh. Springer, Tokyo, pp 107–118. https://doi.org/10.1007/978-4-431-54249-0_7

Islam MN, Tamanna S, van Amstel A, Noman M, Ali MSS, Aparajita DM, Ghosh A (2021) Climate change impact and comprehensive disaster management approach in Bangladesh: A review. In: Islam MN, van Amstel A (eds) Bangladesh II: Climate change impacts, mitigation and adaptation in developing countries. Springer International Publishing, Cham, pp 1–39. https://doi.org/10.1007/978-3-030-71950-0_1

Karim M, Mimura N (2008) Impacts of climate change and sea-level rise on cyclonic storm surge floods in Bangladesh. Glob Environ Chang 18(3):490–500. https://doi.org/10.1016/j.gloenvcha.2008.05.002

Khan SA (2024) The role of innovation and adaptation technologies in climate change-induced disaster risk reduction: Enhancing resilience in Satkhira Communities of Bangladesh. University of Manitoba, Winnipeg. Retrieved from https://umanitoba.ca/environment-earth-resources/sites/environment-earth-resources/files/2024-04/MNRM%20Thesis%20Khan%2C%20Sabbir.pdf

Kitagawa K (2020) Development of disaster risk reduction policy in Thailand. Disaster Prev Manag: Int J, Ahead-Of-Print (Ahead-Of-Print). https://doi.org/10.1108/DPM-08-2019-0244

Kulatunga U, Wedawatta G, Amaratunga DG, Shamsuzzoha (2013) Disaster risk and community vulnerability: The case of Patuakhali, Bangladesh. Retrieved from https://api.semanticscholar.org/CorpusID:15175672

Kumar P, Avtar R, Dasgupta R, Johnson BA, Mukherjee A, Ahsan MN, Mishra BK (2020) Socio-hydrology: A key approach for adaptation to water scarcity and achieving human well-being in large riverine Islands. Prog Disaster Sci, 8. https://doi.org/10.1016/j.pdisas.2020.100134

Macrotrends (2024) Bangladesh GDP 1960–2024. Retrieved 20 August 2024, from https://www.macrotrends.net/global-metrics/countries/BGD/bangladesh/gdp-gross-domestic-product

Mahamud U, Takewaka S (2018) Shoreline change around a river delta on the cox's Bazar coast of Bangladesh. J Mar Sci Eng 6(3):80. https://doi.org/10.3390/jmse6030080

Mahin Uddin M, Al-Amin M, Akter S (2022) Exploring Bangladesh disaster management's institutional and policy challenges: A critical review. Indiana J Econ Bus Manag, 2(5) https://doi.org/10.5281/zenodo.7267722

Mannan S, Haque DME, Sarker NCD (2021) A study on national DRR Policy in alignment with the SFDRR: Identifying the scopes of improvement for Bangladesh. Prog Disaster Science. Elsevier Ltd. https://doi.org/10.1016/j.pdisas.2021.100206

Matin N, Hasan GMJ (2021) A quantitative analysis of shoreline changes along the coast of Bangladesh using remote sensing and GIS techniques. CATENA 201:105185. https://doi.org/10.1016/j.catena.2021.105185

Matin N, Taher M (2001) The changing emphasis of disasters in Bangladesh NGOs. Disasters 25(3):227–239. https://doi.org/10.1111/1467-7717.00174

Mazumdar A, Alharahsheh HH (2020) Digital Bangladesh—vision 2021: What is the digital Bangladesh concept? South Asian Res J Eng Technol 02(01):6–9. https://doi.org/10.36346/sarjet.2020.v02i01.002

Miyaji M, Okazaki K, Ochiai C (2017) A study on the use of cyclone shelters in Bangladesh. J Arch Plan (Trans AIJ) 82(737):1871–1880. https://doi.org/10.3130/aija.82.1871

MoDMR (2010) Comprehensive disaster management programme phase II. Dhaka. Retrieved from https://www.preventionweb.net/files/globalplatform/entry_bg_paper~cdmpphaseii.pdf

MoDMR (2015) National Disaster Management Policy 2015, Pub. L. No. 51.00.0000.322.38.049.15–86. Bangladesh: Disaster Management Knowledge Portal (DMKP), Ministry of Disaster Management and Relief (MoDMR), Government of the People's Republic of Bangladesh. Retrieved from http://203.76.123.197/egls/documents/458/search-detail

MoEFCC (2022a) Climate change initiatives of Bangladesh: Achieving climate resilience. Dhaka. Retrieved from https://doe.portal.gov.bd/sites/default/files/files/doe.portal.gov.bd/npf block/2022-11-02-08-08-ade27c3a48eeeedbf1394e5fa527edd2.pdf

MoEFCC (2022b) Mujib climate prosperity plan 2022–2041. Dhaka. Retrieved from https://moef.gov.bd/site/publications/95fc67a4-9759-479b-8ea7-ad067bfd7ced/Mujib-Climate-Prosperity-Plan-2022-2041

MoF (2020) Bangladesh climate fiscal framework 2020. Dhaka. Retrieved from https://www.undp.org/sites/g/files/zskgke326/files/migration/bd/Updated-Climate-Fiscal-Framework-2020.pdf

Mukherjee M, Wickramasinghe D, Chowdhooree I, Chimi C, Poudel S, Mishra B, Shaw R (2022) Nature-based resilience: Experiences of five cities from South Asia. Int J Environ Res Public Health, 19(19). https://doi.org/10.3390/ijerph191911846

Nasreen M, Hossain KM, Khan MM (2023) Coastal Disaster risk management in Bangladesh. Routledge, London. https://doi.org/10.4324/9781003253495

Neeraj S (2022) A build back better assessment tool for post-disaster Reconstruction and recovery. University of Auckland, Auckland. Retrieved from https://researchspace.auckland.ac.nz/handle/2292/58250

Nepal P, Khanal NR, Sharma BPP (2018) Policies and institutions for disaster risk management in Nepal: A review. Geogr J Nepal 11:1–24. Retrieved from https://www.climatenepal.org.np/sites/default/files/doc_resources/Nepal%20et%20al.%2C%202018%20GJoN.pdf

Van Niekerk D, Nemakonde LD, Kruger L, Forbes-Genade K (2018) Community-based disaster risk management. In: Handbooks of sociology and social research. Springer Science and Business Media B.V., pp 411–429. https://doi.org/10.1007/978-3-319-63254-4_20

NSET (2011) Comprehensive Disaster Management Program (CDMP), Bangladesh. National Society for Earthquake Technology-Nepal (NSET), Kathmandu. Retrieved from https://nset.org.np/nset2012/images/publicationfile/20111220153428.pdf

Ogra A, Donovan A, Adamson G, Viswanathan KR, Budimir M (2021) Exploring the gap between policy and action in disaster risk reduction: A case study from India. Int J Disaster Risk Reduct, 63. https://doi.org/10.1016/j.ijdrr.2021.102428

Pal SK, Sarker PC (2023) Smart Bangladesh vision 2041: Concept of a sustainable developed country. Environ Manag Sustain Dev 12(1):67. https://doi.org/10.5296/emsd.v12i1.20666

Parikh M, Hota DR (2022) Comparative analysis of legal framework for disaster management in Pakistan, India, and Bangladesh. In: Singh A (ed) International handbook of disaster research. Springer Nature Singapore, Singapore, pp 1–12. https://doi.org/10.1007/978-981-16-8800-3_141-1

PCGN (2023) Toponymic fact file: Bangladesh. London. Retrieved from https://assets.publishing.service.gov.uk/government/uploads/system/uploads/attachment_data/file/1185349/Bangladesh_Toponymic_Factfile_July_2023_update2.pdf

Perwaiz A, Parviainen J, Somboon P, Mcdonald A, Kumar A, Wilcox T, Amach CO (2020) Disaster risk reduction in Bangladesh. Bangkok. Retrieved from https://www.undrr.org/media/48524/download?startDownload=20240806

Rahman MS, Gain A (2020) Adaptation to river bank erosion induced displacement in Koyra Upazila of Bangladesh. Prog Disaster Sci 5. https://doi.org/10.1016/j.pdisas.2019.100055

Rahman MH, Rahman MS, Rahman MM (2017) Disasters in Bangladesh: Mitigation and management. Barisal Univ J Part 1, 1(1):139–163. Retrieved from https://bu.ac.bd/uploads/BUJ1V4I1/13.%20Hashinoor.pdf

Rahman S, Tasnim F (2023) The role of NGOs in ensuring local governance in Bangladesh: from the perception of other actors of governance. Asia-Pac J Reg Sci. https://doi.org/10.1007/s41685-023-00283-w

Raj SJ, Ullah MS, Akhter R (2010) From dissemination to response: In search of new strategies for broadcast media in terms of cyclone warnings for Bangladesh. J Sci Commun, 9 (4):A03. https://doi.org/10.22323/2.09040203

Rasheduzzaman M, Shamsuzzoha M, Akter A (2020) Strengthening economic security through social capital: household's adaptation to cyclone risk in Bangladesh. Int J Multidiscip Res Dev:69–75. Retrieved from https://www.allsubjectjournal.com/assets/archives/2020/vol7issue5/7-4-34-390.pdf

Rashida MB, Sheika MR, Haquea AJME, Patwarya MAA (2022) Climate change impact and frequency of cyclone surge in Bangladesh Coast: Proper plan to address sustainably. Int J Econ Environ Geol 13(2):15–20. https://doi.org/10.46660/ijeeg.v13i2.25
Rathore V (2016) Technology in disaster management and disaster risk reduction: A review of applications. 6(4). Retrieved from www.iiste.org
Sadik MS, Nakagawa H, Rahman R, Shaw R, Kawaike K, Fujita K (2018) A Study on cyclone Aila recovery in Koyra, Bangladesh: Evaluating the inclusiveness of recovery with respect to predisaster vulnerability reduction. Int J Disaster Risk Sci 9(1):28–43. https://doi.org/10.1007/s13753-018-0166-9
Salequzzaman M, Gorana RN (2016) Education for Sustainable Development (ESD) initiatives and programmes in Bangladesh. In: Gorana RN, Kanaujia PR (eds) Reorienting educational efforts for sustainable development: Experiences from South Asia, 1st ed. Springer Netherlands, Dordrecht, pp 53–70. https://doi.org/10.1007/978-94-017-7622-6_4
Sattar MA, Biswas AAA, Islam MT, Hossain MA, Siddeqa M, Rahim MA, Aktar S (2020) Disaster vulnerability and mitigation of humanitarian issues in Coastal Bangladesh: Local evidence and knowledge gaps. Prog Disaster Sci, 8. https://doi.org/10.1016/j.pdisas.2020.100138
Shammin MR, Firoz R, Hasan R (2022) Frameworks, stories and learnings from disaster management in Bangladesh. In: Haque AKE, Mukhopadhyay P, Nepal M, Shammin MR (eds) Climate change and community resilience: Insights from South Asia. Springer Nature Singapore, Singapore, pp 239–256. https://doi.org/10.1007/978-981-16-0680-9_16
Shamsuzzoha M, Kormoker T, Ghosh RC (2018a) Implementation of water safety plan considering climatic disaster risk reduction in Bangladesh: A study on patuakhali pourashava water supply system. Procedia Eng 212:583–590. https://doi.org/10.1016/j.proeng.2018.01.075
Shamsuzzoha M, Rasheduzzaman M, Ghosh RC (2018b) Building resilience for drinking water shortages through reverse osmosis technology in Coastal areas of Bangladesh. Procedia Eng 212:559–566. https://doi.org/10.1016/j.proeng.2018.01.072
Shamsuzzoha M, Noguchi R, Ahamed T (2022) Rice yield loss area assessment from satellite-derived NDVI after extreme climatic events using a fuzzy approach. Agric Inf Res 31(1):32–46. https://doi.org/10.3173/air.31.32
Shamsuzzoha M, Shaw R, Ahamed T (2024b) Machine learning system to assess rice crop change detection from satellite-derived RGVI due to tropical cyclones using remote sensing dataset. Remote Sens Appl: Soc Environ 35:101201. https://doi.org/10.1016/j.rsase.2024.101201
Shamsuzzoha M, Ahamed T (2023) Shoreline change assessment in the coastal region of Bangladesh delta using tasseled cap transformation from satellite remote sensing dataset. Remote Sens, 15(2). https://doi.org/10.3390/rs15020295
Shamsuzzoha M, Ahamed T (2024a) Application of introducing damaged area index for cultivated agricultural lands affected by recurrent tropical cyclones Bulbul and Amphan using satellite remote sensing. In: Ahamed T (ed) Remote sensing application II. New frontiers in regional science: Asian perspectives, vol 77. Springer, Singapore, pp 165–178. https://doi.org/10.1007/978-981-97-1188-8_6
Shamsuzzoha M, Ahamed T (2024b) Assessment of shoreline and agricultural land use changes in the onshore Coastal Region of Bangladesh delta using satellite remote sensing and GIS. In: Ahamed T (ed) Remote sensing application II. New frontiers in regional science: Asian perspectives, vol 77. Springer, Singapore, pp 85–119. https://doi.org/10.1007/978-981-97-1188-8_4
Shamsuzzoha M, Al-Maruf A (2012) Post SIDR life strategy: Adaptation scenario of settlements of the South. Inst Bangladesh Stud (IBS) J 19:207–222. Retrieved from https://www.ru.ac.bd/ibs/wp-content/uploads/sites/56/2022/07/JIBS-Bangla-19.pdf
Shamsuzzoha M, Noguchi R, Ahamed T (2021) Damaged area assessment of cultivated agricultural lands affected by Cyclone Bulbul in Coastal region of Bangladesh using landsat 8 OLI and TIRS datasets. Remote Sens Appl: Soc Environ, 23. https://doi.org/10.1016/j.rsase.2021.100523
Shamsuzzoha M, Shaw R, Ahamed T (2024a) Country perspective on disaster risk governance in Bangladesh. In: Rethinking disaster risk management in South Asia, Panel-1: A synergistic

approach to disaster risk governance. Islamabad. Retrieved from https://www.researchgate.net/publication/379946968
Shaw R, Izumi T, Shi P (2016) Perspectives of science and technology in disaster risk reduction of Asia. Int J Disaster Risk Sci 7(4):329–342. https://doi.org/10.1007/s13753-016-0104-7
Shaw R, Takeuchi Y, Rouhban B (2009) Education, capacity building and public awareness for disaster reduction. In: Sassa K, Canuti P (eds) Landslides—disaster risk reduction. Springer Berlin Heidelberg, Berlin, Heidelberg, pp 499–515. https://doi.org/10.1007/978-3-540-69970-5_26
Shaw R, Islam A, Mallick F (2013a) Disaster, risk and evolution of the concept, pp 3–21. https://doi.org/10.1007/978-4-431-54252-0_1
Shaw R, Islam A, Mallick F (2013b) National perspectives of disaster risk reduction in Bangladesh, pp 45–62. https://doi.org/10.1007/978-4-431-54252-0_3
Shaw R, Mallick F, Islam A (2013c) Disaster risk reduction approaches in Bangladesh. In: Shaw RMFIA (ed). Springer, New York. Retrieved from http://www.springer.com/series/11575
Shaw R, Izumi T, Shiwaku K (2018) 1—Science and technology in disaster risk reduction in Asia: Post-sendai developments. In: Shaw R, Shiwaku K, Izumi T (eds) Science and technology in disaster risk reduction in Asia. Academic Press, pp 3–16. https://doi.org/10.1016/B978-0-12-812711-7.00001-8
Shi P, Ye T, Wang Y, Zhou T, Xu W, Du J, Okada N (2020) Disaster risk science: A geographical perspective and a research framework. Int J Disaster Risk Sci 11(4):426–440. https://doi.org/10.1007/s13753-020-00296-5
Shuvo SRS, Islam MdN, Islam ST (2022) Role of information and communication technologies in build back better to post disaster recovery practices: Insights from Bangladesh. Int J Disaster Resil Built Environ 13(2):233–247. https://doi.org/10.1108/IJDRBE-08-2021-0097
Siddeqa M, Islam MT, Hasan MF, Islam R, Rasheduzzaman M (2023) Land use and land cover change detection of Teknaf Upazila due to Rohingya crisis by using GIS, and RS techniques. Asian J Soc Sci Leg Stud 5(6):246–252. https://doi.org/10.34104/ajssls.023.02460252
Su H, Liu C, Dai D, Chen W, Zhang Z, Wang Y (2023) Distribution characteristics and influencing factors of the national comprehensive disaster-reduction demonstration community in China. Land, 12(8). https://doi.org/10.3390/land12081633
Syed MA, Shrestha MS, Khadgi V (2021) Chapter 33—Last mile communication of multihazard early warning: A case study on Bangladesh. In: Pal I, Shaw R, Djalante R, Shrestha S (eds) Disaster resilience and sustainability. Elsevier, pp 725–765. https://doi.org/10.1016/B978-0-323-85195-4.00002-0
Tabassum T, Farzana M, Nahar AN, Araf Y, Ullah MA, Rahaman TI, Hosen MJ (2023) COVID-19 in Bangladesh: Wave-centric assessments and mitigation measures for future pandemics. Heliyon 9(10):e20113. https://doi.org/10.1016/j.heliyon.2023.e20113
Talukder MF, Shamsuzzoha M, Hasan I (2018) Damage and agricultural rehabilitation scenario of post cyclone Mahasen in coastal zone of Bangladesh. J Sociol Anthropol, 2 (1). https://doi.org/10.12691/jsa-2-1-7
Tambe S (2022) Sustainable livelihoods approach. In: Teaching and learning rural livelihoods: A guide for educators, students, and practitioners, 1st ed. Springer Cham, pp 45–56. https://doi.org/10.1007/978-3-030-90491-3_4
Terwisscha Van Scheltinga C (2013) Planning today for delta management of the future: Towards a Bangladesh delta plan 2100. Retrieved from https://www.researchgate.net/publication/260189530
UN (2015) Transforming our world: The 2030 agenda for sustainable development. New York. Retrieved from https://digitallibrary.un.org/record/3923923?v=pdf
UN (2023) Progress towards the sustainable development goals: Towards a rescue plan for people and planet. New York. Retrieved from https://digitallibrary.un.org/record/4014344?v=pdf
UN (2024) The sustainable development goals report 2024. New York. Retrieved from https://unstats.un.org/sdgs/report/2024/The-Sustainable-Development-Goals-Report-2024.pdf

UNDRR (2015) Sendai framework for disaster risk reduction 2015–2030. Geneva. Retrieved from https://www.undrr.org/media/16176/download?startDownload=20240812

Wahiduzzaman M, Paul GC (2023) Assessing the impact of extensive severe cyclonic storm fani on the coastal communities of Bangladesh: A case study. Saudi J Civ Eng 7(03):16–28. https://doi.org/10.36348/sjce.2023.v07i02.002

WB (2024) GDP growth (Annual %)—Bangladesh. World Bank (WB). Retrieved from https://data.worldbank.org/indicator/NY.GDP.MKTP.KD.ZG?locations=BD

Woodruff JD, Irish JL, Camargo SJ (2013) Coastal flooding by tropical cyclones and sea-level rise. Nature 504(7478):44–52. https://doi.org/10.1038/nature12855

Wright H, Fenton A, Huq S, Stott C, Taub J, Chow J (2019) Introduction: Bangladesh responds to climate change. In: Huq S, Chow J, Fenton A, Stott C, Taub J, Wright H (eds) Confronting climate change in Bangladesh: Policy strategies for adaptation and resilience. Springer, London, pp 1–6. https://doi.org/10.1007/978-3-030-05237-9_1

Zhang H, Nakagawa H (2018) Validation of indigenous knowledge for disaster resilience against river flooding and bank erosion. In: Science and technology in disaster risk reduction in Asia. Elsevier, pp 57–76. https://doi.org/10.1016/B978-0-12-812711-7.00005-5

Chapter 20
Echoes of the Indian Ocean Tsunami in Sri Lanka: Two Decades of Lessons Learnt and Future Challenges in Disaster Management

Deepthi Wickramasinghe, Sujeewa Kumari, Rekha Nianthi, and Rajib Shaw

Abstract The Tsunami's impact occurred in the Indian Ocean in 2004 devastating the island of Sri Lanka, bringing significant and cascading environmental, socio-economic and political impacts. However, it also marked a turning point in disaster management on the island. In the two decades since the Tsunami disaster, Sri Lanka has made significant developments in disaster risk reduction (DRR) and building resilience. This chapter examines the evolution of Sri Lanka's DRR policies and legislations, elaborating on advancements in preparedness for disasters, community-based resilience and early warning systems programs. Key initiatives such as the establishment of the Disaster Management Centre (DMC) and the implementation of the "National Policy on Disaster Management" are also highlighted. In addition, the chapter addresses issues and challenges including climate change, resource constraints, and socio-economic inequalities. By assessing both achievements and ongoing obstacles, this chapter presents a comprehensive overview of Sri Lanka's journey towards a more resilient future.

Keywords India Ocean tsunami · Disaster risk governance · DRR policy · Stakeholder engagement · Sri Lanka

---

The original version of the chapter has been revised. A correction to this chapter can be found at https://doi.org/10.1007/978-981-96-2669-4_23

D. Wickramasinghe (✉) · S. Kumari
University of Colombo, Colombo, Sri Lanka
e-mail: deepthi@zoology.cmb.ac.lk

R. Nianthi
University of Peradeniya, Peradeniya, Sri Lanka

R. Shaw
Keio University, Minato, Japan

R. Shaw et al. (eds.), *Two Decades from the Indian Ocean Tsunami*, Disaster Risk Reduction, https://doi.org/10.1007/978-981-96-2669-4_20

20.1 Background

Disaster Management plays a crucial role in handling and preventing emergencies and disasters. It provides a structured framework to assist communities in high-risk situations. The main goal is to protect communities by reducing the possibility of catastrophic events and addressing vulnerabilities. Disaster Management involves various functions such as planning, organizing, and implementing preventive measures, mitigation strategies, and preparedness for emergency response and recovery (Math et al. 2015). Disaster Management provides support to disaster victims and ensures their effective recovery. By employing a well-structured framework, the focus is on restoring normalcy to people's lives and minimizing the impact on lives and properties within a short period. By implementing clearly defined procedures, Disaster Management aims to mitigate the impact of disasters, enabling the prompt rescue of victims and accelerating post-disaster recovery efforts. Mitra and Shaw (2022) suggest fundamental principles and a conceptual framework for Integrated Disaster Resilience (IDR) with the application of a Whole Systems Approach.

Covering an expanse of 65,000 km^2 with a population of 22.156 million as per the Report of the Department of Census and Statistics, 2021, this Island is characterized by a primary topographic feature a mountain massif shaped like an anchor. The island's topography is combined with variations in regional climate (see Fig. 20.1), and serves as a fundamental factor contributing to the diverse contrasts observed in various aspects across the region.

The increase in risk of hazards, disasters, and their impacts on them to Sri Lanka can be attributed to multiple factors. Population rate, growing demands for natural resources such as land, water and uncontrolled and speed urbanization, and infrastructure development that lacks consideration for disaster risk and exposure are among the primary contributors to this trend. According to a study, Sri Lanka's population resides areas around 87% of areas that have been identified as sensible hotspots for disasters (Disaster Management Centre 2024).

In light of this context, this section outlines the legal and policy structure for handling disasters and the responsibilities delegated to crucial organizations for policy and capacity building at various levels. It examines the resources, plans, and methods employed in addressing various risk factors, such as climate change and unsustainable development.

20.2 Disaster History Within Two Decades

Sri Lanka, known for its stunning landscapes and coastline, is highly vulnerable to hazards, disasters and disasters due to the nature of geography, topography, and climate. The diverse landscape, with coastal plains and hilly regions, increases the risk of flooding and landslides during the monsoon season (Rathnayaka and Herath 2005). Its equatorial location also makes it vulnerable to tropical cyclones, which

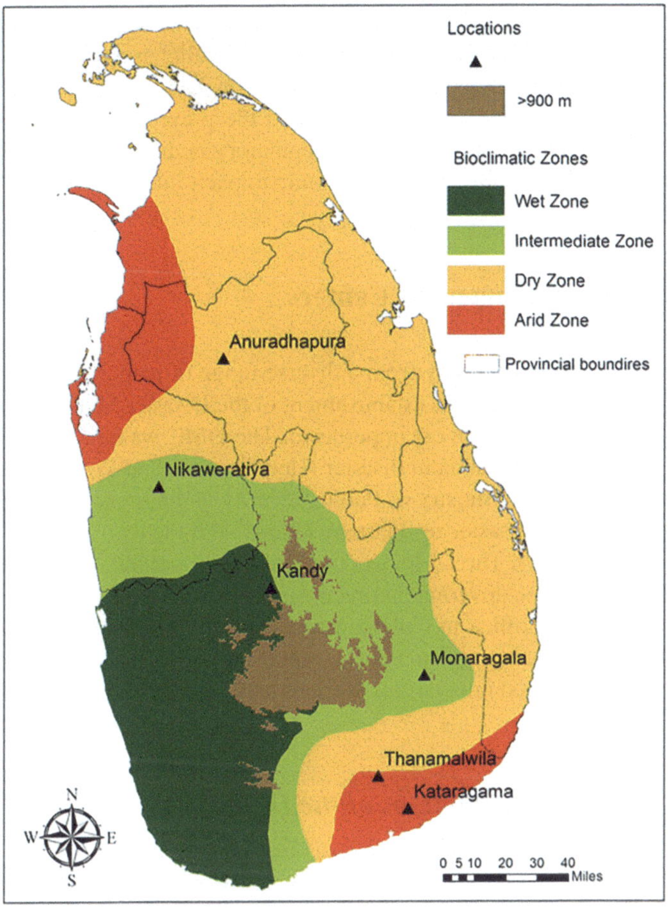

Fig. 20.1 Three major climatic zones of Sri Lanka (*Source* https://www.climatechange.lk/Climate_Profile.html)

bring heavy rainfall, tough winds, and storm floods (Srisangeerthanan et al. 2015). Additionally, changing climatic patterns are leading to more frequent and severe droughts, affecting agriculture and water resources. Common hazards in Sri Lanka include floods, cyclones, landslides, droughts and wildfires (Alahacoon et al. 2018).

Sri Lanka rarely experiences Tsunamis, but the 2004 event left an indelible mark due to its devastating impact. On December 26, powerful Tsunami waves, triggered by a massive underwater earthquake off the Indonesian coast of Sumatra, struck Sri Lanka's shores. Over 30,000 lives were lost, and coastal districts suffered from extensive damages (Rossetto et al. 2006 and Kaplan et al. 2009). The eastern coastal area and southern coastal areas were the hardest affected areas, prompting a significant international humanitarian response. This disaster not only altered Sri Lanka's landscape but also heightened focus on disaster preparedness, legislation, and early warning systems to mitigate future tsunami impacts (Mulligan and Nadarajah 2011).

The 2004 Tsunami event led to the creation of new laws and policies on disaster management. These changes aimed to improve the prevention and mitigation of natural disasters, giving more authority to regulatory bodies to address the social, environmental, cultural, political, and economic impacts effectively.

20.3 Disaster Management Centers

In order to safeguard our country from a diverse range of natural environment and human-affected catastrophes, the establishment of the Disaster Management Centre (DMC) in Sri Lanka holds utmost importance. The DMC was established in 2005 following the devastating Tsunami disaster (Hettiarachchi and Weeresinghe 2014). DMC operates under the Ministry of Public Security and has evolved into a robust institution focusing on disaster response, readiness, and the recovery process (Siriwardana et al. 2018a, b). The DMC works closely with various stakeholders, such as international partners, government organizations, and non-governmental organizations, from its main office in Colombo to ensure the efficient fulfillment of its responsibilities for disaster preparedness, response, and recovery, the DMC of Sri Lanka has been structured into various sections and divisions (Wanninayake 2018).

20.3.1 Powers and Functions of the DMC

Disaster preparedness ranks high among the DMC's core responsibilities. The center is dedicated to developing early warning systems, creating hazard maps, and conducting risk assessments. DMC's primary goal is to ensure that communities are well-informed about approaching disasters, allowing for timely evacuations and preparedness measures to be put in place. To achieve this, the DMC relies on meteorological data and cutting-edge technologies. During times of natural crisis, the DMC performs a vital role in coordinating disaster response determinations. The Operations Division is responsible for managing the logistics of relief operations and deploying emergency response teams and resources. This coordination involves working closely with various government departments, non-governmental organizations, and foreign partners to ensure a unified and effective response.

Moreover, The DMC performs an active part in post-disaster reconstruction with recovery efforts. The center assesses the impacts of natural hazards, coordinates the aspects of distribution of relief supplies, and supports affected areas in rebuilding. Collaboration with local authorities, humanitarian organizations, and other stakeholders is essential to address both immediate and long-term needs and DMC acts accordingly.

The Planning and Mitigation Division of DMC creates strategies and blueprints to minimize vulnerabilities and boost community resilience in disaster risk situations. The DMC carries out educational initiatives, skill-building programs, and community outreach efforts to prepare individuals and organizations to withstand and effectively manage disasters.

The organization serves as a central hub for studying and advancing disaster management. In order to stay ahead of emerging challenges, the Research and Development Division carries out research, collects data on disaster patterns, and explores state-of-the-art technologies. This information plays a crucial role in continually enhancing strategies and policies.

Furthermore, the DMC enhances skills and provide training opportunities for stakeholders. The Training and Capacity Building Division organizes workshops, exercises, and instructional programs aimed at enhancing the capabilities of individuals and groups involved in disaster management. Through these positive efforts, stakeholders can be confident in their preparedness to handle various crisis situations.

20.3.2 Special Measures Taken by DMC

Sri Lanka has adopted a wide array of disaster risk management strategies to facilitate and implement the sections stipulated in the plans and policies related to disaster risk management (Disaster Management Centre 2024). A key strategy amongst them is an institutional arrangement that the government has made with the Provincial Council, District, Divisional, Grama Niladhari and Village levels. To put this into action on the ground, specialized technical committees have been established as outlined in the Act. These committees will consist of experts and representatives from various stakeholder organizations responsible for addressing different natural and human-affected hazards outlined in the DMC Act. Their primary responsibility is to provide guidance on specific disaster risk management activities to the relevant institutions. This decentralized approach allows even village-level institutions to participate in reducing disaster risks (Caruson and MacManus 2012). The endurance of this strategy has been ensured by linking these institutions to the DMC. The situation of Sri Lanka holds horizontal and/or vertical collaboration concepts following the suggestions of Caruson and MacManus (2012) and Bae et al. (2016). Despite this, efficient multi-agency collaboration has been identified and it has been highlighted in relation to handling the natural hazards or disasters in the past in Sri Lanka (Abdeen et al. 2021). Another study conducted by Fernando (2021) draws on examples from the October 2014 landslide in the Badulla area and the 2017 floods highlighted the disquiets over the degree of cooperation that exists between the assistances/agents involved in disaster management.

The Disaster Management Center (DMC) is dedicated to advancing research in order to identify, analyze, and evaluate disaster risk, monitor changes in such risk, forecast future scenarios, and develop strategic measures to mitigate disaster risk. As part of these ongoing efforts, the DMC continuously conducts multi-hazard mapping, vulnerability assessments, and risk evaluations to consolidate all pertinent disaster-related information and hazard maps in a centralized repository. This approach ensures that such information is easily accessible and supports the transfer of technology, particularly in collaborative projects with international agencies. Priority areas identified for the map have been determined based on the rigorousness of vulnerability, and significance of the sector such as health facilities, education, housing, agriculture and ongoing development projects in the area. Integrating disaster risk reduction at the initial stages of development has also been emphasized. The initiative also aims to advance research studies carried out by agencies and universities to improve the disaster information database. This includes a focus on vulnerable areas, societal issues, and specific natural hazards, such as those arising from technological and human-induced risks resulting from unsustainable development. The Sendai Framework builds upon the Hyogo Framework for Action (HFA) by encompassing both natural as well as man-made hazards, expanding the scope of activities related to recovery, rehabilitation, and reconstruction (Disaster Management Centre 2024). Emphasizing various hazard types necessitates a comprehensive understanding of their essential features, including magnitude, duration, and frequency. To achieve this goal, there is a necessity to improve a National Hazard Profile separately developed by the management of DMC too.

The Sri Lankan government has implemented a key strategy involving the establishment of "Early Warning Centers" (EWC). These centers play a crucial role in ensuring effective communication of early warnings during hazardous events (Abdeen et al. 2021). The EW Centers work diligently with the special police emergency units to enhance community and police awareness of early warning signs and immediate actions. These centers aim to establish reliable communication systems, integrate alternatives, and collaborate with community media to safeguard the timely distribution of information (Disaster Management Centre 2024). As part of their role, they must consistently inform relevant administrative bodies and local committees. Additionally, communities and sea-dependent individuals likely to be affected receive prompt notifications from the EW Centers, contributing to effective disaster preparedness and response (Disaster Management Centre 2024).

There are guidelines circulated by DMC to develop a "Disaster Preparedness Plan" and a "Response Plan" by all government units, private sector and educational units can be viewed as another strategy implemented by the government to manage risks of hazards posed due to human and natural activities (Disaster Management Centre 2024). And also the DMC has taken positive steps to promote micro-financing schemes to reduce the exposure of hazard-prone residents, thus addressing disaster resilience within communities. This approach is crucial in breaking the vicious cycle of poverty and disaster, as stated by Abdeen et al. (2021).

DMC utilizes crucial approaches, including training, public awareness, and education, to guarantee the enduring management of disaster risks (Disaster Management Centre 2024). Training initiatives, tailored to the capacities of different authorities, involve identifying target groups, including the government sector, private sector, NGOs, and Community-Based Organizations (CBOs) that work for Disaster Risk Management. The plans encompass every phase of the Disaster Management Cycle taking into account the unique requirements of each district while aligning with ongoing initiatives. The Ministry of Disaster Management carries out public awareness initiatives that emphasize livelihood development and poverty reduction (2018). A project initiated in 2017 integrates global environmental and disaster risk management agreements into Sri Lanka's sustainable development, with completion in December 2018. The project's awareness programs aim to reach wide-ranging individuals, groups, the general public, government officials, political leaders, small and medium-sized business owners, and journalists. These initiatives highlight the significance of utilizing metadata in decision-making processes. Through this well-rounded approach, the project effectively meets the training and awareness requirements of different sectors and communities.

Encouraging different types of stakeholders outside the government organization for "Achieving Maximum Effect" in disaster risk response is a decentralized strategy implemented by the DMC. The aim of this strategy is to improve disaster response and recovery by leveraging the private sector. The private sector is being urged to create alliances and craft specific plans in line with the National Action Plan and the guiding principle of the National Disaster Management Strategy (Ministry of Disaster Management 2018). DMC aims to support the organization and capacity building of these groups, working closely with them to access resources for the post-disaster response and include them in the National Resource database. The DMC also fosters a culture of volunteerism at multiple levels; promoting the registration and training of volunteer groups while acknowledging and rewarding volunteers and employers who champion volunteerism (Disaster Management Centre 2014). Moreover, the DMC acknowledges the vital role of the media, emphasizing their responsibility to deliver accurate and timely information to the general public only to reduce the disaster impacts, support disaster management activities, including the inclusion of their resources in the National Resource database. A framework for the operations of the NGOs and INGOs in Sri Lanka during disasters has also been established to facilitate the role of these organizations and adopt a Code of Conduct for NGOs (Disaster Management Centre 2014).

20.4 Divisions of the DMC

The Disaster Management Centre of Sri Lanka consists of eight divisions for the smooth function of the DMC (Table 20.1).

Table 20.1 Divisions and responsibilities of the DMC

Division	Responsibilities
Operations division	Supervise and manage operational activities, including disaster response, resource allocation, emergency team deployment, and ensuring prompt and efficient emergency response
Planning and mitigation division	Create and implement disaster risk reduction strategies, mitigation plans, and community awareness initiatives; develop policies to improve national resilience against various hazards
Information and Communication Technology (ICT) division	Oversee technology applications for early warning systems, data management, and communication during disaster events
Training and capacity building division	Facilitate training programs, workshops, and drills to boost the skills and preparedness of individuals and organizations in disaster management
Research and development division	Perform research on disaster patterns, vulnerability assessments, and innovative technologies to guide DMC's strategic decisions
Finance and administration division	Administer financial resources, budgeting, financial planning, and general administrative duties to ensure the DMC operates smoothly
Public awareness and education division	Conduct public awareness campaigns, educational initiatives, and outreach programs to inform the public about disaster risks and appropriate response measures
Coordination division	Coordinate the efforts of government agencies, NGOs, and international partners; promote collaboration for a cohesive and unified disaster response

20.5 Laws and Policies

20.5.1 Brief of Disaster Management Act No. 13 of 2005

The Disaster Management Act No. 13 of 2005 of Sri Lanka is the primary national legislation related to disaster risk management (Abdeen et al. 2021). The main goal of this Act is to safeguard the lives of the general public and the natural environment of the country from the negative impacts of natural hazards and/or disasters. This is achieved through the development of National Policies and plans, as well as the establishment of centrally coordinated committees and institutions tasked with implementing these policies and plans (Siriwardana et al. 2018a, b).

Section 2 of the Act provides provisions to ensure the disaster management structure of Sri Lanka which is spearheaded by a National Council for Disaster Management (NCDM) that is carried out through DMC (Sri Lanka Disaster Management Act no 13 of 2005). When contemplating the structure of the NCDM, Section 3 of the Act guarantees the inclusion of the President, prime minister, opposition leader, opposition representatives, ministers covering various subjects, and even provincial representatives as members of the council. This ensures a broad representation across different levels of governance (Sri Lanka Disaster Management Act 2005). This Act ensures the compliance of the 'government representation' principle of the Sendai Framework for Disaster Risk Reduction and establishes a strong and able body to manage disaster situations in Sri Lanka (Abeysinghe and De Mel 2019; Ministry of Disaster Management 2018).

A comprehensive as well as a clear account of the part and responsibilities of the NCDM are provided in Section 4 of the Act. Inter alia, the NCDM is responsible for responding to disaster emergency situations, formulating the National Disaster Management Policy, formulating and implementing the provisions of the National Disaster Management Plan and the provisions of the National Emergency Operation plan and the implementation of disaster management programmes. This Act also provides that the foremost role of the NCDM is to 'take all steps necessary to counter any disaster or impending disaster' and it can be achieved by implementing, coordinating and being accountable for the disaster management initiatives of a variety of institutions that belong to the disaster management framework. The Act further offers provisions for the NCDM with the sole and overarching responsibility and capacity to approve the plans and make decisions in relation to disaster-related management (Abeysinghe and De Mel 2019). Furthermore, NCDM in Sri Lanka, established by the Act, is facing practical challenges due to loopholes in the legislation.

Despite the Act's strong foundation in theory, the NCDM has not been able to fully fulfil its responsibilities. Specifically, the council's irregular meetings, which fall short of the mandated quarterly requirement and the misuse of quorum provisions, have been identified as key issues. For instance, only 13 meetings have been held from 2005 to December 2018, with an average attendance of six members instead of the required 14. Additionally, ministers' absence, substituted by representatives, has hindered timely decision-making at the cabinet level. These loopholes have led to prolonged delays in approving crucial disaster preparation, reduction, and mitigation strategies. Administrative failures, such as the delayed completion of the "Road Map for Disaster Management 2005–2015" and, the non-issuance of gazette notifications on government department disaster management plans, underscore the urgent need for improvements in Sri Lanka's National Legislation on "disaster risk management". This analysis was given based on the studies done by Abeysinghe and De Mel (2019) and Siriwardana et al. (2018a, b). Anyhow, it is evident that the establishment of the NCDM and its role is a major contribution towards a better disaster management framework in Sri Lanka.

20.5.2 Brief of Tsunami (Special Provisions) Act No. 16 of 2005

As a result of the 2004 Tsunami tragedy, significant social and legal changes occurred in Sri Lanka. The Tsunami Act emerged as a crucial piece of legislation, ensuring the rights of victims and their families. According to Section 2 of the Tsunami Act, if individuals were missing as of December 26, 2004, and had not been heard from for six months, the burden of proving their survival falls on those asserting they are alive. This provision addresses the unique challenges posed by the natural disaster and ensures legal recognition of the victims' status. Section 3 of this Act outlines the procedure for obtaining a certificate of death for individuals referred to in Section 2.

Section 5 specifies the offences under this Act, which include:

(a) Knowingly making a false statement in an application submitted under Section 3 or providing false information;
(b) Failing to inform the Registrar-General that a person, whose death has been registered based on an application under Section 3, is actually alive;
(c) Dishonestly or fraudulently using a death certificate issued under the Births and Deaths Registration Act when aware, or having reason to believe, that the person referred to in the certificate is still alive.

Those convicted of these offences by the High Court may face imprisonment for a term ranging from three to five years.

Section 7 clarifies the application of this Act, which includes:

(a) Every child and young person left an orphan;
(b) Every child left with a single parent unable to care for them; and
(c) Every young person left with a single parent who cannot provide necessary care and protection, as a result of the Tsunami that occurred on December 26, 2004.

A foster care order is a vital provision in this Act to safeguard the "best interests" of a child. According to Section 16, upon receiving the Panel's recommendations, and if the applicant is deemed suitable, the Magistrate's Court may issue a Foster Care Order. This order, with the consent of children over ten or young persons, appoints the specified applicants as foster parents. The order must adhere to the format and conditions outlined in the Eighth Schedule of the Act.

After issuing a Foster Care Order, according to Section 17 of the Act, foster parents should provide the child or young person, all the care and protection, exercise relevant control and discharge all the other obligations adequate education and any other special care that might be needed for emotional disturbance suffered by the child/the young person, acting as the natural parent of the foster child.

After the Tsunami hazard, land ownership issues became prominent. Section 30 of this Act addresses land possession affected by the Tsunami. If information about such land is received under Section 66 of the Primary Courts' Procedure Act, No. 44 of 1979, between December 26, 2004, and December 26, 2005, the court will determine possession. The Act stipulates that the person who had the possession on or before December 26, 2004, or their co-possessors if the original possessor is missing or deceased, is deemed to be the rightful possessor, and the judge will issue an order accordingly.

20.5.3 The National Disaster Management Policy

Section 4(a) of the Disaster Management Act No. 13 of 2005 of Sri Lanka issues guidelines to formulate a "National Disaster Management Policy" (SLNDMP) and a "National Disaster Management Plan" (NDMP) to ensure an effective disaster management framework in Sri Lanka. The SLNDMP serves as the main policy for disaster management and "Disaster Risk Reduction" (DRR) and is built on the global standards and the Sendai principles. The "National Policy on Disaster Management" sets out measures aimed at protecting the population, assets, and natural surroundings of Sri Lanka from calamities (Siriwardana et al. 2018a, b). This policy includes measures to ensure that infrastructure and economic development in disaster-prone areas remain strong. It also focuses on efficiently using resources for events related to preparation, response, prevention, relief, rehabilitation, and reconstruction. Furthermore, this policy aims to raise public awareness while providing training to empower individuals to safeguard themselves from natural hazards and/or disasters. It also emphasizes the importance of planning in relation to pre-disaster, preparedness and mitigation of disasters, and also maintaining and improving capabilities for post-disaster relief situations, recovery, and rehabilitation (National Policy on Disaster Management 2014).

Upon reviewing the guidelines for the policy formulation and its content, it is apparent that several shortcomings can be identified. Addressing these issues to ensure the clarity and helpfulness of the policy is also very important (Siriwardana et al. 2018a, b). It has been identified that the guidelines in the Act provide no direction that requires the SLNDMP to be formulated in a manner that encompasses the Sustainable Development Goals (SDG) and its disaster risk reduction strategies (Siriwardana et al. 2018a, b). Consequently, if SDGs like SDG 12 relating to food waste generation were considered in the policy, it is arguable that risks of disasters such as the Meethotamulla dumping site disaster could have been reduced (Ravishan 2017). Secondly, while the Act provides that the policy be formulated to ensure risk management and increase preparedness, the manner in which these requirements must be met is left vague to the viewer (Siriwardana et al. 2018a, b). Nevertheless, if the 'best fit for best practice' principle was included in the guidelines, legal provisions can be made to ensure that the traditional methods that have been effective long before disaster management frameworks, are incorporated with modern twists to reduce risks (Siriwardana et al. 2018a, b). Lastly, in the Act, there is no proper provision to encompass the 'Build back better' principle in the SLNDMP. Hence, it is argued that the policy fails to provide authority for a Post Disaster Assessment which is a vital tool in relation to reducing the risk of disasters and can be used as a measure of the effectiveness of the disaster management strategies in place (Abeysinghe and De Mel 2019; Siriwardana et al. 2018a, b).

20.5.4 The National Disaster Management Plan

Since 2005, two disaster management plans have been formulated and implemented. The Road Map for Disaster Risk Management was "Towards a Safer Sri Lanka" (2006–2016) and the National Disaster Management Plan for 2013–2017 (Disaster Management Centre 2024).

"The National Disaster Management Plan (NDMP) for 2022–2030" sketches a strategic, multi-disciplinary approach to establish a "disaster risk management framework" in Sri Lanka. It emphasizes collaboration across sectors, ministries, and agencies, the plan spans the entire disaster management cycle (Disaster Management Centre 2024). In addressing this plan, it is crucial to emphasize proactive measures before a disaster strikes and to promote a comprehensive understanding of risk within the framework of national policies. This involves investing in both structural and non-structural approaches to reduce disaster risks. Moreover, it is essential to prioritize disaster preparedness to ensure an efficient response and recovery. This includes implementing emergency response strategies during disasters and encompassing relief, recovery, assessment, rehabilitation, relocation, and reconstruction in the post-disaster phase. The long-term perspective includes sustained community resilience development through a "build back better" approach, emphasizing collaboration and participation from multiple stakeholders for effective implementation (Disaster Management Centre 2024). Figure 20.2 displays how the disaster management plans came into force from 2005.

20.6 Case Studies in Relation to Disaster Management in Sri Lanka

20.6.1 Post-Disaster Impact Assessment in Islamabath, Kalmunai—2008

The UNISDR team from the University of Peradeniya conducted a post-disaster impact assessment (PDIA) field survey in Islamabath Kalmunai (M.D), Sri Lanka on 04.10.2008. The objective was to assess the community's perception regarding the Tsunami disaster in the pre situations, during, and post-disaster periods. During the survey, the team visited both the original settlement and resettlement areas, interviewing 19 households and key informants using a direct sampling method. Several key findings emerged from the survey:

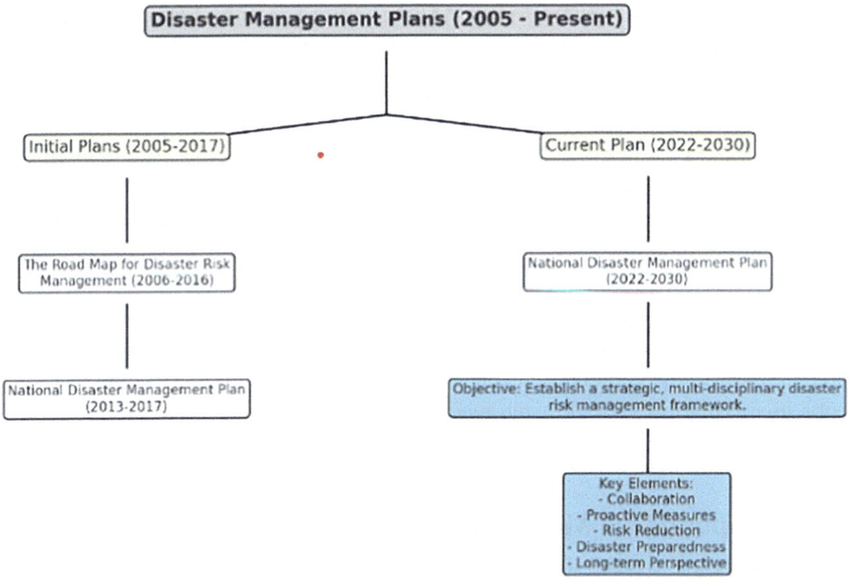

Fig. 20.2 Evolution of disaster management plan from 2005 to present (developed by the author)

1. **Housing and Resettlement**: Diverse housing conditions were observed, with some families living in small huts, while others received improved housing units. However, there were issues such as unclear ownership and improper distribution of houses, irregular and unmetered water and electricity supply, and unsatisfactory waste disposal. Furthermore, roads and drainage were not in a satisfactory condition.
2. **Security and Environment**: Inadequate security was noted in the area, along with environmental issues such as sand mining, garbage dumping, and pollution from construction and industries.
3. **Social and Cultural Aspects**: Lack of privacy and confidentiality in the resettlement areas, families scattered across different locations, presence of a liquor shop near the settlement, and dissatisfaction with the activities of government officials.
4. **Livelihood and Economy**: Lack of livelihood assistance for those in need, and some people who lost their livelihoods were unable to restart.
5. **Health and Education**: Limited healthcare facilities, unsatisfactory services of health officers, and nutritional standards not meeting the required levels. The survey recommended priority areas for interventions, including livelihood improvement, water supply and sanitation, completion of damaged houses, clarification of house ownership, provision of health facilities and security, construction of parapet walls and playgrounds, improvement of the drainage system, and rehabilitation of the handicapped and disabled.

20.6.2 Developing a Training Module for Stakeholders

In 2009, a collaborative project was initiated in relation to the 2004 Indian Ocean Tsunami, involving Kyoto University, University of Madras, University of Peradeniya and Colombo, and Institute of Technology, Bandung. The goal was to develop training modules for integration into university curricula, with input from local NGOs, governments, and school teachers. Refined through workshops and field feedback, eight modules were created. This effort led to the formation of the "Asian University Network of Environment and Disaster Management" (AUEDM). The initiative highlighted the importance of disaster risk reduction education involving diverse stakeholders and continuous curriculum updates, emphasizing field practices and community engagement. As Petal (2009) noted, interdisciplinary knowledge and problem-solving tools should continuously evolve beyond specific projects.

20.7 Proposals for More Effective Disaster Management Plans, Policies and Laws for the Sri Lankan Legal System

Based on the existing data and information, below are the proposals for a better disaster management plan, policy and laws for Sri Lanka (Fig. 20.3).

Inclusive Stakeholder Engagement: Engaging a diverse array of stakeholders is essential for the effective development, review, and refinement of disaster management policies. This approach encompasses government organizations, communities, NGOs, academic institutions, and private sector players. By fostering an inclusive decision-making process, we ensure that policies capture the comprehensive needs and viewpoints of all relevant parties. The rising focus on Community-based strategies at the stage of national disaster risk reduction (DRR) plans highlights the crucial role of communities in both disaster preparedness and recovery (Shaw et al. 2012).

Research on Disaster-Related Issues: There is a notable deficiency in the nation's focus on researching and analyzing critical but often overlooked aspects of disaster management. These include disaster finance, insurance, economic impacts, and psychological effects on victims, as noted by Amarakoon et al. (2024). Furthermore, exploring the interplay between climate change and disasters is imperative for a well-rounded understanding and advancement in disaster management.

Integration of Technology: Leveraging cutting-edge technologies such as artificial intelligence, remote sensing, and geographic information systems (GIS) is crucial for enhancing the accuracy and timeliness of data collection, early warning systems, and risk assessments. The incorporation of advanced technology significantly amplifies the effectiveness of disaster management strategies.

Legislation Review and Update: Regular evaluation of disaster management legislation is vital to identify and address deficiencies and opportunities for improvement. Updating laws to incorporate lessons learned from past events, emerging risks,

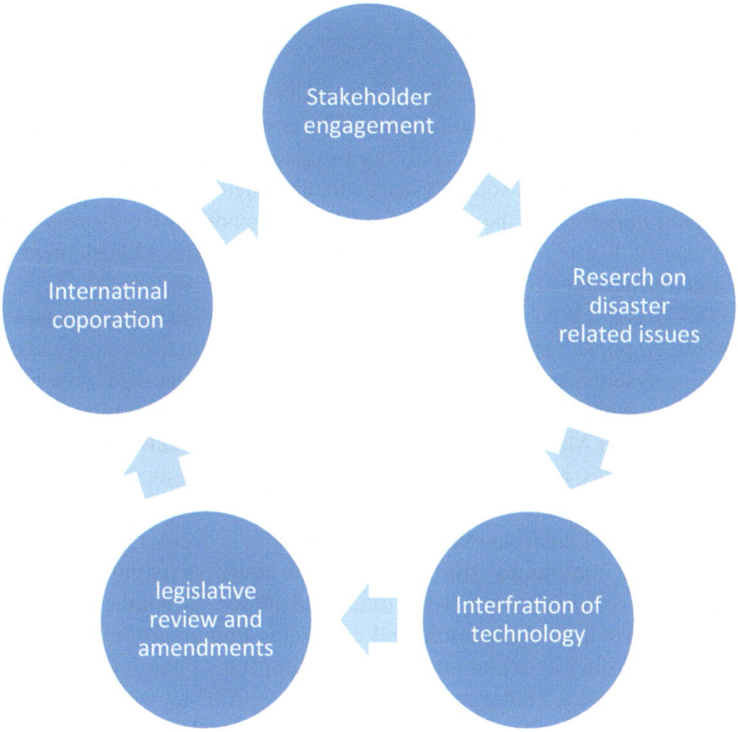

Fig. 20.3 Proposed mechanisms for effective disaster management (developed by the author)

and global best practices ensures that legal frameworks remain robust and adaptable to new challenges.

International Cooperation: Building and maintaining international partnerships is essential for strengthening disaster management policies. Learning from global experiences and best practices, and participating in regional and international networks, enhances a nation's ability to manage large-scale disasters and facilitates mutual support during crises.

By adopting these strategies, governments can significantly enhance their disaster management regulations and legislative frameworks. This proactive approach will not only strengthen their ability to effectively respond to and manage emergencies but also improve their overall readiness to protect lives, safeguard valuable assets, and preserve the environment during and after disaster events. By refining and implementing comprehensive policies, governments will be better equipped to handle the complexities of disaster scenarios, ensuring a more resilient and adaptive response to future challenges.

20.8 Conclusion

In conclusion, Sri Lanka's vulnerability to natural hazards is a pressing concern due to its unique geographical and climatic factors. The country's diverse landscape, exposure to tropical cyclones, and susceptibility to various hazards like floods, droughts and landslides necessitate a robust disaster management strategy. The devastating 2004 Tsunami underscored the urgent need for a comprehensive disaster management framework, leading to the establishment of the Disaster Management Centre (DMC) and amendments for the main Act No. 13 of 2005.

The DMC has since become a pivotal institution in enhancing preparedness for disasters and response capabilities. Its multifaceted approach includes developing early warning systems, coordinating disaster response, and leading recovery efforts. Despite these advancements, challenges persist, particularly concerning the National Council for Disaster Management's (NCDM) operational effectiveness. Issues such as irregular meetings and administrative delays highlight areas for improvement within the legislative framework.

Overall, while Sri Lanka has made significant strides in disaster management since the 2004 Tsunami, continuous efforts are needed to address existing gaps and ensure that the country is better prepared for future calamities. Strengthening legislative compliance, improving coordination, and enhancing community resilience remain crucial to safeguarding lives and minimizing the impacts of natural disasters.

Further, the Disaster Management Act No. 13 of 2005 as well as the Tsunami (Special Provisions) Act No. 16 of 2005 represent crucial legislative measures in Sri Lankan management efforts to enhance disaster risk management and address the aftermath of natural calamities. The Disaster Management Act delivers robust legal coverage for disaster risk management, with a structured approach through the National Council for Disaster Management (NCDM) in Sri Lanka. However, challenges like irregular meetings and administrative inefficiencies highlight the need for legislative and operational improvements. Meanwhile, the Tsunami Act addresses immediate post-disaster needs, including the welfare of victims and their families, and manages land disputes resulting from the disaster. Both Acts demonstrate Sri Lanka's commitment to improving its disaster management systems and providing legal remedies and support to affected individuals. Moving forward, addressing the identified shortcomings and enhancing the implementation of these laws will be crucial in building a more resilient and effective disaster management framework in Sri Lanka.

Moreover, a multifaceted approach to disaster management is essential for building robust and effective systems that can tackle both present and future challenges. By engaging a diverse range of stakeholders, emphasizing research on critical yet often neglected aspects of disaster management, integrating advanced technologies, regularly updating legislation, and fostering international cooperation, governments can enhance their preparedness and response capabilities. Such strategies ensure that disaster management policies are not only comprehensive but also adaptable to evolving risks and conditions. Embracing these practices will ultimately lead to a more resilient society, better equipped to protect lives, safeguard assets, and maintain environmental integrity in the face of disasters.

References

Abdeen FN, Fernando T, Kulatunga U, Hettige S, Ranasinghe KDA (2021) Challenges in multi-agency collaboration in disaster management: A Sri Lankan perspective. Int J Disaster Risk Reduct

Abeysinghe B, Iwao Y, Bandara S (2004) Development of a risk assessment methodology for landslides in Sri Lanka. Nihon Jisuberi Gakkaishi 41(2):131–141. https://doi.org/10.3313/jls.41.131

Abeysinghe S, De Mel N (2019) Disaster management in Sri Lanka: A case study of administrative failures (briefing note), Verite Research

Alahacoon N, Matheswaran K, Pani P, Amarnath G (2018) A decadal historical satellite data and rainfall trend analysis (2001–2016) for flood hazard mapping in Sri Lanka. Remote Sens 10(3):448. https://doi.org/10.3390/rs10030448

Amarakoon V, Trafford J, Udeshika T, Amarasekara DS, Wickramasinghe D (2024) Towards happy relief campers: Surfacing psycho-social issues, conflicts and other problems for flood-affected residents and officials in Kuruwita, Sri Lanka. Int J Disaster Risk Reduct 101. https://doi.org/10.1016/j.ijdrr.2024.104259

Bae Y, Joo YM, Won SY (2016) Decentralization and collaborative disaster governance: Evidence from South Korea. Habitat Int 52:50–56

Bhardwaj P, Singh O (2019) Climatological characteristics of Bay of Bengal tropical cyclones: 1972–2017. Theoret Appl Climatol 139(1–2):615–629. https://doi.org/10.1007/s00704-019-02989-4

Caruson K, MacManus SA (2012) Interlocal emergency management collaboration: vertical and horizontal roadblocks. Publius 42(1):162–187

Chandrasekara SSK, Uranchimeg S, Kwon H, Lee SO (2018) Coastal flood disaster in Sri Lanka-May 2017: Exploring distributional changes in rainfall and their impacts on flood risk. J Coastal Res 85:1476–1480. https://doi.org/10.2112/si85-296.1

Dang K, Sassa K, Konagai K, Karunawardena A, Bandara RMS, Hirota K, Tan Q, Ha ND (2018) Recent rainfall-induced rapid and long-traveling landslide on 17 May 2016 in Aranayaka, Kagelle District, Sri Lanka. Landslides 16(1):155–164. https://doi.org/10.1007/s10346-018-1089-7

De Silva M, Kawasaki A (2018) Socioeconomic Vulnerability to disaster risk: A case study of flood and drought impact in a rural Sri Lankan community. Ecol Econ 152:131–140. https://doi.org/10.1016/j.ecolecon.2018.05.010

Department of Census and Statistics (2021). http://www.statistics.gov.lk/

Disaster Management Centre (2014) National disaster management plan 2013–2017

Disaster Management Centre (2022) National Disaster Management Plan 2022–2030

Disaster Management Centre (2017) Situation report

Disaster Management Centre (2024). http://www.dmc.gov.lk/index.php?option=com_content&view=featured&Itemid=101&lang=en

Government of Sri Lanka (2005) Report of Sri Lankan parliament select committee on natural disasters. https://www.parliament.lk/uploads/comreports/1476092317082517.pdf

Gunda T, Hornberger GM, Gilligan JM (2015) Spatiotemporal patterns of agricultural drought in Sri Lanka: 1881–2010. Int J Climatol 36(2):563–575. https://doi.org/10.1002/joc.4365

Hapuarachchi HAP, Mendıs SS (eds) (2018) Disaster management in Sri Lanka: A comprehensive approach. Springer

Heenatigala M (2021) Influence of climate change on forest fire occurrence and distribution of Sri Lanka and modeling of forest fire. https://doi.org/10.15760/mem.71
Hemachandra D (2010a) Status of disaster management in Sri Lanka, pp 1–210. http://www.parliament.lk/files/pdf/constitution.pdf
Hemasinghe H, Rangali R, Deshapriya NL, Samarakoon L (2018) Landslide susceptibility mapping using logistic regression model (a case study in Badulla District, Sri Lanka). Procedia Eng 212:1046–1053. https://doi.org/10.1016/j.proeng.2018.01.135
Hettiarachchi S, Weeresinghe S (2014) Achieving disaster resilience through the Sri Lankan early warning system: Good practises of disaster risk reduction and management. Procedia Econ Financ 18:789–794. https://doi.org/10.1016/s2212-5671(14)01003-x
International Organization for Migration (2016) Sri Lanka floods and landslides—Situation report
Jayathissa G, Schröder D, Balke K, Fecker E (2019) Landslide hazard zonation in Sri Lanka: An assessment of manual and GIS based automated procedure in preparation of Geology weight map. Int J Eros Control Eng, 11(4):116–123. https://doi.org/10.13101/ijece.11.116
Jayawickrama N (2015) Disaster management in Sri Lanka: Challenges and strategies. Disaster Prev Manag 24(3):401–414. https://doi.org/10.1108/DPM-12-2014-0265
Kaplan M, Renaud FG, Lüchters G (2009) Vulnerability assessment and protective effects of coastal vegetation during the 2004 Tsunami in Sri Lanka. Nat Hazard 9(4):1479–1494. https://doi.org/10.5194/nhess-9-1479-2009
Karunarathne AY, Gress DR (2022) The role of organizational networks in ameliorating flood disaster impacts: A case study of flood inundated rural and urban areas in Sri Lanka. Int J Disaster Risk Reduct 71:102819. https://doi.org/10.1016/j.ijdrr.2022.102819
Kumara SBA, Premasiri RHM (Eds) (2016) Analysis of spatial distribution of landslides in Ratnapura district Sri Lanka. In: Proceedings of the international conference on multidisciplinary approaches. http://dr.lib.sjp.ac.lk/handle/123456789/5919
Kumar SP, Wickramasinghe MS (2019) Climate change and disaster risk management in Sri Lanka. Routledge
Math SB, Nirmala MC, Moirangthem S, Kumar N (2015) Disaster management: Mental health perspective. Indian J Psychol Med 37(3):261–271. https://doi.org/10.4103/0253-7176.162915
Ministry of Disaster Management (2018) Annual performance report
Ministry of Disaster Management (2014) Sri Lanka comprehensive disaster management programme
Ministry of National Policies and Economic Affairs (2017) Sri Lanka rapid post disaster needs assessment. Ministry of Disaster Management. https://lk.one.un.org/wp-content/uploads/2017/11/PDNA-Sri-lanka-2017.pdf
Mitra A, Shaw R (2022) Systemic risk from a disaster management perspective: A review of current research. Soc Sci Res Netw. https://doi.org/10.2139/ssrn.4140988
Mohanty A, Dubey A, Singh RB (2022) The major cyclonic disasters in Sri Lanka. In: Cyclonic disasters and resilience, pp 129–153. https://doi.org/10.1007/978-981-19-1215-3_5
Mulligan M, Nadarajah Y (2011) Rebuilding community in the wake of disaster: lessons from the recovery from the 2004 tsunami in Sri Lanka and India. Community Dev J 47(3):353–368. https://doi.org/10.1093/cdj/bsr025
National Council for Disaster Management (2014) National Policy on Disaster Management
Nisansala WDS, Abeysingha NS, Islam A, Bandara AMKR (2019) Recent rainfall trend over Sri Lanka (1987–2017). Int J Climatol 40(7):3417–3435. https://doi.org/10.1002/joc.6405
Patankar A (2017) Colombo: Exposure, vulnerability, and ability to respond to floods. https://doi.org/10.1596/1813-9450-8084
Petal M (2009) Education in disaster risk reduction, in disaster management: Global challenges and local solutions. In: Shaw R, Krishnamurthy R (eds), pp 285–304
Perera ENC, Jayawardana D, Jayasinghe P, Bandara RMS, Alahakoon N (2018) Direct impacts of landslides on socio-economic systems: a case study from Aranayake, Sri Lanka. Geoenvironmental Disasters 5(1). https://doi.org/10.1186/s40677-018-0104-6

Perera KK (2020) The impact of natural disasters on Sri Lankan communities: Case studies and lessons learned. Springer

Ranasinghe AKRN, Bandara RMS, Puswewala UGA, Dammalage TL (2019) Efficacy of using radar induced factors in landslide susceptibility analysis: Case study of Koslanda, Sri Lanka. Nat Hazards Earth Syst Sciences. https://doi.org/10.5194/nhess-2018-335

Ranatunge E, Malmgren BA, Hayashi Y, Mikami T, Morishima W, Yokozawa M, Nishimori M (2003) Changes in the Southwest Monsoon mean daily rainfall intensity in Sri Lanka: relationship to the El Niño-Southern oscillation. Palaeogeogr Palaeoclim Palaeoecol 197(1–2):1–14. https://doi.org/10.1016/s0031-0182(03)00383-3

Ratnayake U, Herath S (2005) Changing rainfall and its impact on landslides in Sri Lanka. J Mt Sci 2(3):218–224. https://doi.org/10.1007/bf02973195

Ravishan (2017) A brief history of the meethotamulla garbage dump. Roar reports. [Online]. https://roar.media/english/reports/reports/a-brief-history-of-the-meethotamulla-garbage-dump

Shaw R, Takeuchi Y, Krishnamurthy RR, Pribadi KS, Nianthi R (2009) Collaborative post tsunami action research: Lessons and reflections. Asian J Environ Disaster Manag, 1(113–124c). Research Publishing Services. https://doi.org/10.3850/S179392402009000064

Rossetto T, Peiris N, Pomonis A, Wilkinson S, Del Re D, Koo R, Gallocher S (2006) The Indian Ocean tsunami of December 26, 2004: Observations in Sri Lanka and Thailand. Nat Hazards 42(1):105–124. https://doi.org/10.1007/s11069-006-9064-3

Schott F, Reppin J, Fischer J, Quadfasel D (1994) Currents and transports of the monsoon current south of Sri Lanka. J Geophys Res 99(C12):25127–25141. https://doi.org/10.1029/94jc02216

Senaratne RS, Silva SS (2017) Resilience and disaster management in Sri Lanka: Policy and practice. J Disaster Res 12(5):785–797. https://doi.org/10.20965/jdr.2017.p0785

Sri Lanka Disaster Management Act, No. 13 of 2005

Senatilleke U, Abeysiriwardana HD, Makumbura RK, Anwar F, Rathnayake U (2022) Estimation of potential evapotranspiration across Sri Lanka using a distributed dual source evapotranspiration model under data scarcity. Adv Meteorol 2022:1–14. https://doi.org/10.1155/2022/6819539

Shaw R, Takeuchi Y, Krishnamurthy RR, Pereira JJ, Mallick F (2012) Chapter 4 universities and community-based disaster risk reduction. In: Community, environment and disaster risk management, pp 55–66. https://doi.org/10.1108/s2040-7262(2012)0000010010

Shelton S, Ogou FK, Pushpawela BG (2021) Spatial-temporal variability of droughts during two cropping seasons in Sri Lanka and its possible mechanisms. Asia-Pac J Atmos Sci 58(1):127–144. https://doi.org/10.1007/s13143-021-00239-0

Shibayama T, Ashton MS, Singhakumara B, Griscom HP, Ediriweera S, Griscom BW (2006) Effects of fire on the recruitment of rain forest vegetation beneath Pinus caribaea plantations. Sri Lanka for Ecol Manag 226(1–3):357–363. https://doi.org/10.1016/j.foreco.2006.01.016

Siriwardana CSA, Jayasiri GP, Hettiarachchi SSL (2018a) Investigation of efficiency and effectiveness of the existing disaster management frameworks in Sri Lanka. Procedia Eng:1091–1098

Siriwardana C, Jayasiri G, Hettiarachchi S (2018b) Investigation of efficiency and effectiveness of the existing disaster management frameworks in Sri Lanka. Procedia Eng 212:1091–1098. https://doi.org/10.1016/j.proeng.2018.01.141

Sri Lanka: Drought—Aug 2023 (2023) ReliefWeb. https://reliefweb.int/disaster/dr-2023-000151-lka

Srisangeerthanan S, Lewangamage C, Wickramasuriya SS (2015) Tropical cyclone damages in Sri Lanka. Wind Eng, JAWE 40(3):294–302. https://doi.org/10.5359/jawe.40.294

The World Bank Group (2021) Sri Lanka—vulnerability. Climate change knowledge portal. Retrieved June 13, 2022, from https://climateknowledgeportal.worldbank.org/country/sri-lanka/vulnerability

The World Bank (2021) FROM crisis to green, resilient, AND inclusive recovery. Jurnal Economi 114

United Nations Office for Disaster Risk Reduction (2015) Sendai framework for disaster risk reduction 2015–2030. https://www.unisdr.org/files/43291_sendaiframeworkfordrren.pdf

Wanninayake WMSB (2018) Roles and responsibilities of government stakeholder agencies in managing disasters–reviewing Sri Lankan experience. J Trop Environ 1(1)

Chapter 21
Determination of Constraints in Existing DRR Coordination Mechanisms Among Different Humanitarian Actors of Coastal Bangladesh

Md. Shamsuzzoha, A. K. M. Abdul Ahad Biswas, Md. Mahmudul Hasan, Md. Faisal, Papri Hazra, Rajan Chandra Ghosh, Md. Rasheduzzaman, Promi Debnath, Md. Abdur Rahman, and Rajib Shaw

Abstract Bangladesh is placed fourth in the Asia–Pacific region for risk factors and has been identified as one of the most disaster-prone nations in terms of extreme climate change occurrences. The study was undertaken in the coastal Barguna district of Bangladesh, to identify different disaster risk reduction (DRR) actors in humanitarian sectors; existing constraints of DRR coordinating mechanisms among different stakeholders; to formulate and suggest a framework to resolve the problems related to DRR coordination mechanisms. Results revealed that different humanitarian actors (HA) in DRR-based active working sectors of Bangladesh. The selected

Md. Shamsuzzoha (✉) · R. C. Ghosh · Md. Rasheduzzaman
Department of Emergency Management, Faculty of Environmental Science and Disaster Management, Patuakhali Science and Technology University, Dumki, Bangladesh
e-mail: zohageo@pstu.ac.bd

A. K. M. A. A. Biswas · P. Debnath
Department of Disaster Risk Management, Faculty of Environmental Science and Disaster Management, Patuakhali Science and Technology University, Dumki, Bangladesh

Md. M. Hasan
Department of Geo-Information Science and Earth Observation, Faculty of Environmental Science and Disaster Management, Patuakhali Science and Technology University, Dumki, Bangladesh

Md. Faisal
Department of Disaster Resilience and Engineering, Faculty of Environmental Science and Disaster Management, Patuakhali Science and Technology University, Dumki, Bangladesh

P. Hazra
Department of Environmental Science, Faculty of Environmental Science and Disaster Management, Patuakhali Science and Technology University, Dumki, Bangladesh

Md. A. Rahman
Office of the Senior Consultant, Knowledge Craft Bangladesh (KCB), Dhaka, Bangladesh

R. Shaw
Graduate School of Media and Governance, Keio University, Fujisawa, Japan

R. Shaw et al. (eds.), *Two Decades from the Indian Ocean Tsunami*, Disaster Risk Reduction, https://doi.org/10.1007/978-981-96-2669-4_21

DRR actors were the Department of Agricultural Extension (DAE), Department of Fisheries (DoF), Department of Livestock Services (DLS), Bangladesh Rural Development Board (BRDB), Directorate General of Health Services (DGHS), Department of Public Health Engineering (DPHE), Directorate General of Family Planning (DGFP), different Non-governmental Organization (NGO), and International non-governmental organizations (INGOs). Existing gaps found among the acting humanitarian actors related to DRR coordination mechanisms made it difficult to find strong coordination and relationship practices among the different actors; no master plan was prepared to perform DRR activities; no established structure had been found to coordinate among the DRR actors; more emphasis on own reputation rather than community services and lower practices of database preservation related to DRR practices by humanitarian actors. Existing gaps found among the different humanitarian actors and community people related to DRR coordinating mechanisms were: the respondents of M. Balia Toli, Bodor Khali, and Noltona *Union* opined that the community people were not selected as a representative of any group or committee; community people not received sufficient seeds, fertilizer and other agricultural inputs for emergency recovery due to their vulnerability reasons; inappropriate dissemination of early warning through locally activated humanitarian organization before disaster and the humanitarian actors provided insufficient evacuation assistance in taking rescue shelters. The constraints faced by the community people related to DRR coordination were a lack of opportunities for DRR-related training; early warning practices, participation in community awareness programs; unwillingness to share knowledge and skills regarding DRR with community group meetings; lack of access to digital media (radio and television) and mistrust on warning and forecasting with showing no interest for moving to shelter. Considering the different limitations in practicing DRR activity found in the current study, an appropriate selection of specific problems, limitations, and actors-based coordination mechanisms might be the best approach to solve existing problems faced by the different HA in the district disaster management committee (DDMC) and District Relief and Rehabilitation Office (DRRO). A draft framework was proposed to get the basic guidelines to reduce gaps in coordination mechanisms among the different actors to achieve minimum humanitarian standards in disaster response.

Keywords Disaster risk communication · Disaster risk reduction · Humanitarian actors · Humanitarian standards · Coordination mechanisms · Coastal Bangladesh

21.1 Introduction

21.1.1 Background of the Study

Humanitarian actors for disaster response are non-profit civilian organizations, whether national or international, United Nations (UN)-based organizations, that commit to humanitarian standards, protection principles, humanitarian charters, and sustainable development activities (Sphere Association 2011, 2018). Humanitarian actors share a commitment to working according to the disaster risk reduction (DRR) four priorities, the sphere project's humanitarian charter, protection, minimum standards in disaster response, and other recognized humanitarian guidelines and actions (CPHA 2019; Ersel 2015; Sphere Association 1998, 2004; UNDRR 2015).

The methodical application of policy tools to provide humanitarian aid in a coordinated and efficient manner is known as humanitarian coordination (Pothiawala 2015). Such policy tools include (a) planning strategically; (b) collecting information and handling data; (c) assembling assets and guaranteeing responsibility; (d) putting up an effective division of labour in the industry; (e) establishing a workable framework and negotiating with the political authorities; and (f) contributing leadership to provide humanitarian and emergency response. These policy tools, when used sensibly and delicately, will add a measure of discipline without overly restricting behaviour in DRR (CPWG 2013).

Several structures play a key role in humanitarian coordination in Bangladesh, but there is currently no clear or comprehensive vision for how these fit together (Biswas et al. 2015; Krishnan et al. 2022; Sattar et al. 2020). There are many parts of the system, and each set of humanitarian actors (Government, UN, donors, local NGOs, international NGOs, sector groups/clusters) needs to be internally well-coordinated (Matin and Taher 2001; Rahman and Tasnim 2023), and also strong mechanisms that bring those parts of the system together for policy dialogue, information sharing, and coordinate plans and responses to disasters are needed (Ray-Bennett et al. 2020).

The humanitarian response group for emergencies has played a central coordinating role in DRR (Feng and Cui 2021). The response group works as the key mechanism stakeholders of the humanitarian charter and minimum standards of humanitarian action (Sphere Association 2000). While the DRR group was the primary place where stakeholders looked for coordination between the government and the international community, other groups mentioned related to coordination were (i) the Government of Bangladesh (GoB), (ii) the Ministry of Disaster Management and Relief (MoDMR), (iii) the National Disaster Management Team (NDMT), (iv) related sector coordination/clusters, (v) NGO Coordination Meeting (NGOCM) at district level, (vi) NGO and INGO forums (politically aligned with the major political parties), (vii) other INGO associations (GoB 2010, 2017, 2020a, b; Islam et al. 2013; Planitz 1997).

A humanitarian crisis (or 'humanitarian disaster') is defined as a singular event or a series of events that are threatening in terms of the health, safety, or well-being of a community or large group of people (Ahmed 2018; Cook and Ne 2018). Figure 21.1

shows the institutional actors and stakeholders in the DRR and crisis management. It may be an internal or external conflict and usually occurs throughout a large land area (Perwaiz et al. 2020). Local, national, and international responses are necessary for such events (Alam and Hoque 2022; Houston et al. 2015; Raj, Ullah, & Akhter, 2010). The DRR program for Bangladesh as one of the signatory nations (Zaman et al. 2020) of Sendai Framework for Disaster Risk Reduction (SFDRR) 2015–2030, aims at creating a value-added by filling some of the identified gaps and responding to needs in the field of awareness building on the local level (Azad et al. 2019), capacity development, and mainstreaming DRR into development projects and programs (R. Shaw et al. 2009).

By doing the responsibilities of DRR, some gaps were found regarding the coordination mechanisms of multiple humanitarian actors like governmental, nongovernmental organizations, international organizations, private companies, and others (Basher 2013; Kabra et al. 2015; Masoomi et al. 2023; United Nations Resident Coordinator Office 2022). Additionally, each operational actor had distinctive characteristics relating to existing political and socioeconomic factors within socio-religious and cultural conditions (Chakrabarty 2022; Kabra et al. 2015; Kulatunga et al. 2014; Li et al. 2024).

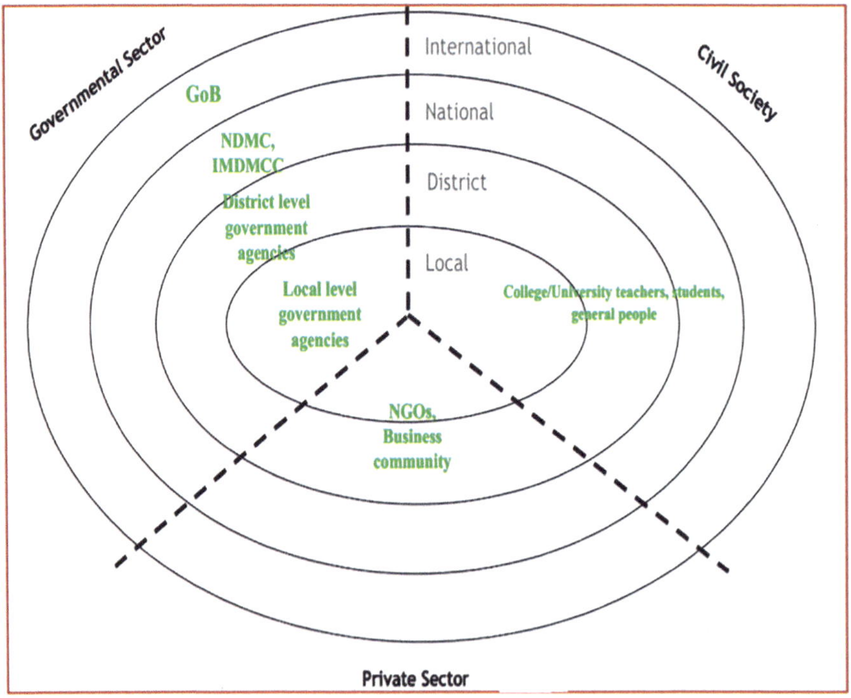

Fig. 21.1 Institutional actors and stakeholders in DRR (Zimmermann 2010)

21.1.2 Justification of the Proposal

Over the last two decades, Bangladesh has shifted from reactive interventions to proactive disaster risk reduction approaches during its disaster risk management activities (Azad et al. 2019; Chaiya et al. 2022; Islam 2023a, b; Naim et al. 2024; Nasreen 2021; Shamsuzzoha et al. 2024a). The results were apparent in the minimum loss of lives and livelihoods destroyed by natural disasters (Faisal et al. 2021; Matsuda 1993; Saha et al. 2024; Shamsuzzoha et al. 2022). Bangladesh is now a global model in disaster management and DRR practices in disaster risk and emergency management (Rahman 2013). Meanwhile, there were vast knowledge and coordination gaps in humanitarian intervention during any phase of a disaster situation such as (i) preparedness/readiness, (ii) response, (iii) recovery and (iv) mitigation/reduction (Feng and Cui 2021; Gaillard and Mercer 2013; Guo et al. 2024; Xie and Chen 2024). Inadequate studies have been conducted on the DRR coordination mechanisms of humanitarian actors. The DRR-based humanitarian actors ultimately aimed to fulfill the principles of the Sendai Framework of DRR, and the Sphere Projects approaches on coordination mechanisms within and across sectors and with relevant stakeholders at all levels (Sphere Association 2018; van Boetzelaer et al. 2024).

21.1.3 Objectives of the Research

21.1.3.1 General Objectives

The main goal of this study was to develop a framework for guidance and recommendations for resolving the current humanitarian issues with DRR coordinating mechanisms among various humanitarian actors at the sub-district level in Bangladesh, including GOs, NGOs, and INGOs.

21.1.3.2 Specific Objectives

i. To identify different humanitarian actors in the DRR coordinating mechanisms of the study area.
ii. To identify the existing constraints of DRR coordinating mechanisms among different humanitarian stakeholders and,
iii. To formulate and suggest a DRR-based framework to solve the existing problems of DRR coordination mechanisms.

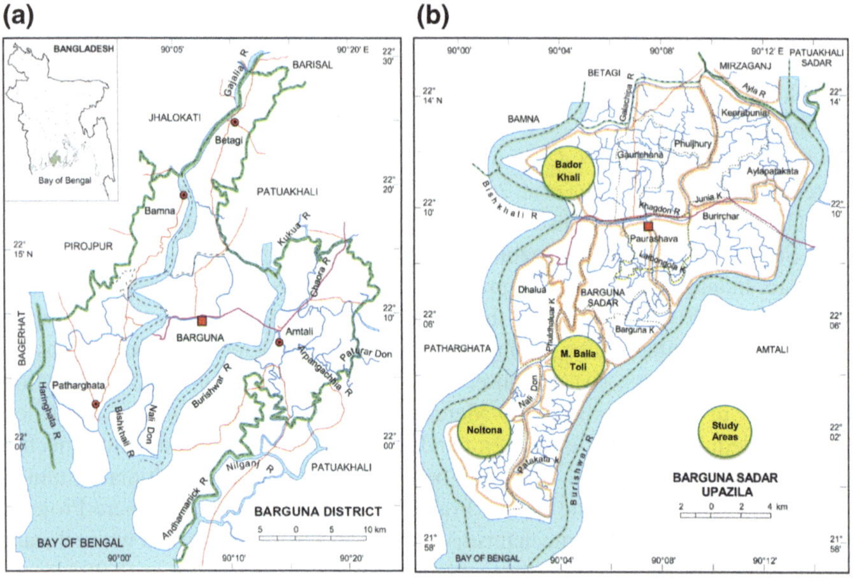

Fig. 21.2 Study area map: Barguna District (GoB 2024a). **b** Study area map: Barguna *Sadar Upazila* (GoB 2024b)

21.2 Materials and Method

21.2.1 Site Selection

Barguna district is situated between 22°29′ N and 22°36′ N latitudes and between 89°52′ E and 90°22′ E longitudes (GoB 2024a) (Fig. 21.2a). The Barguna *Sadar Upazila* (sub-district) of the study district was selected as the study area of the study (GoB 2024b) (Fig. 21.2b). This coastal sub-district is prone to natural climate extremes (Shamsuzzoha and Ahamed 2023, 2024; Shamsuzzoha and Al-Maruf 2012). Therefore, this area was suitable for mapping out the determination of constraints in existing DRR coordination mechanisms among different humanitarian actors (Shaw et al. 2013a).

21.2.2 Rationale for Selecting the Research Site

The Barguna *Sadar Upazila* was selected as one of the most affected sub-districts due to the severity of climatic hazards. Barguna *Upazila* comprises 10 *Unions* (BBS 2024; GoB 2024b). The study was conducted in three local administrative units (known as *Unions*) of the sub-district, and they were (i) M. Balia Toli, (ii) Bodor Khali, and (iii) Noltona (Fig. 21.2b). Natural disasters like tropical cyclones, storm surges, tidal floods, salinity incursions, and so forth happen violently nearly every year, which is why these unions were selected. Before the selection of these administrative *Unions*, a face-to-face discussion with the concerned GO, NGO personnel, and community leaders was conducted by the data collection team of the study. Subsequently, those three *Unions* as the lower administrative units of the country were selected for the research. Barguna *Upazila* is one of the most disaster-affected coastal regions of the Barishal Division in Bangladesh. People of this locality face different kinds of natural calamities every year (Rasheduzzaman et al. 2020; Shamsuzzoha et al. 2021; Shamsuzzoha et al. 2024b; Talukder et al. 2018). Therefore, various humanitarian issues are raised, and the local humanitarian actors of the sub-district play a vital role in managing the humanitarian response issues. However, there are vast knowledge gaps among those actors. But Barguna *Sadar Upazila* is geographically vulnerable to climatic disasters, with high risks to the local community, and has limitations for coping with climatic triggering extreme events. The disaster risks and emergency response-oriented knowledge with experiences of humanitarian issues of the selected three *Unions* were also explored. Therefore, the research team selected the research site for the study (Hasan et al. 2021).

21.2.3 Sample Size Determination

The sample size for this research was determined using the following formula (Yamane 1967).

$$\mathbf{n} = \mathbf{N}/\left(\mathbf{1} + \mathbf{N}\mathbf{e}^{2}\right)$$

where n = Sample size

N = Total number of populations.

e = Level of precision (5%).

N = Population.

21.2.4 Research Methods

The study applied a mixed approach combining qualitative and quantitative methods for data collection, analysis, and discussions (Mukherjee et al. 2022; Shaw et al. 2013a, b, c, 2016). Necessary primary data was collected through an extensive questionnaire survey at the household and institutional levels in the three coastal areas of the Barguna district. Face-to-face interviews were conducted with various key informants to learn about the study area, constraints in existing DRR coordination mechanisms, different humanitarian actors in vulnerabilities, adaptation strategies, and the socio-economic conditions of households (Habiba et al. 2013; Shaw et al. 2013b). Before the questionnaire survey, field observation was conducted with a scientific and analytical mindset to comprehend the overall state and real-world circumstances of the study area for the purpose of choosing households according to their various occupational statuses (e.g., laborers, farmers, fishermen, businessmen, service holders from NGOs and GOs, etc.). This involved gathering pertinent information and data from all accessible published and unpublished research works, books, significant office records, governmental documents, web publications, etc.

Therefore, primary data was collected by three research associates from 150 respondents of separate households of the three study *Unions,* respectively, through a structured questionnaire. To fulfill the second and third objectives, this study conducted five focus group discussions (FGDs) and five (5) key informant interviews (KIIs). Key informant interview (KII) was conducted with related humanitarian stakeholders such as local community peoples, Disaster Management Coordination Committee (DMCC), government officials, community leaders, and different national and international non-governmental organization officers to make it more authentic (Sadik et al. 2018; Shaw et al. 2018). The study was also justified by the collected secondary data and literature review (Arab et al. 2022; Kulatunga et al. 2013; Shamsuzzoha et al. 2018). At the local community level, focus group discussions were conducted with two groups from the local community; one comprising only women and another containing only men; with a checklist about their predisposed vulnerability, risks, assets, livelihoods, opportunities, relationships of trust, membership of groups, bonding, bridging, and linking or networks, access to wider institutions and obstacle's towards DRR coordination mechanisms of the coastal region of Bangladesh (Ali 1996; Rasheduzzaman et al. 2020; Salman 2023).

21.2.5 Data Analysis

Numerous diagrams, charts, and tables were used for the research to present significant data analysis results (Anson et al. 2017). The Microsoft Office Excel 10 and SPSS 19 software programs were used for the required statistical analysis and assessments. Meanwhile, a descriptive discussion was made to elucidate qualitative data and answer the research question. Finally, research results with a summary and conclusion were presented in existing DRR coordination mechanisms among different humanitarian actors of Barguna *Sadar Upazila* (GoB 2024b).

21.3 Results and Discussion

21.3.1 Identification of Different Humanitarian Actors of the Barguna Sadar Upazila

This study was performed to identify different humanitarian actors in the DRR coordinating mechanisms of the study Barguna *Sadar Upazila* area. Different humanitarian actors have performed their work toward Disaster Risk Reduction (DRR) in the coastal zone of Bangladesh. The identified actors were functional in emergency preparedness, response, mitigation, and adaptation-based DRR practices to save local people by operating their sub-district branches (*Upazila* offices) through their administrative hierarchy (Table 21.1).

21.3.2 Existing Constraints of DRR Coordinating Mechanisms Among Different Stakeholders

The study found the existing constraints of DRR coordinating mechanisms among different stakeholders in three aspects; the first aspect was on gaps identification between DRR coordinating mechanisms among the different stakeholders; the second aspect was on gaps minimization between DRR coordinating mechanisms among various stakeholders and community people; and third aspect was for finding constraints of community people in DRR coordinating mechanisms.

21.3.2.1 Gaps Between DRR Coordinating Mechanisms Among Different Humanitarian Stakeholders

The study found that stakeholders do not maintain good relationships among them. There was no master plan to perform DRR works and show a tendency to highlight them. There was no third party or structure to coordinate DRR actors. Organizations think only of their reputation rather than community service. Humanitarian actors had a shortage of databases that work with DRR.

21.3.2.2 Gaps Between DRR Coordinating Mechanisms Among Different Humanitarian Stakeholders and Community People

The organization failed to properly disseminate early warning prior to the disaster; individuals in the community were not elected or chosen as representatives of any group committee; they were not given enough loans, seeds, fertilizer, and other

Table 21.1 Different humanitarian actors and their sub-district branches (*Upazila* offices) at the Barguna Sadar *Upazila*, Barguna District, Bangladesh

Serial no	DRR actors	Authority/Origin
i	District Relief and Rehabilitation Office (DRRO)	Ministry of Disaster Management and Relief, GoB
ii	Department of Agricultural Extension (DAE)	Ministry of Agriculture, GoB
iii	Department of Fisheries (DoF)	Ministry of Fisheries and Livestock, GoB
iv	Department of Livestock Services (DLS)	Ministry of Fisheries and Livestock, GoB
v	Bangladesh Rural Development Board (BRDB)	Rural Development and Co-operatives Divisions, GoB
vi	*Union Parishad* (UP, the Smallest Administrative Unit of Bangladesh)	Local Government Division, GoB
vii	Directorate General of Health Services (DGHS)	Ministry of Health & Family Welfare, GoB
viii	Dept. of Public Health Engineering (DPHE)	Ministry of Health & Family Welfare, Bangladesh
ix	Bangladesh *Krishi* (Agriculture) Bank (BKB)	A Government-owned Bank, GoB
x	Directorate of Family Planning (DFP)	Medical Education and Family Welfare division, Ministry of Health & Family Welfare, GoB
xi	Department of Social Services (DSS)	Ministry of Social Welfare, GoB
xii	World Food Program (WFP)	Intergovernmental organization of the UN
xiii	Bangladesh Red Crescent Society (BDRCS)	The largest humanitarian non-governmental organization in the world, affiliated with the International Red Cross and Red Crescent Movement
xiv	Grameen Bank (Village Bank)	Statutory public authority, Bangladesh
xv	Bangladesh Rural Advancement Committee (BRAC)	A National level Non-profit NGO in Bangladesh
xvi	Community Development Centre (CODEC)	A National level Non-profit NGO in Bangladesh
xvii	The Voluntary Organization for Social Development (VOSD)	A National level Non-profit NGO in Bangladesh
xviii	Sangathita Gramunnyan Karmasuchi (Sangram)	A Local-level NGO of Barishal Division, Bangladesh
xix	Nazrul Smriti Sangsad-NSS	A Local-level NGO of Barguna District, Bangladesh

(continued)

Table 21.1 (continued)

Serial no	DRR actors	Authority/Origin
xx	Resource Development Foundation (RDF)	A Local-level NGO of Barguna District, Bangladesh
xxi	Islamic Relief, Bangladesh	International NGO
xxii	Caritas, Bangladesh	The Bangladeshi Catholic Bishops' Conference oversees this Caritas Internationalis member
xxii	Health Education and Economic Development (Heed), Bangladesh	A Christian faith-based, not-for-profit, non-political, and NGO in Bangladesh
xxiii	Cooperative for Assistance and Relief Everywhere (CARE), Bangladesh	International NGO
xxiv	Oxford Committee for Famine Relief (OXFAM), Great Britain (GB)	International NGO
xxv	Save the Children	International NGO

agricultural inputs for emergency recovery; and they did not receive enough aid in finding shelter.

Representative of Any Humanitarian Response Group or Group Committee

There was a large gap between the representatives and stakeholders due to an inadequate election system. The respondents were elected or voted for as a representative of any group committee at one time and the percentages were found, respectively: M. Balia Toli 8%, Bodor Khali 8%, and Noltona 4%. The percentages were M. Balia Toli 92%, Bodor Khali 92%, and Noltona 96%, respectively, and as a result, they were neither elected nor voted on as representatives of any group committee (Fig. 21.3a).

Loan for Recovery

From the field survey, it was observed that 20% of respondents of the M. Balia Toli *Union* got loans for recovery, and 80% did not get loans from any organizations to overcome their recovery after disasters. On the other hand, 24% of respondents of both Bodor Khali and Noltona *Unions* got loans for recovery. So, 76% of respondents had not received any loan support from any humanitarian organization (Fig. 21.3b).

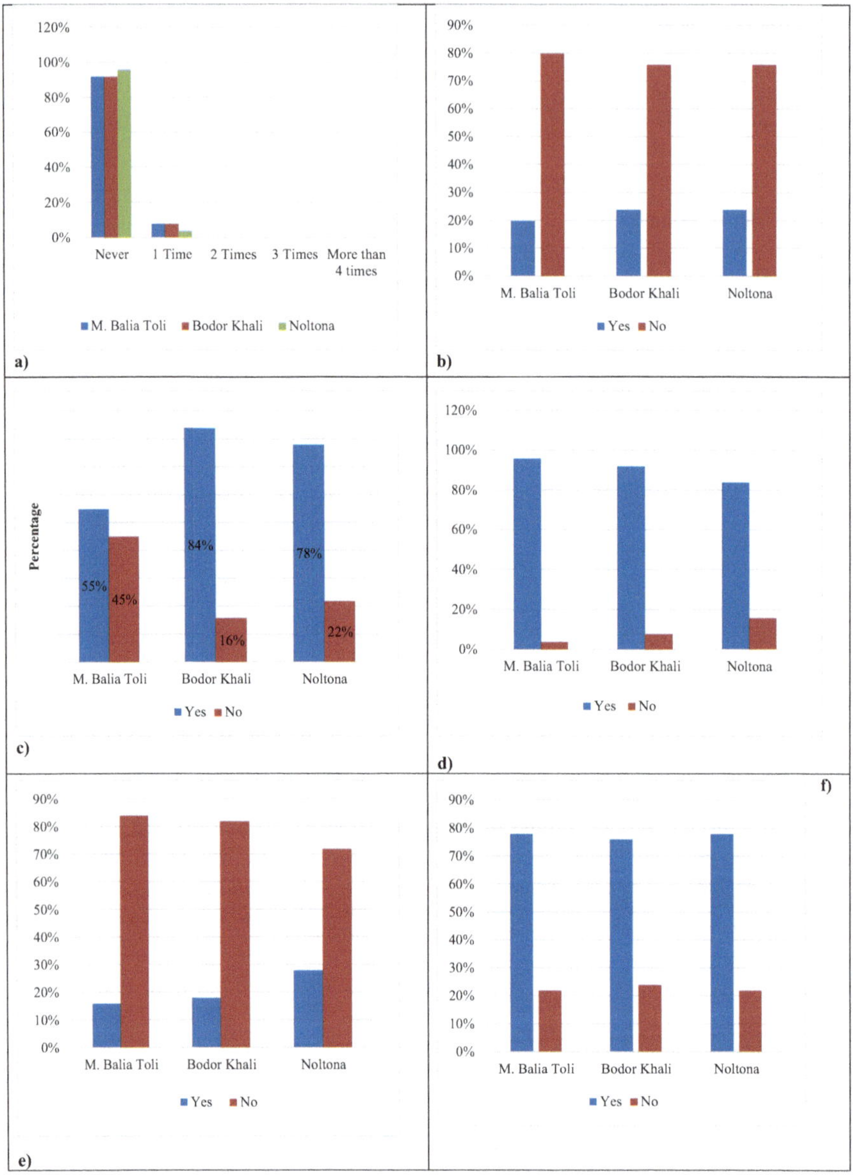

Fig. 21.3 Bar graphs: **a** Elected or voted as a representative of any group committee; **b** Provided any loan for recovery; **c** Provided seed, fertilizer, and other agricultural inputs after a disaster; **d** Dissemination of early warning before the disaster; **e** Organizations assistance in taking shelter; **f** Got relief materials after a disaster

Seed, Fertilizer, and Other Agricultural Reliefs

Respondents of M. Balia Toli, Bodor Khali, and Noltona *Unions* got seeds, fertilizers, and humanitarian subsidies support as relief inputs after the disasters and it was 55%, 84%, and 78%, respectively. On the other hand, 45, 16 and 22% of respondents of those *Unions,* respectively, did not get any subsidies and reliefs (Fig. 21.3c).

Dissemination of Early Warning

The study found that about 96, 92 and 84% of respondents of M. Balia Toli, Bodor Khali, and Noltona unions, respectively, opined that a few organizations disseminate early warning messages before disasters (e.g., floods and cyclones mainly). The rest of the respondents disagreed with the opinion circulated as the early warning, hazard prediction, and forecasting (Fig. 21.3d).

Humanitarian Assistance in Taking Shelter

Most of the respondents opined that they did not get this facility. Only 16, 18 and 28% of respondents of M. Balia Toli, Bodor Khali, and Noltona union, respectively, got shelter from a small number of organizations (Fig. 21.3e).

Relief Materials Before, During and After Disaster

The study found that 78, 76 and 78% of respondents of M. Balia Toli, Bodor Khali, and Noltona Union, respectively, got relief material after a disaster and the rest of the respondents (M. Balia Toli 22%, Bodor Khali 24%, and Noltona 22%) did not get any relief material after a disaster (Fig. 21.3f). Respondents also revealed that they got relief materials from the Union *parishad* and different types of NGOs like BRAC, CODEC, Grameen Bank, Islamic Relief, OXFAM-GB, etc. Government and non-government organizations supply mixed food and non-food items like tins, clothes, dry food, rice, and medicines. In special cases, cash support in Bangladesh Taka (BDT) was given to the remotely affected people. They got insufficient relief materials and those did not arrive at the proper time (Fig. 21.3f). It was also observed that only 2, 4 and 4% of respondents of M. Balia Toli, Bodor Khali, and Noltona union, respectively, agreed that they get sufficient relief (Fig. 21.4a).

Satisfaction Level of Services and Information

Three study unions (M. Balia Toli 90%, Bodor Khali 94%, Noltona 62%) were dissatisfied with humanitarian services and information from different organizations.

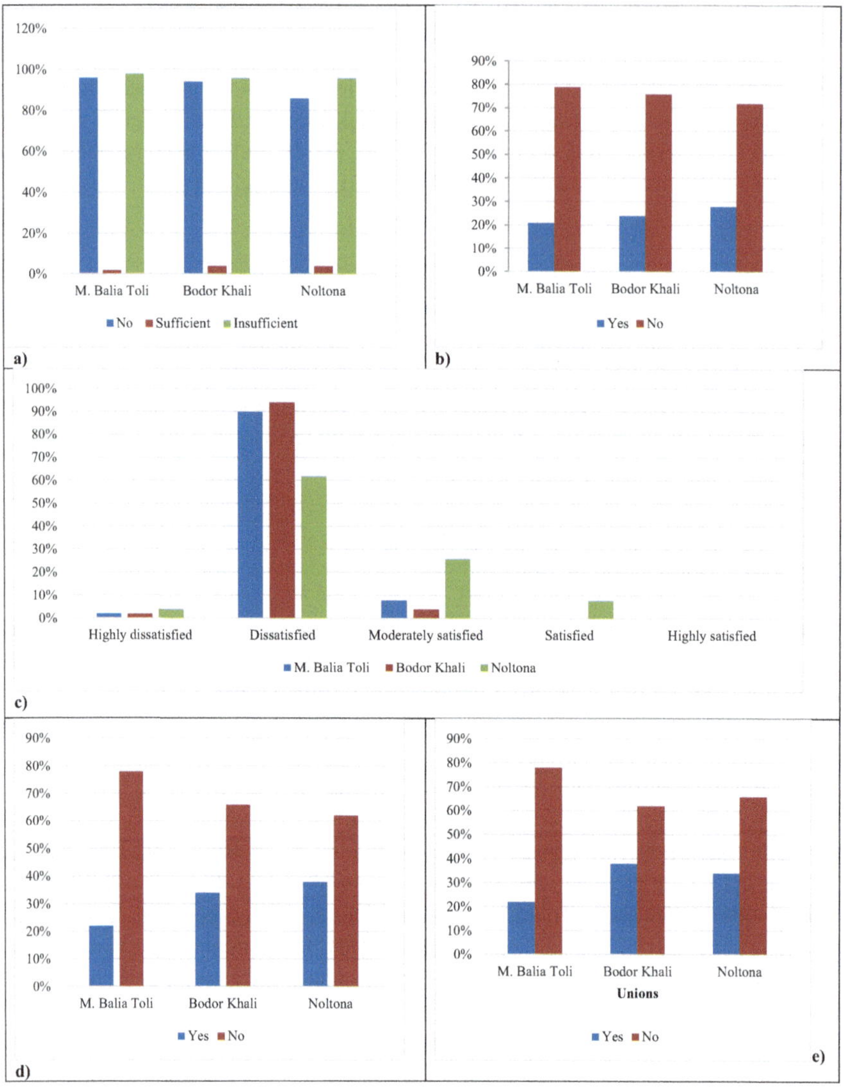

Fig. 21.4 Bar graphs: **a** Condition of getting relief materials; **b** Participation in Agricultural extension training program; **c** Satisfaction level with information and services offered by various organizations; **d** Participation in a training program on livestock and poultry management; **e** Getting a loan to overcome their situation

It was also noted that only respondents of the Noltona union opined moderately satisfied (26%) in this regard (Fig. 21.4c).

Role of Support Services and Their Influences

An activity required for the successful execution of a humanitarian program or process was known as departmental support service. The support service of related organizations needed good structural services to succeed in their influencing role. Therefore, there were many humanitarian support services in the study areas, among agricultural, livestock, poultry, and fisheries extensions were counted for surveying.

Participation in the Agricultural Extension Training Program

Respondents among M. Balia Toli (79%), Bodor Khali (76%), and Noltona (72%) unions lacked an agricultural extension training program in the past 12 months by 21%, 24%, and 28% respondents, respectively, got the extension-based training program (Fig. 21.4b).

Livestock and Poultry-Related Services

Respondents of M. Balia Toli, Bodor Khali, and Noltona unions, respectively, participated in training programs on livestock and poultry management in the past 12 months (22, 34 and 38%). It was a worrying observation that most respondents did not attend the provided training programs (Fig. 21.4d).

Loan to Overcome Their Situation

The study picked that about 22, 38 and 34% of respondents of M. Balia Toli, Bodor Khali, and Noltona unions, respectively, got loans for necessities. However, the rest of them (M. Balia Toli 78%, Bodor Khali 62%, and Noltona 66%) were not getting any loans from different organizations to overcome their emergencies (Fig. 21.4e).

21.3.2.3 Constraints of Community People in DRR Coordinating Mechanisms

DRR-Related Training Program

People among the three unions (M. Balia Toli, Bodor Khali, and Noltona) got very low (36%) low (30%) moderate (25%), and high (9%) DRR-related training programs which highlighted that they have a small amount of knowledge in DRR-based humanitarian responses (Fig. 21.5a).

Problem Solved by Household Members

Respondents of M. Balia Toil (64%), Bodor Khali (34%), and Noltona (44%) union did not solve their problem without any help from others. However, some of them opined that sometimes they could have solved it without assistance (Fig. 21.5b).

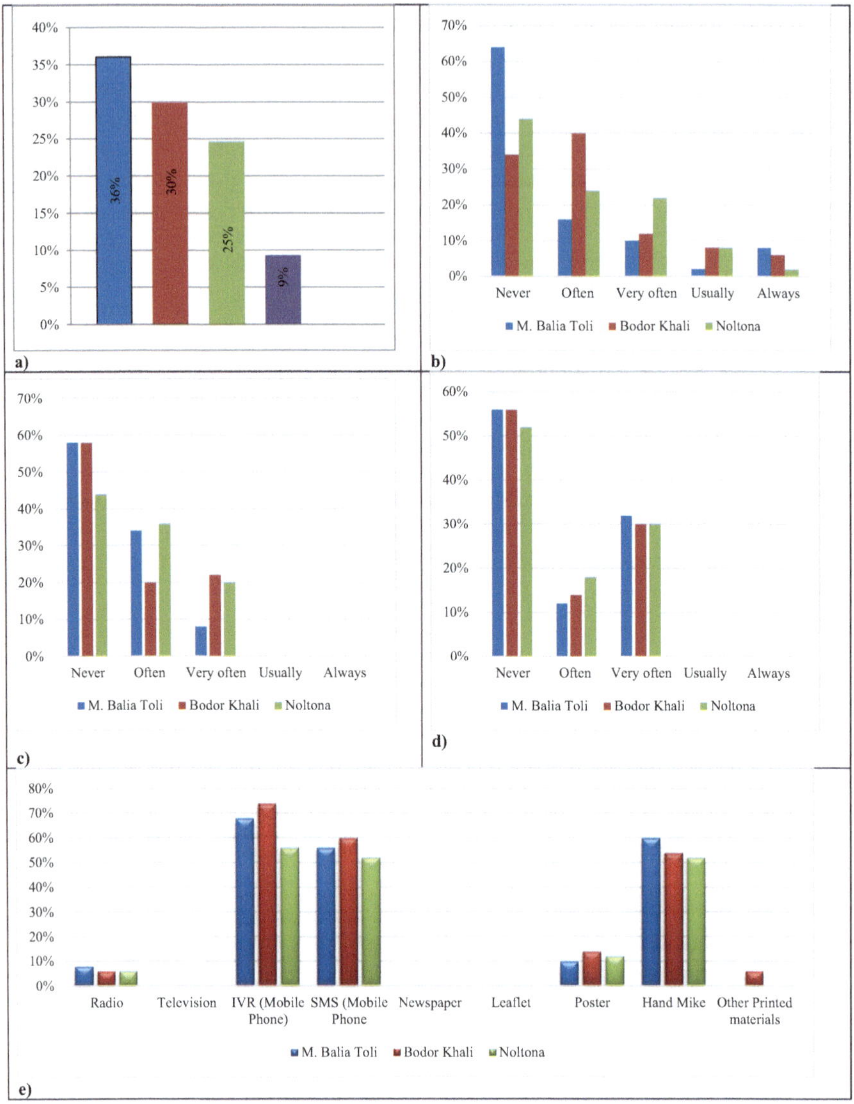

Fig. 21.5 Bar graphs: **a** Experience in DRR-related training program; **b** Problem-solving by household members; **c** Exposing ideas in group meetings; **d** Connection between community members with government and non-government agencies; **e** Disaster information dissemination by mass communication

Exposing Ideas in Group Meeting

The study revealed that respondents of M. Balia Toil, Bodor Khali, and Noltona are never (58%, 58%, 44%, respectively), often (34%, 20%, and 36%, respectively) and very often (8%, 22%, and 20%, respectively) exposed their ideas, problems, and other activities in the group meeting (Fig. 21.5c).

Connection Between Community Members with Government and Non-Government Agencies

Most of the respondents of M. Balia Toli, Bodor Khali, and Noltona union have never had a connection between community members with government and non-government agencies. The rest of the respondents also opined that they had often and very often connections with the GO and NGO-based agencies (Fig. 21.5d).

Assessing Disaster Information Dissemination by Mass Communication and Individual Communication

Disaster Information Dissemination by Mass Communication

The study perceived that respondents of M. Balia Toli, Bodor Khali, and Noltona union got disaster information by IVR (mobile phone) (68, 74 and 56%, respectively), SMS (mobile phone) (56, 60, and 52%) and Hand Mike (60, 64, and 52%), respectively (Fig. 21.5e).

Disaster Information Dissemination by Individual Communication

Field survey revealed that the respondents of M. Balia Toli, Bodor Khali, and Noltona *Unions* got any signal or other information about the disaster from their *Union Parishad* (74, 88 and 82%), *Upazila Parishad* (54, 40 and 54%), and NGOs/ Community-based Organizations (CBOs) (22, 22 and 36%), respectively (Fig. 21.6a).

Connection and Participation

Community Awareness Program for Disaster Mitigation

Respondents of M. Balia Toli, Bodor Khali, and Noltona union participated never (46, 35 and 38%), rarely (34, 44 and 28%), and sometimes (18 19 and 20%), respectively, in any types of community awareness program for disaster mitigation in the past 10 years (Fig. 21.6b).

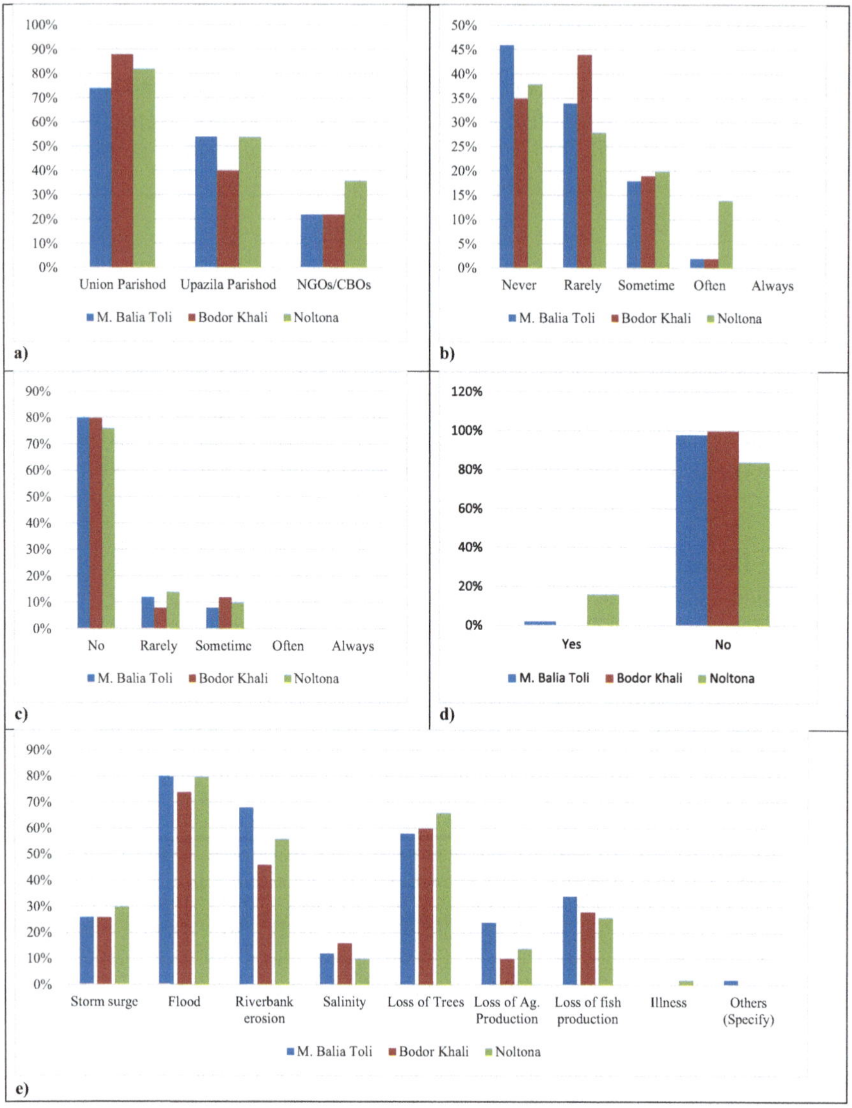

Fig. 21.6 Bar graphs: **a** Disaster information dissemination by individual communication; **b** Connection and participation of community awareness program; **c** Communication or network with NGOs and other voluntary organizations; **d** Household's existing on Radio and Television; and **e** Livelihood risk has experienced

Communication or Network with NGOs and Other Voluntary Organizations

The study demonstrated that the respondents had no communication with NGOs and other voluntary organizations. It was observed that a few percentages were not remarkable but surprising among these unions often or always communicated with NGOs and other voluntary organizations (Fig. 21.6c).

Constraints and Risk in the Livelihood Sector

The study found that the local community has faced many disasters in the last 10 years. Respondents of M. Balia Toli, Bodor Khali and Noltona union have faced so many risks like storm surges (26, 26 and 20%), floods (80, 74 and 80%), riverbank erosions (68, 46 and 56%), salinity intrusions (12, 16 and 10%), deforestation (58, 60 and 66%), loss of agriculture production (24, 10 and 14%), loss of fish production (34, 28 and 26%), respectively (Fig. 21.6e).

Emergency Preparedness Strategies

Households Exist on Radio and Television

Respondents of M. Balia Toli (98%), Bodor Khali (100%), and Noltona (84%) unions had no radio and television to get information on emergencies (Fig. 21.6d).

Level of Listening Early Warning

Field survey revealed that respondents had listened never (12, 18 and 12%), rarely (20, 42 and 24%), and sometimes (54, 30 and 58%), respectively, in any type of early warning (Fig. 21.7a).

Level of Understanding Early Warning

Study area people understood all signals (24, 18 and 16%), and some signals (76%, 82%, and 84%), respectively, about the early warning. However, there were no respondents who superficially and did not understand disaster signals (Fig. 21.7b).

Believe in Existing Disaster Forecasting

From the study, it was found that 100% of respondents (Fig. 21.7c). believe in existing disaster forecasting systems.

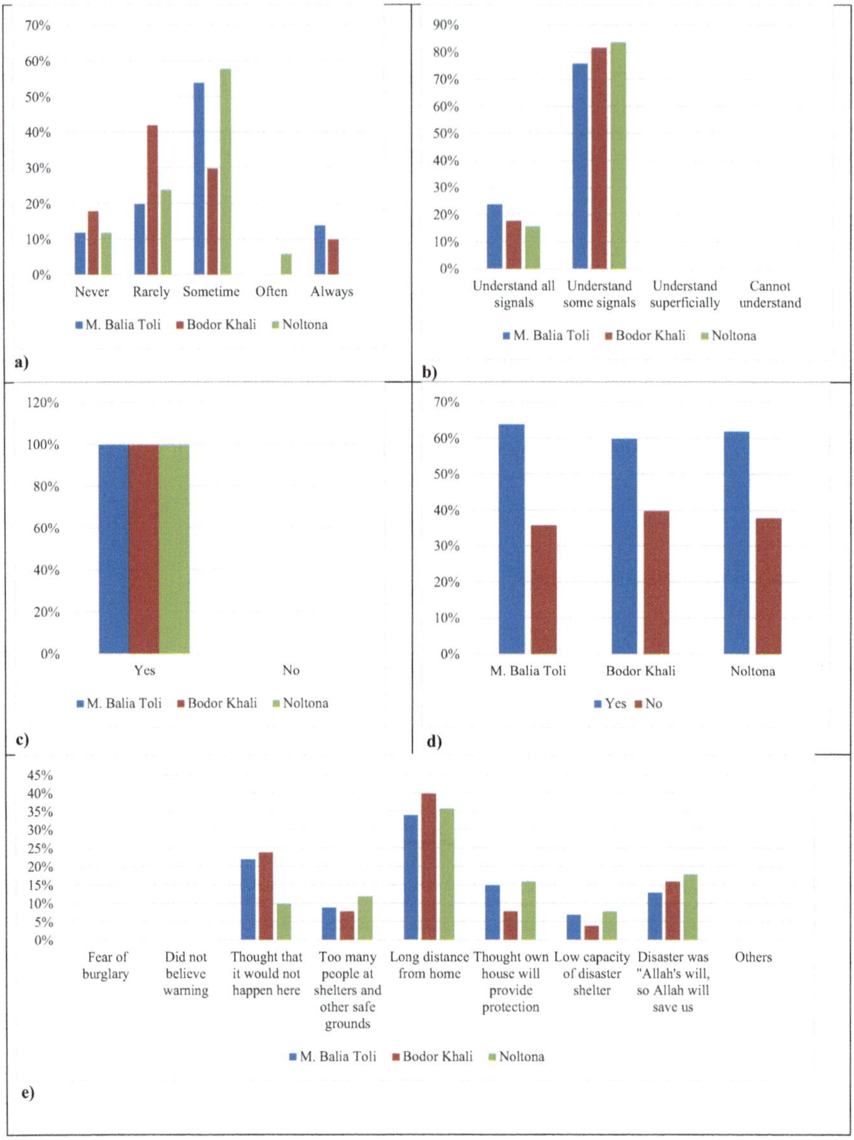

Fig. 21.7 Bar graphs: **a** Level of listening for early warning; **b** Level of understanding of early warning; **c** Belief in existing disaster forecasting; **d** Taking shelter in the last disaster; and **e** Reasons behind not taking shelter during a disaster

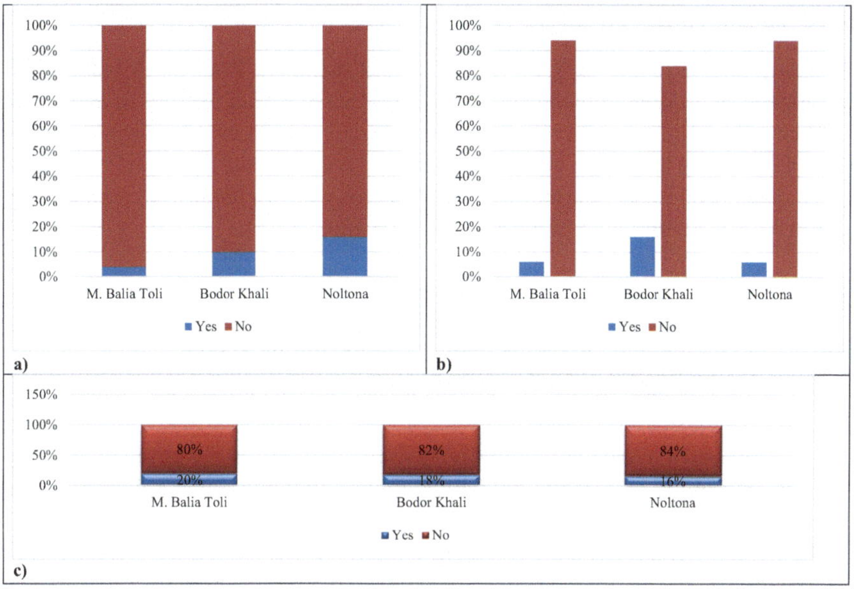

Fig. 21.8 Bar graphs: **a** Diversified income source; **b** Temporary migration due to disaster for earning income; and **c** Household (HH) members permanently migrated due to disaster for earning income

Take Shelter in the Last Disaster

Study observed that the respondents of M. Balia Toli (64%), Bodor Khali (60%), and Noltona (62%) union took shelter (64, 60 and 62%) respectively in the last disaster (Fig. 21.7d).

Reasons Behind not Taking Shelter During a Disaster

Reasons behind not taking shelter during a disaster are not happening; overloaded and other safe grounds; distance from home; owning a house will provide better protection; low capacity; and believing in religion (Fig. 21.7e).

Diversified Income Source

Fisheries, agriculture, livestock, and some other sectors were observed as the key sources of income from the respondents. So, it was found that 96, 90 and 84% of respondents respectively had no other income source (Fig. 21.8a). However, few respondents had alternative income sources for balancing their financial condition at any crisis moment.

Risk Reduction Strategies

The study analyzed that only 6% of household members from the M. Balia Toli *Union*, 16% of household members of the Bodor Khali *Union*, and 6% of household members of the Noltona *Union* migrated temporarily to earn their household income (Fig. 21.8b). They left their vulnerable places and went to Dhaka, Khulna to improve their income level to support their family. Whereas a few of them were permanently migrated for the betterment of their life (Fig. 21.8c).

21.3.3 Framework to Explain the Problems of DRR Coordination Instruments

The study offered a framework for resolving the issues with DRR coordination procedures by improving leadership, predictability, accountability, and collaboration in order to improve coordination and the delivery of humanitarian aid to HA.

The humanitarian organization would continue to serve as a forum for policy discussions related to national development strategies, addressing the strategic aspects of planning a coordinated response when necessary. Its goal is to guarantee the effectiveness of aid through operational arrangements of mutual accountability in order to make tangible and quantifiable progress in DDMC and DRRO (Fig. 21.9).

The DRR's humanitarian subgroup that focusses on coordinated response readiness (shown in the diagram as the "DRR sub-group on coordination") would be the

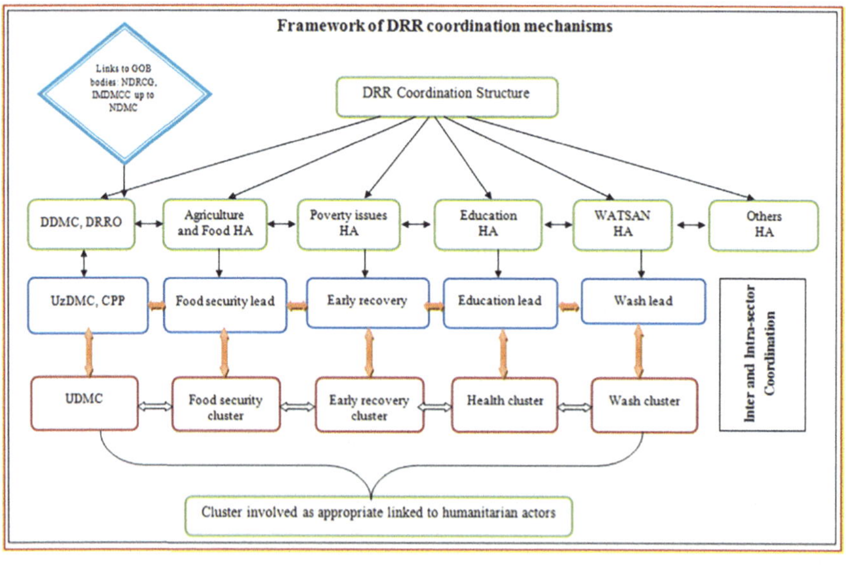

Fig. 21.9 DRR coordination mechanisms

one tasked with ensuring readiness by carrying out tasks related to choosing the right clusters needed in Bangladesh and cluster leadership roles and responsibilities. Additionally, this group would be responsible for elucidating the connections between the international community, humanitarian actors, and important government entities, such as the National Disaster Response Coordination Group (NDRCG), the National Disaster Management Council (NDMC), the Inter-ministerial Disaster Management Coordination Committee (IMDMCC), and the NGO Coordination Committee (NGOCC). During the response, this group would turn into an active inter-cluster coordinating mechanism. The group should consist of humanitarian response DRR members and cluster lead agencies, and it should be co-chaired by a MoDMR representative (e.g., the DG, DDM, and Officer-in-Charge, DRRO). As "humanitarian sub-groups" or clusters at the community level, several clusters that were deemed suitable for the national context should be established; that is, persons with a development agenda unrelated to humanitarian response would not voluntarily join the cluster, whereas those with an interest in humanitarian response would. Between the leadership cluster or subgroup and the humanitarian actors, there are vital connections. The humanitarian organization should have a cluster lead. Other members may not be affiliated with the humanitarian organization (e.g., INGOs), but will be the entities involved in the humanitarian response in the sector.

Because it seeks to guarantee the relationship between humanitarian response, recovery, and the longer-term development agenda, the leadership link to the DRR/humanitarian actors is significant. The sub-group on coordination serves as the primary operational coordinating connection between the clusters. To create readiness plans, the DRR sub-group will convene as needed; cluster leads for humanitarian response may be invited to attend. Clusters are identified based on necessity. These clusters should be linked to the humanitarian sub-group on coordination, as they lack a defined position within an existing DRR/humanitarian actor. Leadership among the humanitarian actors should be decided upon. WASH, Education, and Child Protection are examples of clusters that should, if possible, be housed inside this structure without a change in membership or leadership. For objectives, additional subgroups could be formed in addition to the humanitarian subgroup on coordination. One such group might be a global humanitarian stakeholders' group that the RC leads in its capacity.

21.4 Conclusion and Recommendation

The goal of efficient humanitarian coordination is to make the most use of available resources to provide the most pertinent and suitable response to the needs of those impacted by a catastrophe. The coordinated response must be implemented in various situations to be included in the country's broader development framework and catastrophe risk reduction context. The goal of efficient humanitarian coordination is to make the most use of available resources to provide the most pertinent and suitable response to the needs of those impacted by a catastrophe. The coordinated response must be implemented in various situations to be included in the country's broader development framework and catastrophe risk reduction context.

The humanitarian actors in DRR segments of the study area are: Agricultural Extension; Fisheries; Livestock; BRDB; Bangladesh Krishi Bank; Public Health; Family planning; Social welfare; *Union Parishad*; and NGOs are Red Crescent Society; OXFAM-GB; Save the Children; WFP; CODEC; Grameen Bank; BRAC; CARITAS; CARE; Heed Bangladesh; VOSD; and Sangram. The existing constraints (of DRR coordination) are described through different aspects. The gaps between DRR coordinating mechanisms among different humanitarian stakeholders are not maintaining good relationships among actor organizations; have not had any master plan to perform DRR works; the tendency to highlight themselves; there is no third party or structure to coordinating DRR actors; organizations think only their reputation rather than community service; lack of database that humanitarian actors are working with DRR.

The gaps between DRR coordinating mechanisms among different stakeholders and community people are Respondents are selected or voted as representatives of any group committee at one time and the percentage is (M. Balia Toli 8%, Bodor Khali 8%, and Noltona 4%). Most of the time they are not voted on or elected as a humanitarian representative of any DRR group or group's committee. Humanitarian actors did not provide sufficient loans, seeds, fertilizers, and other agricultural inputs for disaster damage recovery. According to the survey, it was summarized that respondents of M. Balia Toli, Bodor Khali, and Noltona union got loans (20, 24 and 24%), respectively, for recovery and 80, 76 and 76% of respondents didn't get the loan. Only 55, 84 and 78% of respondents, respectively, get seed, fertilizer, and other agricultural inputs after the disaster. It was shown that about 96, 92 and 84% of respondents of M. Balia Toli, Bodor Khali, and Noltona unions respectively agreed that a few organizations disseminate early warning messages before disaster. Respondents of M. Balia Toli, Bodor Khali, and Noltona *Unions* got shelter from a small number of organizations and the rest of the respondents (M. Balia Toli 84%, Bodor Khali 82%, and Noltona 72%) opined that they did not get shelter facilities.

The constraints faced among the community people in DRR coordinating mechanisms are not having enough training about DRR. There is no experienced or knowledgeable person in DRR and warning systems in this area and the percentage were (18, 19 and 20%) respectively in M. Balia Toli, Bodor Khali, and Noltona *Unions* that participated in community awareness programs for disaster mitigation in the past 10 years. They had no radio or television to get information on emergencies in M. Balia Toli (98%), Bodor Khali (100%), and Noltona (84%) *Unions*, respectively.

The concerning issue is that the respondents of M. Balia Toli, Bodor Khali, and Noltona union did not take shelter during the disaster, the reasons behind are not happening possibility (22, 24 and 10%), respectively; overloaded; distance from home. They thought their own house would provide better protection (15, 8 and 16%), and belief in religion (13, 16 and 18%), respectively.

The study developed a framework that helps to overcome the problems of DRR coordination mechanisms among different stakeholders. The following guidelines are thought to be essential to bolstering initiatives for improved humanitarian coordination and connections to DRR, according to the evaluation:

i. Establish attainable and reasonable objectives to enhance coordination.
ii. Reduce the number of new or parallel structures created and maximize the use of already-existing structures.
iii. Complement national emergency and hazard risk management systems.
iv. Boost confidence in humanitarian actions.
v. Ascertain the capacity of humanitarian leadership.
vi. Encourage better collaboration with INGOs.
vii. Encourage coordination at the local level to implement efficient mitigation and adaptation plans.
viii. Analyze DRR case studies (function, links, and membership).
ix. Describe the DRR sub-group on coordination's inter-cluster coordination mechanism.
x. Outline the DRR subgroup's coordination strategy for catastrophes of varying sizes.
xi. Decide which major clusters are necessary for Bangladesh, and where appropriate, think about creating these as LCG-WG subgroups with a humanitarian focus.
xii. Draw important connections between the DRR programs and important GoB, UN, and INGO entities.
xiii. For DRR tasks, use "subgroups" (e.g., international humanitarian community coordination).

Improving coordination in humanitarian response aims to allocate a few resources in the most efficient manner to address needs. These sectors' coordination needs:

i. Policy discussion on the most effective DRR, response, and recovery strategies that lead to established best practices.
ii. Coordinated emergency preparation.
iii. Needs within and between sectors are identified and prioritized (by means of a coordinated approach to evaluation).
iv. Data management and information exchange on who is doing what during a response (which should also minimize overlaps and highlight gaps).
v. Monitoring and assessment using Bangladesh's national disaster management plan.

Acknowledgements The research team would like to express their sincere gratitude to Oxfam-GB and the Community Development Center's (CODEC) Patuakhali and Barguna District Offices in Bangladesh for their cooperative and combined research support. Our special thanks and gratitude to Patuakhali Science and Technology University (PSTU), Bangladesh's administration for signing a MoU with CODEC which makes it possible to conduct this joint research. The research team also highly appreciates the efforts of the academicians and researchers of the Faculty of Environmental Science and Disaster Management (ESDM), PSTU, for their enormous precious efforts to make this research successful. The research team expresses humble gratitude also to all the respondent stakeholders of different government and non-government organizations from District to *Unions* (smallest administrative unit) levels of Barguna district of Bangladesh for their endless support, cooperation, and patience during the data collection.

Author Contributions Principal Investigator (PI), Literature Review, Writing of Original Draft and Corresponding Author: M.S.; Research Coordinator (as the Dean of the Faculty of Environmental Science and Disaster Management, Patuakhali Science and Technology University, Bangladesh) and Writing of Original Draft: A.K.M.A.A.B., Maps, and Figures Preparation and Writing of Original Draft: M.M.H.; Table Preparation, Reports Collection and References: M.F. and M.R.; Field Survey, Report Compilation, and Data Analysis: P.H., R.C.G., P.D., and M.A.R.; Editing, and Supervision: A.K.M.A.A.B., M.S., and R.S. After reading the final draft of the document, all authors gave their approval for publication of the manuscript.

Data Availability Statement Upon a reasonable request, the corresponding author will make the dataset created and analyzed for this study public; however, there are restrictions on the data's repeatability for reuse and commercially reliable details.

Conflicts of Interest Regarding the research, the authors say they have no conflicts of interest.

References

Ahmed KM (2018) Humanitarian response coordination architecture in Bangladesh Rohingya refugee crisis operation: Practice, challenges and prospects. Dhaka. Retrieved from https://dspace.bracu.ac.bd/xmlui/bitstream/handle/10361/11537/13168011_ARC.pdf?sequence=1&isAllowed=y

Alam A, Hoque MM (2022) Local level collaborative governance for pandemic responses: Unpacking a case in Bangladesh. Policy & Gov Rev 6(3):207–228. https://doi.org/10.30589/pgr

Ali, A. (1996). Vulnerability of Bangladesh to climate change and sea level rise through tropical cyclones and storm surges. In: Erda, L., et al. Climate Change Vulnerability and Adaptation in Asia and the Pacific. Springer, Dordrecht. https://doi.org/10.1007/978-94-017-1053-4_16

Anson S, Watson H, Wadhwa K, Metz K (2017) Analysing social media data for disaster preparedness: Understanding the opportunities and barriers faced by humanitarian actors. Int J Disaster Risk Reduct 21:131–139. https://doi.org/10.1016/j.ijdrr.2016.11.014

Arab ST, Islam MM, Shamsuzzoha M, Alam KF, Muhsin N, Noguchi R, Ahamed T (2022) A review of remote sensing applications in agriculture and forestry to establish big data analytics. In: Ahamed T (ed) Remote sensing application. New frontiers in regional science: Asian perspectives, vol 59. Springer, Singapore, pp 1–24. https://doi.org/10.1007/978-981-19-0213-0_1

Azad MAK, Uddin MS, Zaman S, Ashraf MA (2019) Community-based disaster management and its salient features: A policy approach to people-centred risk reduction in Bangladesh. Asia-Pac J Rural Dev 29(2):135–160. https://doi.org/10.1177/1018529119898036

Basher R (2013) Science and technology for disaster risk reduction: A review of application and coordination needs. Geneva. Retrieved from https://www.caneus.org/undrr/resources/3.Science-and-Technology-for-Disaster-Risk-Reduction.pdf

BBS (2024) Statistical yearbook Bangladesh 2023. Dhaka. Retrieved from http://nsds.bbs.gov.bd/storage/files/1/Stat_Yearbook-23.pdf

Biswas AAA, Tariqul Islam M, Abdus Sattar M, Nasrin Mili S, Jahan T (2015) Community based risk assessment of agriculture sector in Sreerampur union of Bangladesh. J Food Secur 3(5):125–136. https://doi.org/10.12691/jfs-3-5-2

Chaiya C, Pal I, Pinthong P (2022) Paradigm of Disaster Risk Reduction: A Comparative Study of Five Asian Countries. Soc Sci Asia, Off J Natl Res Counc Thail Conjunction J Polit GovAnce 8(3):1–33. https://doi.org/10.14456/ssa.2022.17

Chakrabarty S (2022) Factors affecting governance aspect of disaster management: Comparative study of the Sundarbans in India and Bangladesh. In: Chatterjee U, Akanwa AO, Kumar S, Singh SK, Dutta Roy A (eds) Ecological footprints of climate change: Adaptive approaches and sustainability. Springer International Publishing, Cham, pp 219–245. https://doi.org/10.1007/978-3-031-15501-7_9

Cook ADB, Ne FY (2018) Complex humanitarian emergencies and disaster management in Bangladesh: The 2017 Rohingya exodus. Retrieved from https://www.rsis.edu.sg/wp-content/uploads/2018/07/NTS-Report11-Bangladesh-HADR.pdf

CPHA (2019) Minimum standards for child protection in humanitarian action, 2nd ed. Alliance for Child Protection in Humanitarian Action (CPHA), Sphere Project, Geneva. Retrieved from https://spherestandards.org/wp-content/uploads/CPMS-2019-EN.pdf

CPWG (2013) Minimum standards for child protection in humanitarian action, 1st ed. Child Protection Working Group (CPWG), Sphere Project, Geneva. Retrieved from https://spherestandards.org/wp-content/uploads/CPMS-2013-EN.pdf

Ersel M (2015) Water and Sanitation Standards in Humanitarian Action. Turk J Emerg Med. https://doi.org/10.5505/1304.7361.2015.48753

Faisal M, Saha MK, Sattar MA, Biswas AKMAA, Hossain MA (2021) Evaluation of climate induced hazards risk for coastal Bangladesh: A participatory approach-based assessment. Geomat Nat Haz Risk 12(1):2477–2499. https://doi.org/10.1080/19475705.2021.1967203

Feng Y, Cui S (2021) A review of emergency response in disasters: Present and future perspectives. Natural Hazards. Springer Science and Business Media B.V. https://doi.org/10.1007/s11069-020-04297-x

Gaillard JC, Mercer J (2013) From knowledge to action: Bridging gaps in disaster risk reduction. Prog Hum Geogr 37(1):93–114. https://doi.org/10.1177/0309132512446717

GoB (2024a) Barguna district. Retrieved August 24, 2024, from https://www.barguna.gov.bd/en

GoB (2024b) Barguna sadar upazila. Retrieved August 24, 2024, from https://sadar.barguna.gov.bd/en

GoB (2010) Standing orders on disaster 2010. Dhaka. Retrieved from https://ddm.portal.gov.bd/sites/default/files/files/ddm.portal.gov.bd/policies/11a33a04_f11d_44fa_a235_48d2f47521b7/2020-01-13-12-23-a4a05f9a58a819021a6d591667fa4ad6.pdf

GoB (2017) National Plan for Disaster Management (2016–2020): Building resilience for sustainable human development. Dhaka. Retrieved from https://modmr.portal.gov.bd/sites/default/files/files/modmr.portal.gov.bd/policies/0a654dce_9456_46ad_b5c4_15ddfd8c4c0d/NPDM(2016-2020)%20-Final.pdf

GoB (2020a) National plan for disaster management (2021–2025): Action for disaster risk management towards resilient nation. Dhaka. Retrieved from https://modmr.portal.gov.bd/sites/default/files/files/modmr.portal.gov.bd/page/a7c2b9e1_6c9d_4ecf_bb53_ec74653e6d05/NPDM2021-25%20DraftVer5_23032020.pdf

GoB (2020b) Standing orders on disaster 2019. Dhaka. Retrieved from https://reliefweb.int/report/bangladesh/government-people-s-republic-bangladesh-standing-orders-disaster-2019

Guo Y, Song Y, Zhang M (2024) Value realization of intelligent emergency management: Research framework from technology enabling to value creation. Data Sci Manag. https://doi.org/10.1016/j.dsm.2024.06.001

Habiba U, Shaw R, Abedin MA (2013) Community-based disaster risk reduction approaches in Bangladesh. In: Shaw R, Mallick F, Islam A (eds) Disaster risk reduction approaches in Bangladesh. Springer, Tokyo, pp 259–279. https://doi.org/10.1007/978-4-431-54252-0_12

Hasan MM, Islam R, Rahman MdS, Ibrahim M, Shamsuzzoha M, Khanam R, Zaman AKMM (2021) Analysis of land use and land cover changing patterns of Bangladesh using remote sensing technology. Am J Environ Sci 17(3):64–74. https://doi.org/10.3844/ajessp.2021.64.74

Houston JB, Hawthorne J, Perreault MF, Park EH, Goldstein Hode M, Halliwell MR, Griffith SA (2015) Social media and disasters: a functional framework for social media use in disaster planning, response, and research. Disasters 39(1):1–22. https://doi.org/10.1111/disa.12092

Islam A, Shaw R, Mallick F (2013) Bangladesh climate change strategy and action plans. In: Shaw R, Mallick F, Islam A (eds) Climate change adaptation actions in Bangladesh. Springer, Tokyo, pp 107–118. https://doi.org/10.1007/978-4-431-54249-0_7

Islam MT (2023a) Concept of disaster, governance, and development. In: Islam MT (ed) Disaster, governance and development: Perspectives from Bangladesh. Springer, Singapore, pp 15–32. https://doi.org/10.1007/978-981-99-1631-3_2

Islam MT (2023b) Disaster management in Bangladesh: Historical perspective. In: Islam MT (ed) Disaster, governance and development: Perspectives from Bangladesh. Springer, Singapore, pp 33–59. https://doi.org/10.1007/978-981-99-1631-3_3

Kabra G, Ramesh A, Arshinder K (2015) Identification and prioritization of coordination barriers in humanitarian supply chain management. Int J Disaster Risk Reduct 13:128–138. https://doi.org/10.1016/j.ijdrr.2015.01.011

Krishnan S, Zaman S, Ferdaus M, Kabir MH, Khatun H, Rahman SMS, Howard N (2022) How can humanitarian services provision during mass displacement better support health systems? An exploratory qualitative study of humanitarian service provider perspectives in Cox's Bazar, Bangladesh. J Migr Health 6:100132. https://doi.org/10.1016/j.jmh.2022.100132

Kulatunga U, Wedawatta G, Amaratunga D, Haigh R (2014) Evaluation of vulnerability factors for cyclones: The case of Patuakhali, Bangladesh. Int J Disaster Risk Reduct 9:204–211. https://doi.org/10.1016/j.ijdrr.2014.05.011

Kulatunga U, Wedawatta G, Amaratunga DG, Shamsuzzoha (2013) Disaster risk and community vulnerability: The case of Patuakhali, Bangladesh. Retrieved from https://api.semanticscholar.org/CorpusID:15175672

Li Z, She J, Guo Z, Du J, Zhou Y (2024) An evaluation of factors influencing the community emergency management under compounding risks perspective. Int J Disaster Risk Reduct 100:104179. https://doi.org/10.1016/j.ijdrr.2023.104179

Masoomi B, Aghajani H, Jafarnejad A, Movahedi MM (2023) Expert oriented hybrid approach for modeling coordination motivators in humanitarian supply chain: Managerial implications to actors' coordination enhancement. Int J Disaster Risk Reduct 97:104013. https://doi.org/10.1016/j.ijdrr.2023.104013

Matin N, Taher M (2001) The changing emphasis of disasters in Bangladesh NGOs. Disasters 25(3):227–239. https://doi.org/10.1111/1467-7717.00174

Matsuda I (1993) Loss of human lives induced by the cyclone in Bangladesh, pp 319–325. https://doi.org/10.1007/BF00812781

Mukherjee M, Wickramasinghe D, Chowdhooree I, Chimi C, Poudel S, Mishra B, Shaw R (2022) Nature-based resilience: Experiences of five cities from South Asia. Int J Environ Res Public Health 19(19). https://doi.org/10.3390/ijerph191911846

Naim C, Pokorny J, Uyen A, Shortall C, Farra A, Moussally K (2024) The role of humanitarian actors in global governance for AMR. Lancet Glob Health. https://doi.org/10.1016/S2214-109X(24)00319-X

Nasreen M (2021) Fifty years of disaster risk governance in Bangladesh. The Daily Star. Retrieved from https://www.thedailystar.net/supplements/celebrating-50-years-bangladesh/news/fifty-years-disaster-risk-governance-bangladesh-2067109

Perwaiz A, Parviainen J, Somboon P, Mcdonald A, Kumar A, Wilcox T, Amach CO (2020) Disaster risk reduction in Bangladesh. Bangkok. Retrieved from https://reliefweb.int/report/bangladesh/disaster-risk-reduction-bangladesh-status-report-2020

Planitz A (1997) A guide to successful damage and needs assessment. South Pacific Disaster Reduction Programme (SPDRP). Retrieved from https://reliefweb.int/report/world/guide-successful-damage-and-needs-assessment

Pothiawala S (2015) Food and shelter standards in humanitarian action. Turk J Emerg Med. https://doi.org/10.5505/1304.7361.2015.98360

Rahman MM (2013) Assessing natural disaster preparedness and climate change mitigation strategies in the coastal areas of Bangladesh. University of Hong Kong, Pokfulam Road, Hong Kong SAR. https://doi.org/10.5353/th_b5131687

Rahman S, Tasnim F (2023) The role of NGOs in ensuring local governance in Bangladesh: From the perception of other actors of governance. Asia-Pac J Reg Sci. https://doi.org/10.1007/s41685-023-00283-w

Raj SJ, Ullah MS, Akhter R (2010) From dissemination to response: In: Search of new strategies for broadcast media in terms of cyclone warnings for Bangladesh. J Sci Commun 9(4):A03. https://doi.org/10.22323/2.09040203

Rasheduzzaman M, Shamsuzzoha M, Akter A (2020) Strengthening economic security through social capital: Household's adaptation to cyclone risk in Bangladesh. Int J Multidiscip Res Dev:69–75. Retrieved from https://www.allsubjectjournal.com/assets/archives/2020/vol7issue5/7-4-34-390.pdf

Ray-Bennett N, Mendez D, Alam E, Morgner C (2020). Inter-agency collaboration for natural hazard management in developed countries. https://doi.org/10.1093/acrefore/9780199389407.013.176

Sadik MS, Nakagawa H, Rahman R, Shaw R, Kawaike K, Fujita K (2018) A study on cyclone Aila recovery in Koyra, Bangladesh: Evaluating the inclusiveness of recovery with respect to predisaster vulnerability reduction. Int J Disaster Risk Sci 9(1):28–43. https://doi.org/10.1007/s13753-018-0166-9

Saha MK, Biswas AAA, Faisal M (2024) Livelihood vulnerability of coastal communities in context of the climate change: A index-based assessment. World Dev Sustain, 4. https://doi.org/10.1016/j.wds.2024.100152

Salman MA (2023) Assessment of vulnerability and capacity to the cyclone 'Amphan' impacts of the southwestern coastal part of Bangladesh: an empirical contextual investigation. Nat Hazards 115(2):1715–1742. https://doi.org/10.1007/s11069-022-05615-1

Sattar MA, Biswas AAA, Islam MT, Hossain MA, Siddeqa M, Rahim MA, Aktar S (2020) Disaster vulnerability and mitigation of humanitarian issues in coastal bangladesh: local evidence and knowledge gaps. Prog Disaster Sci, 8. https://doi.org/10.1016/j.pdisas.2020.100138

Shamsuzzoha M, Rasheduzzaman M, Ghosh RC (2018) Building resilience for drinking water shortages through reverse osmosis technology in coastal areas of Bangladesh. Procedia Eng 212:559–566. https://doi.org/10.1016/j.proeng.2018.01.072

Shamsuzzoha M, Noguchi R, Ahamed T (2022) Rice yield loss area assessment from satellite-derived NDVI after extreme climatic events using a fuzzy approach. Agric Inf Res 31(1):32–46. https://doi.org/10.3173/air.31.32

Shamsuzzoha M, Shaw R, Ahamed T (2024b) Machine learning system to assess rice crop change detection from satellite-derived RGVI due to tropical cyclones using remote sensing dataset. Remote Sens Appl: Soc Environ 35:101201. https://doi.org/10.1016/j.rsase.2024.101201

Shamsuzzoha M, Ahamed T (2023) Shoreline change assessment in the coastal region of Bangladesh delta using Tasseled Cap transformation from satellite remote sensing dataset. Remote Sens, 15(2). https://doi.org/10.3390/rs15020295

Shamsuzzoha M, Ahamed T (2024) Assessment of Shoreline and agricultural land use changes in the Onshore Coastal Region of Bangladesh delta using satellite remote sensing and GIS. In: Ahamed T (ed) Remote sensing application II. New frontiers in regional science: Asian perspectives, vol 77. Springer, Singapore, pp 85–119. https://doi.org/10.1007/978-981-97-1188-8_4

Shamsuzzoha M, Al-Maruf A (2012) Post SIDR life strategy: Adaptation scenario of settlements of the South. Inst Bangladesh Stud (IBS) J 19:207–222. Retrieved from https://www.ru.ac.bd/ibs/wp-content/uploads/sites/56/2022/07/JIBS-Bangla-19.pdf

Shamsuzzoha M, Noguchi R, Ahamed T (2021) Damaged area assessment of cultivated agricultural lands affected by Cyclone Bulbul in Coastal Region of Bangladesh using Landsat 8 OLI and TIRS datasets. Remote Sens Appl: Soc Environ, 23. https://doi.org/10.1016/j.rsase.2021.100523

Shamsuzzoha M, Shaw R, Ahamed T (2024a) Country perspective on disaster risk governance in Bangladesh. In: Rethinking disaster risk management in South Asia, panel-1: A synergistic approach to disaster risk governance. Islamabad. Retrieved from https://www.researchgate.net/publication/379946968

Shaw R, Izumi T, Shi P (2016) Perspectives of science and technology in disaster risk reduction of Asia. Int J Disaster Risk Sci 7(4):329–342. https://doi.org/10.1007/s13753-016-0104-7

Shaw R, Takeuchi Y, Rouhban B (2009) Education, capacity building and public awareness for disaster reduction. In Sassa K, Canuti P (eds) Landslides—disaster risk reduction, pp 499–515. Springer Berlin Heidelberg, Berlin, Heidelberg. https://doi.org/10.1007/978-3-540-69970-5_26

Shaw R, Islam A, Mallick F (2013a) Disaster, risk and evolution of the concept, pp 3–21. https://doi.org/10.1007/978-4-431-54252-0_1

Shaw R, Islam A, Mallick F (2013b) National perspectives of disaster risk reduction in Bangladesh, pp 45–62. https://doi.org/10.1007/978-4-431-54252-0_3

Shaw R, Mallick F, Islam A (2013c) Disaster risk reduction approaches in Bangladesh. In: Shaw RMFIA (ed). Springer, New York. Retrieved from http://www.springer.com/series/11575

Shaw R, Izumi T, Shiwaku K (2018) 1—Science and technology in disaster risk reduction in Asia: Post-Sendai developments. In: Shaw R, Shiwaku K, Izumi T (eds) science and technology in disaster risk reduction in Asia. Academic Press, pp 3–16. https://doi.org/10.1016/B978-0-12-812711-7.00001-8

Sphere Association (1998) Humanitarian charter and minimum standards in disaster response. Sphere Project. Retrieved from https://spherestandards.org/wp-content/uploads/SphereHandbook-1998.pdf

Sphere Association (2000) Humanitarian charter and minimum standards in disaster relief response, 1st ed. Sphere Project, Geneva. Retrieved from https://spherestandards.org/wp-content/uploads/Sphere-Handbook-2000-English.pdf

Sphere Association (2004) Humanitarian charter and minimum standards in disaster response, 2nd ed. Sphere Project, Geneva. Retrieved from https://spherestandards.org/wp-content/uploads/Sphere-Handbook-2004-English.pdf

Sphere Association (2011) Humanitarian charter and minimum standards in humanitarian response, 3rd ed. Sphere Project, Geneva. Retrieved from https://spherestandards.org/wp-content/uploads/2018/06/Sphere_Handbook_2011_English.pdf

Sphere Association (2018) The sphere handbook: Humanitarian charter and minimum standards in humanitarian response, 4th ed. Sphere Project, Geneva. Retrieved from https://spherestandards.org/wp-content/uploads/Sphere-Handbook-2018-EN.pdf

Talukder MF, Shamsuzzoha M, Hasan I (2018) Damage and agricultural rehabilitation scenario of post cyclone Mahasen in Coastal Zone of Bangladesh. J Sociol Anthropol, 2 (1) https://doi.org/10.12691/jsa-2-1-7

UNDRR (2015) Sendai framework for disaster risk reduction 2015–2030. Geneva. Retrieved from https://www.undrr.org/media/16176/download?startDownload=20240812

United Nations Resident Coordinator Office (2022) Bangladesh information management working group: Cyclone Sitrang Humanitarian Snapshot (26 October 2022). Retrieved April 5, 2023, from https://reliefweb.int/map/bangladesh/bangladesh-information-management-working-group-cyclone-sitrang-humanitarian-snapshot-26-october-2022

van Boetzelaer E, van de Kamp J, Keating P, Sharma S, Pellecchia U, Browne JL, Franco OH (2024) Involving older people in the preparedness, response, and recovery phases in humanitarian emergencies: A theoretical framework on ageism, epistemic injustice, and participation. Lancet Healthy Longev 5(1):e76–e82. https://doi.org/10.1016/S2666-7568(23)00244-1

Xie M, Chen L (2024) Turning every disaster into an opportunity: An exploratory study on refugees' perceived emergency management capacity and community resilience. Int J Disaster Risk Reduct 111:104727. https://doi.org/10.1016/j.ijdrr.2024.104727

Yamane T (1967) Statistics: An introductory analysis. Can J Econ Polit Sci (2014/11/7). Harper and Row, New York. Retrieved from https://archive.org/details/statisticsanintr0000taro

Zaman S, Sammonds P, Ahmed B, Rahman T (2020) Disaster risk reduction in conflict contexts: lessons learned from the lived experiences of Rohingya refugees in Cox's Bazar, Bangladesh. Int J Disaster Risk Reduct 50:101694. https://doi.org/10.1016/j.ijdrr.2020.101694

Zimmermann M, Glombitza KF, Rothenberger B (2010) Disaster risk reduction programme for Bangladesh. The Swiss Agency for Development Cooperation (SDC), Directorate of Humanitarian Aid and SHA: Bern, Switzerland

Chapter 22
20 Years After: A Snapshot of Affected Communities in the Aftermath of the 2004 Indian Ocean Tsunami in Sri Lanka

Nishara Fernando, Dilanthi Amaratunga, Richard Haigh, and Yeshani Fernando

Abstract The 2004 Boxing Day Tsunami was one of the most devastating natural hazards ever recorded in Sri Lankan history, taking the lives of thousands of people while displacing millions. In the aftermath, the Sri Lankan Government, in collaboration with many international and local non-governmental organisations, initiated massive relocation efforts, including declaring a 'No Construction Zone' along the coastline and resettling displaced communities away from the Coastline. Despite these efforts, subsequent studies immediately after the relocation efforts have raised questions regarding its efficacy in the long term. In line with the 20th anniversary of the Tsunami, the study at hand discusses the impact of Tsunami relocation efforts on affected communities a long time after relocation by taking Galle and Batticaloa as case studies, which were heavily affected by the 2004 Tsunami. The study explored (a) the issues and challenges that have emerged from the relocation process 'a long time' after relocation, (b) livelihoods and social infrastructure restoration initiatives, (c) the gendered impacts of post-tsunami relocation, (d) the impact exerted by the relocation process on vulnerable groups such as children, the disabled and the elderly, (e) the current status of Tsunami Early Warning Mechanisms and preparedness in Sri Lanka (f) long-term impacts of post-Indian Ocean Tsunami reconstruction, (e) the status of community formation and social cohesion amongst relocated communities, and (f) coping and adaptation strategies employed by the relocated communities to overcome the challenges. These objectives were explored among 4 groups: original tsunami resettlers, original resettlers who moved back to the buffer zone,

N. Fernando (✉)
Department of Sociology, University of Colombo, Colombo, Sri Lanka
e-mail: nishara.fernando@soc.cmb.ac.lk

D. Amaratunga · R. Haigh
Global Disaster Resilience Centre, School of Applied Sciences, University of Huddersfield, Queensgate, Huddersfield HD1 3DH, UK

Y. Fernando
Federation of Sri Lankan Local Government Authorities, Colombo, Sri Lanka

R. Shaw et al. (eds.), *Two Decades from the Indian Ocean Tsunami*, Disaster Risk Reduction, https://doi.org/10.1007/978-981-96-2669-4_22

new settlers, and the host community. In addition to data gathered from 650 respondents using questionnaires and structured interview schedules, 30 in-depth interviews were conducted with key informants in the 4 relocation sites. The study's findings revealed many negative impacts that relocated populations face, such as increased distance to the city, which has reduced accessibility to educational and employment opportunities. The study also revealed a propensity for resettlers to return to their previous places in the Buffer Zone, revealing that relocation was unsuccessful in the long term. Furthermore, the relocation process disrupted the social relations of the elderly and impacted the educational opportunities of children, with a significant impact on poorer communities. Hence, the relocation process has increased the vulnerability of populations affected by the Tsunami and has been unsuccessful at mitigating the socio-economic challenges of the affected populations following the relocation. Hence relocation has to be considered as a long-term process in a way that vulnerabilities of affected populations are mitigated.

Keywords Post tsunami relocation · Long-term impacts · Coping strategies · Poverty · New settlers

22.1 Introduction

The 2004 Indian Ocean Tsunami left a permanent mark in world history as one of the most devastating natural hazards ever recorded. The tidal waves crashed into the coastal regions of Indonesia, India, Sri Lanka and many other countries, killing thousands and displacing millions in the region. The scale of the damage in human and physical terms has been well documented. In Sri Lanka, the Southern, Northern, Eastern and Western provinces were badly affected, rendering 889,175 families homeless (DCS 2005). Over a hundred and fifty thousand individuals lost their main source of income (DCS 2005). Given its sheer scale, the effort required to manage its aftermath in terms of relief, recovery, reconstruction, resettlement and rehabilitation has naturally been unprecedented. The situation worsened with the reimposement of the "No Construction Zone" declared by the Government of Sri Lanka (GoSL) for residential purposes which designated selected strips of land along the coastal regions of the country as inhabitable to counter the threat of future Tsunamis and other coastal hazards (Boano 2009). As a result, more families had to leave their homes and settle in temporary camps and transitory shelters set up by the government, inter-governmental organisations, and international and local charity organisations until they permanently received houses from new settlements. Considering the large number of missing and dead people reported, the government reimposed the buffer zone regulation as a disaster risk reduction strategy to protect their lives from future disasters. Despite resistance from various groups, the government saw it as an opportunity to reimpose the buffer zone regulation soon after the tsunami to change settlements close to the sea and build more resilient communities.

The GoSL had to grapple with the challenges of providing housing to displaced families, restoring livelihoods and the built environment, etc. With the support of international aid agencies and intergovernmental organisations the government initiated the relocation of displaced communities to locations at a safe distance away from the sea. The relocation process was plagued with accusations of mishandling aid, corruption allegations, low quality of building infrastructure, lack of community-centric approaches, failure to address the true needs of the displaced communities, appropriateness of the relocation sites, etc. (Ahmed 2020; Arunatilake 2018; Bittel 2019; Fernando et al. 2021; Oliver-Smith 2018). In the initial years, the impacts of the tsunami and the challenges brought forth by the displacement and relocation process attracted the attention of academia and scholars around the world. Researchers studied topics including the aptness of relocation sites, socio-economic and psychological challenges experienced by relocates, livelihoods, and vulnerable groups in detail, creating a wealth of knowledge.

An examination of the status of relocated communities in 2014, 10 years after relocation, revealed a great deal about the impact of the recovery process on different population groups (Hettige et al. 2014). The relocated communities experienced both positive and negative impacts related to the recovery process, and these impacts highlighted several key factors affecting recovery outcomes in Sri Lanka. Haigh et al. (2015) identified them in seven categories, namely, '(1) The nature and extent of settlement planning; (2) The role of local institutions including local government; (3) Socio-economic characteristics of victims; (4) The role of donor agencies; (5) Distance between the original and new settlement; (6) Quality of construction; (7) The nature and extent of social infrastructure; and (8) Beneficiary participation and community mobilisation'.

Hence, the experiences of affected communities regarding personal well-being, access to livelihoods and other services, physical and social infrastructure, have varied widely owing to a range of personal and structural circumstances. Therefore, identifying these factors was considered to be vital to preventing these inequities from arising in future recovery efforts, related to tsunamis or other hazards, in Sri Lanka or elsewhere. These research efforts also informed Sri Lanka's Action Plans to address the targets and priorities set out in the Sendai Framework for Action on Disaster Risk Reduction (2015–2030). However, since then the enthusiasm surrounding relocation and the plight of relocatees have diminished.

22.2 Rationale of the Study

26th of December 2024 marks the 20th anniversary of the Indian Ocean Tsunami. Since 2004, relocated communities have lived their lives in relocation sites and have encountered and coped with many socio-economic, political, and psychological challenges. Studies revealed that some have managed to flourish in the new settlements by employing various enhancement strategies while some have failed to survive, prompting them to leave the relocation settlements in search of a better life (Fernando 2012; Muggah 2008; Weerasena et al. 2018; Hettige et al. 2018). Nevertheless, longitudinal studies regarding the relocated communities are limited. As a result, there exists a significant gap in knowledge surrounding the relocated communities, especially regarding the way their lives have progressed 'a long time' after relocation. Therefore, no other time is more appropriate to delve into the lives of relocated communities and to explore the status of their lives than the 20th anniversary of the Tsunami Disaster. Within this backdrop, this study collected data from three relocation settlements in Akmeemana Galle in the Southern Province and one resettlement in Batticaloa City in the Eastern Province to identify the achievements and persisting issues and highlight the key lessons learned and the recommendations that can be made to improve disaster preparedness and recovery processes in Sri Lanka and elsewhere. The study paid specific attention to the status of relocation settlements including the quality of the built environment, status of the infrastructure, livelihoods, social relations including host and relocated community relations, vulnerable groups such as children and the elderly, challenges and impacts created by the relocation process and by the location of the resettlements along with coping strategies and adaptation measures employed by the communities.

22.3 Conceptual Framework: Displacement, Forced Relocation/Resettlement

According to Turton, forced relocation occurs when "Persons are forced to leave their homes or homelands for whatever reason and have been allocated specific areas for them to settle down in their own country with at least minimum resources and services to re-establish their lives" (Turton 2006: 13–14). Therefore, forced resettlement changes people's socio-spatial settings and often requires the social reconstruction of resettled communities in new socio-spatial settings to be successful. This is a huge challenge especially considering the fact that, oftentimes, forced resettlers belong to the poorest, most politically marginalised members of society who are at risk of ending up alienated from their government (Turton 2006).

Michael Cernea (2000) in his IRR model, identifies 8 important risks faced by displaced people, namely, landlessness, joblessness, homelessness, marginalisation, food insecurity, increased morbidity and mortality, loss of access to common property resources, and social disarticulation, which influence one another. He emphasises that planning and monitoring relocation processes is crucial in minimising these risks (Cernea 2000). Anthony Oliver-Smith (2009) identifies those shortcomings in planning, financing, implementing, and administering relocation projects lead to relocation/resettlement failures. Chris de Wet (2009) states that the inadequacy of input-related factors such as national legal framework and policies, political will, funding, pre-displacement research, careful implementation and monitoring are contributing to relocation failures. He further states that the impact of relocation on people is further determined by their health, gender, and economic position (de Wet

2009). Therefore, the resettlement process comes out of the complex interaction of all these factors leading to unpredictability in the linear rational planning approach. According to Koenig (2009), planning needs to consider the three main rights of relocated people—the rights to livelihoods, housing, and improved living conditions. However, most relocation projects, as in the case of the Tsunami relocation, are defined only in terms of housing, neglecting economic and social reconstruction in the long run. Taking a broader view beyond the norm, this study explored the following:

- The status of community formation and social cohesion amongst the relocated communities;
- The status of livelihoods and social infrastructure restoration initiatives;
- The gender-related impacts of post-tsunami relocation;
- The impact exerted by the relocation process on vulnerable groups such as children and elderly;
- The status of Tsunami Early Warning Mechanisms and preparedness in Sri Lanka;
- Long-term impacts of post-Indian Ocean Tsunami reconstruction;
- The issues and challenges that have emerged from the relocation process a long time after relocation;
- Coping and adaptation strategies employed by the relocated communities to overcome the challenges.

The study collected data from four groups of individuals involved in the relocation process, namely: Original Tsunami resettlers, Original resettlers who moved back to the Buffer zone, New settlers, and the Host Community.

(1) **Original Tsunami Resettlers**—The term "Original Tsunami Resettlers" refers to individuals who are victims of the Indian Ocean Tsunami and were relocated by the government to their current settlement. This study explored their lived experiences in the community and the social, political and economic challenges they faced, the experiences of vulnerable groups such as children and the elderly in the settlement, and the successes and failures of the relocation process.

(2) **Original Settlers who moved back to the Buffer Zone**—This group of individuals moved back permanently to the locations of their original places or houses by either closing, selling, or renting their houses in the relocated settlements. The study explored the factors that motivated them to move back to their original location, the challenges encountered when moving back, their life in the current location, their preparedness to face another Tsunami threat and their comprehension of Emergency Warning mechanisms, among others.

(3) **New Settlers**—"New Settlers" refers to individuals who have moved into the relocation settlement by either illegally purchasing or renting households within the settlements that were once inhabited by originally relocated communities. The present study explored the factors that prompted or motivated them to move into the community, their experiences in relocation sites, as well as challenges and issues they had to grapple with in the settlement.

(4) **Host Community**—The "Host Community" refers to individuals who had lived in relocation sites or closer to sites prior to the settlement of relocated communities. This research explored the nature of the relationship between the relocated and the host communities, whether the relocated communities have successfully integrated into the host community by entrenching behavioural, social, political, and economic norms of the host communities after 20 years as some theoretical approaches have suggested.

22.4 Research Methodology

22.4.1 Selection of the Relocation Sites

The study explored the status of several relocation sites in the Southern and Eastern provinces of Sri Lanka. The rationale behind this selection criterion is the availability of a significant body of secondary data that would enable the current study to build a comprehensive and holistic picture of the relocation sites. This enabled the study to demonstrate the progress of relocation activities and to identify the challenges and issues that arise in certain phases of the relocation process. These phases that were identified include: (a) immediately after relocation, (b) sometime after relocation, and (c) long time after relocation.

Based on this rationale, the study selected Salzburg, Kadiragamar, and China Friendship relocation sites from the Akmeemana Divisional Secretariat in Galle (Southern Province) and Swiss Village (Block J, K, and H) of Manmunai North Divisional Secretariat in Batticaloa (Eastern Province), Sri Lanka. People who lived in the buffer zone area of the Galle Municipal Council before the tsunami were relocated into new large settlements (with 50 or more housing units) constructed in the Akmeemana Pradeshiya Sabha, which is a Rural Council area. These three settlements were situated 10–15 km away from the Galle city area. We also purposively selected a Swiss resettlement village in Batticaloa Municipal Council (Eastern Province) as another location to conduct fieldwork. People who stayed in the buffer zone areas of Dutch Bar, Navaladi, and Kallady were relocated into the new settlement situated in the same Municipal Council area, but 7 km away from the old settlements.

22.4.2 Sampling and Data Collection

A random sample of re-settlers (95 from Batticaloa and 188 from Galle), new settlers (47 from Batticaloa and 75 from Galle), host community members (47 from Batticaloa and 60 from Galle) and those who had returned to the buffer zone by the time of the research (53 from Batticaloa and 50 from Galle) were selected to conduct one to one interview. Separate interview schedules were constructed to interview each group to collect both qualitative and quantitative data for the study. Different data collection tools that were used to collect data from respondents of different main groups are briefly discussed:

(1) **A structured questionnaire with original tsunami settlers**

A structured questionnaire was administered by data collection enumerators, who were trained graduates, for a selected sample of 283 original Tsunami settlers from both locations. The interview schedule consisted of both open and closed questions. The questions covered socio-demographic details of the selected households and social, economic, and political challenges that original resettlers had to grapple with during and after the relocation period. Moreover, the questions explored issues and problems that the communities had to tackle immediately after, sometime after and a long time after the relocation. The interview schedule was initially developed in English and later translated into Sinhala and Tamil languages.

(2) **Semi-structured interviews with original resettlers who moved back to the buffer zone**

Semi-structured interviews were conducted with a sample of 103 purposively selected resettlers who had moved back to the location of their original houses either by selling, renting, or closing their houses in the resettlements. The interviews were conducted using a semi-structured interview schedule, developed initially in English and later translated into vernacular languages. The interview schedule consists of open-ended questions seeking to understand the reasons that prompted individuals to resettle in their previous location, the social, economic, and political benefits gained from the resettlement, their opinions regarding both the positive and negative aspects of the relocation process, their readiness to face another Tsunami threat, and awareness of early warning measures, among others.

(3) **Semi-structured interviews with new settlers**

Semi-structured interviews with new settlers will be carried out with 122 purposively selected individuals who had moved into the relocation site by renting or purchasing houses from original resettlers. The interviews, carried out using a semi-structured interview schedule, included broad questions on the reasons behind moving into the relocation site, reasons for selecting the relocation site, challenges experienced when moving to the location and their future prospects. The interview schedule was initially developed in English and then translated into Sinhala and Tamil languages.

(4) **Semi-structured interviews with members of the host community**

The study conducted 107 semi-structured interviews with members of the host community to understand the level of integration of relocated communities in the local community. In addition, the interviews explored the opinions of host community members regarding the relocation process and whether relocation has created additional issues for the community sometime or a long time after relocation. Like the semi-structured interviews conducted with new settlers and the resettlers who had moved out, an interview schedule with open-ended questions was utilized to conduct interviews. The interview schedule was developed in English and translated into vernacular languages.

(5) **Key informant interviews with national and local government officials**

Key informant interviews were conducted with selected national and local government officials including the divisional secretaries of the respective districts, officials of the land registry, disaster management officers, relief services officers, GN officers of the selected settlements, economic development officers, SAMURDHI officers, officials responsible for child protection and education, officials responsible for the care of the elderly and disabled individuals, land officers and other elected officials. A semi-structured interview schedule was utilized to conduct the interview which included questions on the status of the relocation sites, the livelihoods and well-being of community members, challenges and issues they experience, land rights and the provision of deeds to community members, status of vulnerable groups, and interactions with the host community, among others.

Data collection spanned nearly 9 months and a further 3 months were spent on data entry and analysis. The collected survey data were entered into the Statistical Package of Social Sciences (SPSS) and conducted relevant univariate, bivariate, and multivariate analyses. The findings of the study were presented, first, to local-level officials in two separate meetings, held in Akmeemana on the 26th of February 2024 and in Manmunai North Divisional Secretariats on the 11th of June 2024.

22.5 Results and Discussions: The Issues and Challenges that Have Emerged from the Relocation Process a Long Time After Relocation

The aftermath of the Indian Ocean tsunami provided valuable context for examining the real-life experiences of resilience and vulnerability in a diverse group of communities. The study utilizes this opportunity to examine how different groups of displaced populations have coped with its impacts over 20 years since the tsunami, based on the most recent fieldwork. Since the tsunami, all the displaced communities have been resettled in new settlements, either away from or near the buffer zone. State agencies facilitated the relocation process by allocating land, and providing infrastructural facilities such as electricity, water, and access roads, while most of the new settlements were built with either financial support or the direct involvement of various non-governmental organisations. The fieldwork covers only two areas affected by the tsunami-induced displacement and resettlement, yet it facilitated researchers to draw significant conclusions on several issues in the recovery process, pertaining to the following areas:

22.5.1 Poverty Among Resettlers

Interviewed resettlers were requested to identify whether their households belonged to 'very poor', 'poor', or 'non-poor' categories, considering the socio-economic condition of their households at the time of the interview. Results show that 58% of Batticaloa resettlers belong to very poor (16%) or poor (42%) households while 66% of Galle resettlers belong to very poor (11%) or poor (55%) households. 42% of Batticaloa resettlers and 34% of Galle resettlers identify their households as non-poor households. A prominent characteristic of the poor and very poor households, in both Galle and Batticaloa, is the fact that they do not have a permanent income source in the household. However, at least one household member is casually employed with some members looking for employment. This income is insufficient for household survival, and they suffer from a lack of savings. Some household members are engaged in vulnerable lifestyles as drug addicts and alcoholics and have even fallen into debt traps. There are also chronically ill household members. Most of the households are beneficiaries of the Samurdhi social security program and receive other government assistance for the elderly and the disabled. On the other hand, non-poor households have at least one or two permanent sources of income in the government or private sector or are engaged in businesses as trawler boat owners, normal boat owners, or even running a grocery shop. Some of these households have at least one member working in the Middle East or South Korea.

22.5.2 Satisfaction with Housing and Place of Residence

Loss of house and place of residence due to relocation closely relates to other risks such as joblessness and marginalization (Cernea 2000). It's vital to avert these risks by releasing sufficient funds to construct durable housing close to the previous settlement without disrupting the existing social capital and livelihoods. It is against this background researchers inquire about satisfaction with the present housing unit and place of residence from respondents in Galle and Batticaloa, 61% of respondents from Batticaloa stated that their present housing unit and place are better than the previous one. This is mainly due to reasons such as the quality of housing, homeownership, and housing units with more space. However, 74% of Galle respondents are of the view that their previous housing unit is better than the present one, with only 20% stating the opposite. This is mainly due to reasons such as distance from the city, lack of services in the new place, and poor-quality housing. Most of the resettlers who lived in the city of Galle before the tsunami aspire to move into new settlements constructed in the city or closer to the city. This is because relocating to the Akmeemana area, after the buffer zone regulation was re-imposed following the tsunami, disrupted their income-earning activities and social relationships. However, this was not possible as only very few resettlements were constructed in Galle city due to land scarcity. Most large resettlements were constructed in the Akmeemana area about 10–15 km away from the city of Galle. In this context, Galle city dwellers converted into rural dwellers because of the relocation, which reduced access to various common quality services and other employment opportunities available to city dwellers.

Since then, various politicians and officials in the Akmeemana area have pledged to develop Akmeemana as a regional city, following the construction of the Southern Highway in General and the Pinnaduwa exit point, which is especially close to the Akmeemana area. However, it is evident from the responses of resettlers in the Akmeemana area that they still depend on the city of Galle to obtain various services even after residing in the new settlements for sixteen or more years. They also believed that their previous place and housing unit were better than the present one. Comparatively, Batticaloa resettlers were resettled in places well within the city area after displacement.

22.5.3 The Status of Livelihoods and Social Infrastructure Restoration Initiatives

The risk of losing wage employment because of relocation is very high among both rural and urban populations (Cernea 2000). Those who work in the industry and service sectors lost their jobs in the urban sector while, shopkeepers, craftsmen and others lost their small businesses. As a result, some have changed their income-earning activities. When exploring the status of livelihoods of resettlers, nearly 48% of Galle resettlers, compared to only 15% in Batticaloa, stated that they had to change or quit income-earning activities soon after relocating to the new settlements, mainly due to the distance to the city and transport difficulties. When further exploring resettlers' household economic situation after relocation and staying 16 years or more in resettlements, only one-third of respondents both in Galle and Batticaloa stated that their household economic situation got better. Two-thirds stated that the situation got worse due to multiple reasons such as increased expenses, decreased income, and lack of employment opportunities. When further questioned about their outlook on their household economic situation in the future, a significant proportion of resettlers in Batticaloa (80%) were pessimistic compared to resettlers in Galle (52%). This could be mainly due to the economic situation of the country at present, the high cost of living, and the lack of employment opportunities in the area.

This pattern was further confirmed when a significant proportion of respondents answered negatively (73% in Batticaloa and 86% in Galle) when inquired from resettled families whether their household income is generally sufficient to meet household expenses. As a result, they have employed various strategies to cope with the situation such as borrowing money from friends and relatives, cutting down expenses, and withdrawing savings. It is evident from these findings that a significant proportion of the sample employs available social capital for their survival. On the other hand, one-third of respondents from both locations employ vulnerable strategies such as pawning valuables, depending on money lenders, and buying essential goods on credit, indicating dependency on external sources for their survival. This in turn will force them to stay in a vicious cycle of poverty.

Social infrastructure is comprised of the facilities, spaces, services, and networks that support the Communities' quality of life and well-being. It helps people to be happy, safe and healthy. People use these services daily which in turn will increase their quality of life. Resettlers have received necessary social infrastructure-related facilities over the past 20 years. However, resettlers from both locations requested the development of proper solid waste management and wastewater disposal systems in the settlements to improve their living conditions. Improper wastewater and solid waste management systems are havens as dengue breeding spots which could be a reason why dengue fever is prevalent in these locations. They also requested to build factories in the vicinity of the settlements to create employment opportunities for casually employed or unemployed youth in the area.

22.5.4 The Gender-Related Impacts of Post-Tsunami Relocation

Unemployment and underemployment can often be visible among resettlers (Cernea 2000). In this context, the study examined the gender variation of unemployed household members. There were more unemployed females (84%) compared to their male counterparts (16%). Unemployment was also prevalent among those with education only up to post-primary level (26%) or Ordinary Level (44%). 72% of all respondents interviewed stated that there was at least one member without employment and in search of employment, at the time of the interview. Regarding the difference between research locations, a slightly higher proportion of households with unemployment was reported from Batticaloa (nearly 73%) compared to Galle resettlements (71%). Households with unemployed members were higher in the 'poor' and 'very poor' categories both in Batticaloa (60 and 77%, respectively) and Galle (65 and 73%, respectively) resettlements. It became evident from the interviews conducted with the main respondents that it is difficult to find employment in the new settlement areas compared to before. It is important to create more opportunities for them to acquire more skills and financial assistance to start new income-earning activities using these skills.

Unemployed women aspired to get training from the government or any other institution, to build skills in cookery or sewing, so that they could engage in self-employment or work in a garment facility. These institutions could fulfil a crucial need of the relocated communities by assisting women in getting an additional income source for the family which in turn assists families in the 'poor' and 'very poor' categories to move out of the cycle of poverty in the long run.

22.5.5 The Impact on Vulnerable Groups Such as Children and the Elderly

The study also explored issues and problems that school children and the elderly had to face after relocation. Overall, nearly 16% of resettled households reported that their children dropped out of school after relocation (21% in Batticaloa and 14% in Galle). This is mainly due to the loss of income of the household head, increased distance to the previous school, and increased expenditure on transportation. This has disproportionately affected the very poor households with 50% in Batticaloa and 36% in Galle reporting school dropouts. At present, most of these children have grown into adults and are engaged in casual employment or household activities. Nearly 38% of households, 30% in Batticaloa and 56% in Galle have moved their children to schools closer to the new settlements to cope with increased transportation expenditure. Although some are concerned about the quality of education, which the children received from popular schools in the city before relocation. There is a small proportion of parents in Galle who still send their children to popular schools in the city irrespective of the distance and increased transport expenditure.

Nearly 48% of respondents in Galle stated that at least one household member suffers from chronic disease/s while Batticaloa households reported only 18%. It is evident from the interviews conducted with elderly people in both relocated settlements who are suffering from chronic illnesses that they have difficulties accessing services in the city due to transport difficulties. The elderly in general also raised their concern about being away from the social environment (city, sea, burial grounds, neighbourhood, etc.) that they had before. They had lived in their previous places for two or three generations. Relocation has uprooted them from these familiar places, and as a result, they have lost neighbours, friends, and relatives who lived with them before relocation, and they do not have the time to reconstruct such a friendly social environment in the new settlements. Hence, they feel isolated more than ever before. This could be the reason why most elderly people are completely opposed to relocation (Gamburd 2014).

22.5.6 The Status of Community Formation and Social Relationships Amongst the Relocated Communities

Resettlement fragments communities dismantle patterns of social organization, scatter interpersonal ties with kinsmen and disrupt informal networks with friends and neighbours (Fernando 2012). The creation of community structures and community-owned resources is a complex process that cannot be accomplished overnight (Cernea 2000). Creating a community in resettlement comes with many challenges as members in the resettled community need to overcome trauma, adjust to the new environment, build new social networks, and collaborate with external organizations for support in rebuilding their lives, without becoming dependent (Alaniz 2017). The success of creating a community can be measured by the way people treat each other and their perceptions about their neighbours, host community, and their community.

According to 99% of Batticaloa respondents, their neighbourhoods and resettlements are peaceful. The percentage is much less in Galle, where nearly 77% of respondents consider their neighbourhoods and settlements mostly peaceful, with some reporting quarrels and tensions caused by alcoholics, personal jealousies, land disputes, and extramarital relationships. In both locations interviewed respondents know their neighbours and have reciprocal relationships such as helping each other in difficult situations either by giving money, advice, or labour. It is important to note that 98% of Batticaloa respondents know at least one or two of their present neighbours from the previous settlement. This percentage is slightly low in Galle (95%). Similarly, respondents in both Galle and Batticaloa have at least one known friend from the previous settlement residing in the present settlement. Known friends assist each other in difficult situations by giving money, advice, and labour. Tsunami-related resettlement disrupted long-standing and meaningful relationships between people and place and reconstructed relationships with friends, neighbours and kinsmen which took considerable time. New settlers reportedly have a very peaceful relationship with both host and resettled communities where they actively engage in organizing religious activities and other community activities like New Year celebrations with host and resettled communities.

22.5.7 Coping and Adaptation Strategies Employed by the Relocated Communities to Overcome the Challenges

Migration could be identified as a last resort enhancement strategy employed by those who are unable to successfully cope with the combination of stresses they are exposed to because of forced relocation. In this background, the study explored whether resettlers have any plans of leaving the resettlement. A significant proportion of the interviewed resettlers (nearly 90%) from Batticaloa stated that they intended to live in the settlement, while only 10% stated they wanted to leave. Those who wanted to leave would like to move back to their previous place or somewhere closer to the city, mainly by selling their housing unit. A significant proportion of the interviewed resettlers from Galle (nearly 80%) intended to live in the settlement, while only 20% wanted to leave. Those who wanted to leave stated that they would like to move back to their previous place or somewhere close to the city, mainly by selling their housing unit. While only a small portion of interviewed resettlers wanted to leave their respective resettlements at the time of the interview, data from the Divisional Secretariat show that more than half of the resettlers in the settlements chosen for the previous study had gone back to their previous place before the tsunami by selling, renting, or closing their housing units.

The researchers also explore the views of the people who have moved into the buffer zone regarding whether they will be safe from future tsunamis and other coastal hazards. Most people are of the view that they now know what a tsunami is, unlike in 2004, and where to run to protect their lives when they get the evacuation call. When asked further they also stated that they did not know what would happen to their movable and immovable property if a Tsunami strikes. Government officials are of the view that they may have to expect more casualties and damages in future tsunamis especially in the Galle area as now there are more families and various businesses that operate in the buffer zone area. Coast Conservation Department officials have the authority to demolish new constructions in the buffer zone area, constructed without their permission (Coast Conservation Department 1997, 2005). According to officials, they refrain from doing so mainly due to political pressure.

Resettlers who have sold their lands have done so by signing a transfer letter in front of a lawyer as they did not receive the original deed at that time. However, according to the land officer at the Divisional Secretariat, this practice has not been accepted as legally valid. As a result, land deeds were only handed over to original settler families. As such, a new issue has already emerged for people who had purchased the land from original resettlers, who had sold their houses before obtaining the deed. As original deeds were not given to those who purchased lands from original resettlers, new settlers are now unable to obtain the property deeds under their name until they pay the government-valued amount for the property.

22.5.8 New Settlers (Purchased or Rented)

Some settlers have gone back to the buffer zone or places close to the city by renting or selling their houses which is a new development 20 years after relocation. People who have purchased or rented such houses are categorized as new settlers. Some salient features of new settlers are discussed below.

47 respondents from Batticaloa who have either rented (29) or purchased (17) houses were interviewed. Most people who have purchased housing units have done so by paying Rs.500,000 to Rs.1,000,000 and those who have sold their housing units illegally have done so by signing an agreement in front of a lawyer (88%). Those who have rented the house have done so by obtaining a monthly rental of Rs. 5000 or below. Many renters and new settlers who have purchased or rented housing units have done so mainly due to the low price. However, most of them have identified the distance to the city as one of the main disadvantages. Nearly 94% of new settlers intend to stay in this settlement.

In Galle, 74 respondents who have either purchased (63) or rented (11) houses in resettlements were interviewed. People who have purchased housing units have done so by paying Rs.500,000 to Rs.1,500,000 and those who have sold their housing units illegally have done so by signing an agreement in front of a lawyer (85%). Those who have rented the house have done so by obtaining a monthly rental between Rs.5000 and Rs.10,000. Most renters and new settlers who have purchased or rented housing units have done so mainly due to the low price and the proximity to relatives, who are in either the settlement or the neighbouring village. Most new settlers have identified an unfavourable social environment (conflicts, drug addiction, alcoholism, and being labelled as tsunami resettlers) as one of the main disadvantages. Generally, new settlers have mostly a peaceful relationship with both host and resettled communities with some quarrels and tensions in between. New settlers actively engage in organizing religious activities and other community activities like New Year celebrations along with host and resettled communities and nearly 67% intend to stay in this settlement.

22.5.9 Tsunami Early Warning Systems

Tsunami early warning systems were introduced in Sri Lanka as a formal response to increasing the resilience of coastal communities from tsunamis. An end-to-end tsunami early warning and mitigation system was established in 2008, and this system has been fully operational since 2013 in all the affected countries (IOC/UNESCO 2015). Tsunami early warning is a complex system. It involves multiple components which are supported by various scientific bodies, experts, and stakeholders. A regional early warning system is divided into two phases: upstream, interface and downstream. Detection of the earthquake, prediction of the tsunami-by-tsunami service providers, and delivery of the warning to the national centre for further analysis are mainly conducted in the upstream phase (Sakalasuriya et al. 2021). Interface is where the warning is received by the national focal point and processed, and the decision to evacuate is taken before disseminating it to the public. This involves the three principal actions of issuing the warning, conveying the warning and ordering for evacuation (Sakalasuriya et al. 2018). Evaluation of a national-level warning, country-specific impacts of the tsunami, disseminating the message to the public for evacuation and handling the evacuation are the main activities conducted in the downstream phase (Haigh et al. 2020). Disaster Management officials in Sri Lanka have conducted various activities aimed at tsunami preparedness in coastal regions including evacuation drills, construction of evacuation plans, development of different dissemination media for warning at local levels and activities raising awareness (Birkmann et al. 2010; Lovholt et al. 2014), and development of the Standard Operating Procedures for Sri Lanka's Tsunami Early Warning System (Disaster Management Centre 2020).

Sri Lanka has also participated in Exercise IOWave. The Tsunami Exercise IOWave is a key exercise that evaluates the Indian Ocean Tsunami Warning and Mitigation System (IOTWMS) in responding to a potentially destructive tsunami. The exercise aims to improve tsunami preparedness at every level by evaluating and improving the operational effectiveness of Tsunami Service Providers (TSPs), National Tsunami Warning Centres (NTWCs), National Disaster Management Organisations (NDMOs), and Local Disaster Management Organisations (LDMOs). Exercises IOWave 18 and IOWave 20, conducted in 2018 and 2020 respectively, simulated a tsunami situation in Indian Ocean countries, including Sri Lanka, which prompted NTWCs and NDMOs to activate their Standard Operating Procedures (SOPs). The exercises stimulated the development, training, testing, and evaluation of Disaster Plans and SOPs, with the aim of understanding the governmental-level readiness for communication and decision-making during a disaster without disrupting the public. Accordingly, these exercises help to understand the shortcomings in the tsunami warning and response chain as well as country-level tsunami-capacity-building needs (Amaratunga et al. 2020; Haigh and Amaratunga 2018).

It is against this background that researchers explored the awareness of tsunami early warning systems among people who resided in the buffer zone or close to it, as they are the ones who live very close to the sea and are most vulnerable due to exposure to the sea. A significant proportion (88%) of respondents from both locations answered positively. When further examining the type of early warning system they were aware of, a majority of respondents who lived in the buffer zone in Batticaloa were aware of the siren system (nearly 77%), while respondents in the buffer zone area in Galle were aware of the public announcement system administered by the police. Several tsunami warnings were issued after the 2004 tsunami, and these were the channels through which interviewed respondents had received the message to evacuate in time during the past incidences. It is vital to note that overall, 92% of respondents stated that they would first contact the village-level administrative officer (GN) after the tsunami and even the relief officer at the Divisional Secretariat if they need any further assistance. However, only a very small proportion of respondents (18%) have participated in tsunami evacuation drills such as IOWAVE exercises, and those who participated in these drills have stressed the importance of such exercises as they give a clear picture to the people on where to evacuate and what to take with them.

22.6 Conclusion

This study has explored the long-term impact of tsunami relocations in purposively selected locations in Galle and Batticaloa over nearly a year interviewing four groups of settlers, namely, old settlers, new settlers, the host community, and those who moved back to the buffer zone. The tsunami relocation process has had many negative impacts on affected communities, further increasing their social and economic vulnerability at the household level. Initially, the distance from the city to their resettlement compared to their previous places created transportation difficulties which adversely affected people's livelihoods with the impacts exacerbated for 'poor' and 'very poor' households as most of the income is generated in these households through casual employment. Only one-third of resettlers belong to the non-poor category from both locations and two-thirds of resettled households belonging to poor and very poor households need various interventions in the short and long run to break the cycle of poverty. Many of the children from poorer households had dropped out of school, due to increased transportation costs and decreased household income. The lack of education in turn contributed to unemployment and economic vulnerability among these households. Women were disproportionately affected by unemployment, compared to their male counterparts. As a coping strategy, women requested that vocational training institutions and factories be built near their settlements so that they may increase their skills and ability to find employment.

The social relationships of the elderly have also been negatively impacted and they feel no sense of belonging in their new settlements. The distance to hospitals from resettlements has contributed to reduced accessibility to proper healthcare facilities for disabled communities and individuals suffering from diseases. New settlers, who have moved into relocation settlements after purchasing or renting houses from original Tsunami resettlers, faced a new issue related to home ownership. House ownership of most of the new settlers is deemed illegal due to government regulations, and they are unable to get a deed in their names, despite purchasing the houses from previous resettlers.

Social relationships among the relocated communities have improved in the past 20 years, with many Tsunami resettlers and new settlers claiming that they plan to continue living in the settlements. However, at the time of the study, many original resettlers had moved back to the Buffer Zone as an enhancement strategy against the negative impacts caused by the relocation process. Hence, 20 years after the Tsunami, there are now more families and businesses in the Buffer Zone and an increased risk of property damage and casualties in case of a Tsunami. The awareness of Tsunami early warning systems installed after the 2004 Tsunami was relatively high among the communities in both Galle and Batticaloa, especially the sirens and the public announcement systems. However, participation in evacuation drills was relatively low, which proved to be an issue, considering that these drills were the most effective in training vulnerable communities on evacuating effectively and efficiently. As such the findings of the study highlight the negative impacts of not considering relocation as a long-term process. It stresses the importance of considering the long-term impact of relocation projects to avoid further increasing the vulnerability of the affected population.

References

Ahmed I (2020) Sustainable development through post-disaster reconstruction: A unique example in Sri Lanka. Extern Interv Disaster Risk Reduct: Impacts Local Communities:65–79. https://doi.org/10.1007/978-981-15-4948-9_4

Alaniz R (2017) From strangers to neighbours: Post-disaster resettlement and community building in Honduras. Texas University Press, Texas. https://doi.org/10.7560/313831-007

Amaratunga D, High R, Premalal S, Siriwardana C (2020) Observer report on exercise Indian Ocean wave 2020: An Indian Ocean-wide tsunami warning and communications exercise, 13th October 2020 in Sri Lanka

Arunatilake N (2018) Post-disaster housing: Lessons learnt from the 2004 Tsunami of Sri Lanka. Asian Tsunami Post-Disaster Aid:211–232. https://doi.org/10.1007/978-981-13-0182-7_11

Birkmann J, Buckle P, Jaeger J, Pelling M, Setiadi N, Garschagen M, Fernando N, Kropp J (2010) Extreme events and disasters: a window of opportunity for change? Analysis of organizational, institutional and political changes, formal and informal responses after mega-disasters. Nat Hazards 55:637–655. https://doi.org/10.1007/s11069-008-9319-2
Bittel EP (2019) Disaster recovery, social capital, and the Sri Lankan context: A comparative study of two communities in Batticaloa (Doctoral dissertation, University of Colorado)
Boano C (2009) Housing anxiety and multiple geographies in post-tsunami Sri Lanka. Disasters 33(4):762–785. https://doi.org/10.1111/j.1467-7717.2009.01108.x
Cernea M (2000) Risks, safeguards and reconstruction. A model for population displacement and resettlement. Econ Polit Wkly, 35(41):3659–3678
Coast Conservation Department (1997) Coast conservation management plan (Revised) 1997. Coast Conservation Department, Colombo
Coast Conservation Department (2005) Developers guide and procedures for coastal development. Coast Conservation Department, Colombo. Available from https://www.coastal.gov.lk/images/pdf/Permit_Guidline.pdf
DCS (2005) Impact of tsunami 2004 on Sri Lanka. Department of Census & Statistics, Colombo
De Wet C (2009) Does development displace ethics? The challenge of forced resettlement. In: Oliver-Smith A (ed) Development and dispossession: The crisis of forced displacement and resettlement, 77, 96. School of Advanced Research Press, New Mexico. UNU-EHS, Bonn
Disaster Management Centre (2020) Standard operating procedures for Sri Lanka's tsunami early warning system: Ministry of Defence, Government of Sri Lanka
Fernando N (2012) Forced relocation after the Indian Ocean Tsunami 2004: Case study of vulnerable populations. UNU-EHS, Bonn
Fernando N, Amaratunga D, Haigh R, Wise B, Prasanna J, Jayasinghe N (2021) Life two years after relocation: Status quo of natural hazard-induced displacement and relocation in Kegalle, Sri Lanka. In: Multi-hazard early warning and disaster risks. Springer International Publishing, pp 3–25
Gamburd MR (2014) The golden wave. Culture and politics after Sri Lanka's Tsunami disaster. Indiana University Press, Indiana
Haigh R, Amaratunga D (2018) Observer report on exercise Indian Ocean Wave 2018: An Indian Ocean-wide tsunami warning and communications exercise, 5th September 2018 in Sri Lanka
Haigh R, Amaratunga D, Hettige S (2015) Briefing paper on taking stock of the tsunami recovery process
Haigh R, Sakalasuriya MM, Amaratunga D, Basnayake S, Hettige S, Premala S, Arachchi AJ (2020) The upstream-downstream interface of Sri Lanka's tsunami early warning system. Int J Disaster Resil Built Environ, 11(2):219–240
Hettige S, Amaratunga D, Hettiarachchi S, Haigh R (2014) Proceedings of the tsunami recovery in Sri Lanka: Ten years on. Social Policy Analysis and Research Centre (SPARC), University of Colombo Sri Lanka
Hettige R, Haigh R, Amaratunga D (2018) Community level indicators of long-term disaster recovery. J Eng Procedia, Elsevier, 212:1287–1294. SCOPUS Indexed
IOC/UNESCO (2015) IOTWMS factsheet. https://unesdoc.unesco.org/ark:/48223/pf0000233627
Koenig D (2009) Urban relocation and resettlement: distinctive problems and distinctive opportunities. In: Oliver-Smith A (ed) Development and Dispossession: The crisis of forced displacement and resettlement. School of Advanced Research Press, New Mexico, pp 119–139
Løvholt F, Setiadi NJ, Birkmann J, Harbitz CB, Bach C, Fernando N et al (2014) Tsunami risk reduction–are we better prepared today than in 2004? Int J Disaster Risk Reduct 10:127–142. https://doi.org/10.1016/j.ijdrr.2014.07.008
Muggah R (2008) Relocation failures in Sri Lanka. A short history of internal displacement and resettlement. Zed books, London
Oliver-Smith A (2009) Introduction: Development-forced displacement and resettlement: A global human rights crisis. In: Development and dispossession—the crisis of forced displacement and resettlement, pp 3–23. School of Advanced Research Press, New Mexico

Oliver-Smith A (2018) Disasters and large-scale population dislocations: International and national responses. In: Oxford research encyclopaedia of natural hazard science. https://doi.org/10.1093/acrefore/9780199389407.013.224

Sakalasuriya M, Amaratunga D, Haigh R, Hettige S (2018) A study of the upstream-downstream interface in end-to-end tsunami early warning and mitigation systems. Int J Adv Sci Eng Inf Technol, 8(6). ISSN: 2088–5334. Indexed in SCOPUS

Sakalasuriya MM, Haigh R, Amaratunga D, Hettige S, Weerasena N (2021) An analysis of the downstream operationalisation of the end-to-end tsunami warning and mitigation system in Sri Lanka. In: Amaratunga D, Haigh R, Dias N (eds) Multi-hazard early warning and disaster risks. Springer, Cham. https://doi.org/10.1007/978-3-030-73003-1_45

Thurnheer K (2014) Life beyond survival: Social forms of coping after the tsunami in war-affected Eastern Sri Lanka. Medienproduktion, Wetzlar. https://doi.org/10.1515/transcript.9783839426012

Turton D (2006) Who is a forced migrant? In: De Wet CJ (ed) Development-induced displacement: Problems, policies and people. Berghahn Books, Oxford, pp 13–27

Weerasena N, Amaratunga D, Hettige S, Haigh R, Sridarran P (2018) Provision of social infrastructure to resettled victims of the tsunami: Evidence from the grass roots. J Eng Procedia, Elsevier 212:212, 379–386. https://doi.org/10.1016/j.proeng.2018.01.049, SCOPUS Indexed

Correction to: Echoes of the Indian Ocean Tsunami in Sri Lanka: Two Decades of Lessons Learnt and Future Challenges in Disaster Management

Deepthi Wickramasinghe, Sujeewa Kumari, Rekha Nianthi, and Rajib Shaw

Correction to:
Chapter 20 in: R. Shaw et al. (eds.), *Two Decades from the Indian Ocean Tsunami*, Disaster Risk Reduction, https://doi.org/10.1007/978-981-96-2669-4_20

In the original version of the book, the author name in Chapter 20 has been updated as Rekha Nianthi. The book and the chapter have been updated with the changes.

The updated version of this chapter can be found at
https://doi.org/10.1007/978-981-96-2669-4_20

R. Shaw et al. (eds.), *Two Decades from the Indian Ocean Tsunami*, Disaster Risk Reduction, https://doi.org/10.1007/978-981-96-2669-4_23

The manufacturer's authorised representative in the EU is Springer Nature Customer Service Centre GmbH, Europaplatz 3, 69115 Heidelberg, Germany. If you have any concerns regarding our products, please contact ProductSafety@springernature.com

Printed and bound by CPI Group (UK) Ltd, Croydon, CR0 4YY
01/10/2026
02225913-0017